COST EFFECTIVE DESIGN/BUILD CONSTRUCTION

ANTHONY J. BRANCA, P.E.

R.S. MEANS COMPANY, INC.
CONSTRUCTION CONSULTANTS & PUBLISHERS
100 Construction Plaza
P.O. Box 800
Kingston, Ma 02364-0800
(617) 747-1270

Printed in the United States of America

10 9 8 7 6 5 4 3 2 1

Library of Congress Cataloging in Publication Data

ISBN 0-87629-088-8

COST EFFECTIVE DESIGN/BUILD CONSTRUCTION

ANTHONY J. BRANCA, P.E.

Illustrated by
Carl W. Linde

"Success by credibility:
Succeeding at what you are doing because you are good at it."

To Betty, Elizabeth and Jeannette
To Joseph and Gussie
To Olga

I had a need, you understood.

TABLE OF CONTENTS

FOREWORD

The sequence of *Cost Effective Design/Build Construction* is the same as that followed by an actual design/build project. The chapter headings specifically define their content. The text begins with the start-up requirements of the company and then proceeds through the different stages of project development.

In covering the steps associated with design/build construction, this book outlines techniques that have been or can be used to successfully perform all of the tasks. Absolute conformance to the specific methods outlined in this book is not necessary to achieve success in design/build contracting. Success lies within the individual and his approach. In the end, what works is correct, and what does not is not.

To find one's way in unknown territory, there are two options: one can ask questions and grope for directions, or one can follow a road map. Both methods may get the traveler to his destination, but the map is a faster and surer way. For those who are entering the realm of design/build construction, this book is intended to serve as such a map.

INTRODUCTION

The roots of design/build can be traced to the early history of building construction. Design/build may be seen as an outgrowth of the Master Builder method, an approach that predominated until early in the 20th century.

Financing the work and securing the work force for early construction was achieved primarily through conquest. The plunder of one's neighbor's wealth and the subjugation of his populace was the standard method of acquiring resources. The owner then hired the master builder who acted as the architect, engineer, and contractor for the project. This policy of conquer and build was practiced to perfection by the greatest conquering administrators of all time, the Romans.

With the dissolution of conquering forces such as those of the Roman Empire, patrons no longer had an unlimited supply of materials and labor with which to build. The spoils from invasions into neighboring territories became scarcer as those neighbors became either poorer or stronger. Those who sought resources in the form of plunder had to travel far with no assurance of success. An alternative was to subjugate one's own people and to levy taxes. The people would not stand for servitude, however, and taxes had limitations. Unlike their predecessors who had the means to build on a grand scale, the master builders now had to forecast costs. This was a very difficult task for projects spanning decades and longer.

As project durations became more reasonable due to better techniques and more modest ambitions, the costs became more predictable. More master builders gained experience and entered the marketplace. Given these factors, competition was inevitable, and a new construction method evolved – lump sum bidding. To establish standard value and further control, a designer was commissioned independently to design the project.

The competitive and capitalist American economy was well suited to the lump sum bid method, which became the standard used until the late 1950's. By this time, certain shortcomings in this method had begun to appear. Rapidly rising construction costs were a major factor. As inflation forced costs up, time became a valuable commodity and the inefficiencies of the lump sum bid method grew more expensive. The stage was set for a new construction method that would be less complicated and time-consuming: the design/build method. With design/build, complete and detailed working drawings were not needed before construction could start, and thus valuable time (and money) could be saved. The project's designer could work with the design/build staff as a team, performing the same functions that were once the domain of the Master Builder. Any changes made in the course of the project could be carried out much more quickly and easily – without the delays that arise when outside agencies and traditional bidding procedures are involved.

Design/build solves many problems posed by the lump sum bid method, but it, too, is subject to social and economic pressures. That is why success in design/build depends upon effective corporate and project management, and the mutual understanding and cooperation of all of the parties involved in the project.

Part One

DEFINING DESIGN/BUILD

Chapter One

Comparison of Methods

Design/build is a unique form of building construction. To understand the design/build method, we should first discuss and illustrate other methods currently practiced in the construction industry.

The Traditional Bid Method

Traditional bid work involves independent activities of the architect/engineers and the contractor. The architect, together with the owner, develops the concept of the product, his building. That concept is then given a specific identity, described by a complete set of plans and specifications. These documents are examined, quantified, and priced to result in a bid, or offer amount, to produce the building. The low qualified bidder is then awarded the job. The contractor's responsibility is to construct the building. The relationship between the owner, the designer, and the contractor may be illustrated in two ways: contractual and operational.

Figure 1.1 shows that although there are individual contracts between the owner and the contractor and between the architect/engineer and the owner, there exists no contractual relationship between the architect and the contractor. The operational relationship between these parties can be illustrated as shown in Figure 1.2.

The American Institute of Architects (AIA) General Conditions of the Contract for Construction usually identifies the designer (architect/engineer) as the review and approval entity. The architect is also assigned some responsibilities as the contract or project administrator. This is not a complete representation of the architect's role, but it does clarify one of his duties. Certain responsibilities are willingly deferred by the architect; among them are the owner's approval of the contractor, the bid, the duration of the project, the progress payments, and the extra work and changes to the contract. In this way, the architect's function is limited to an advisory role with the approval of the owner.

Examination of the contractual relationship clearly shows that responsibility for the project is fragmented and shared (not necessarily equally) by the contractor, the architect, and also the owner who remains the eventual recipient of disputes and claims. The operational illustration (Figure 1.2), may show a linear relationship, but it does not identify the architect as the responsible entity. This illustration simply demonstrates the architect's (A/E) role as a reviewing conduit for contract information, interpreter of design questions, and arbitrator of possible disputes.

The Construction Management Method

Construction management is another method of building. In the middle to late sixties, and again in the early seventies, construction management became a desired approach to producing a building. The two main reasons for the development of the construction management technique were scarce money and high interest rates, and the consequential need to produce the product quickly.

As money available for construction became less abundant and high interest rates became the norm, many lending institutions grew increasingly concerned with the risks involved should a developer be unable to complete a project. The lending institutions had a number of concerns; they wanted the authority to remove an inexperienced or

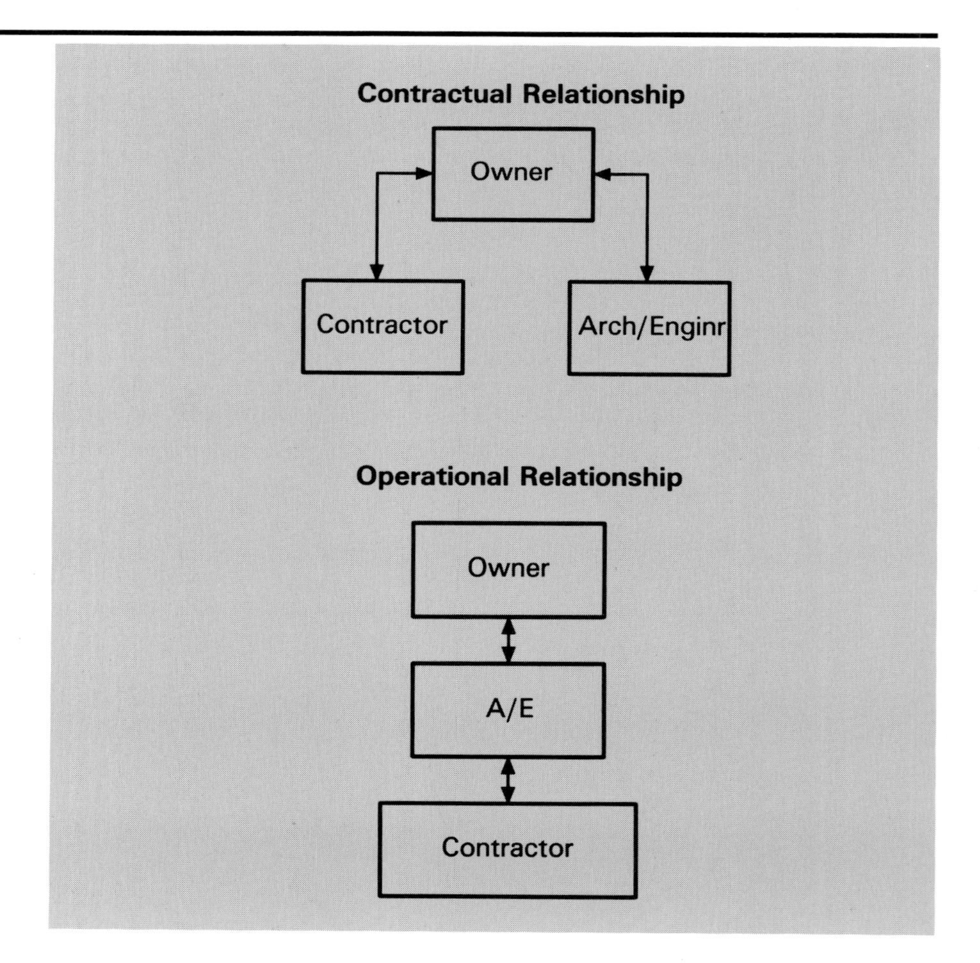

Figure 1.1

Figure 1.2

inept project manager; and they sought to achieve and maintain better methods to control costs and changes to the work. Solutions could not be found in the traditional bid method, which prohibited that kind of control. To reduce the risk, the lending institutions insisted, as a condition of the loan, on professional construction management throughout the process – from inception to completion.

As a consequence of the high construction interest rates, developers began to demand that the building be completed as quickly as possible so as to become an income producer, converted to the permanent mortgage. At the same time, there were regional demands for useable space. All of these factors led to the conclusion that the traditional bid method would not suffice. The introduction of *fast track* construction provided a new approach, which could be used in concert with construction management. With the fast track method, construction is started before the design is complete. The fast track approach could only be utilized under contractual arrangements different from the traditional bid method.

As is often the case, the solution to a problem becomes an established procedure, remaining long after the problem goes away. The process of construction management became popular to all parties involved for different reasons. The owner liked the method because it could produce the building and generate revenue more quickly. The architect found that construction management allowed him to test the cost of the building elements in advance of the final design drawings and thereby avoid potential redesign. It also required fewer drawings than the number needed for a bid because certain details could be worked out later on sketches, or resolved on shop drawings. On the other hand, the architect also saw some disadvantages in the construction management method. It required him to design certain elements or trade work out of normal sequence, thereby disrupting the flow of the drawings. The necessity for out-of-sequence design was to accommodate items of work that were in progress and critical components that had to be ordered in advance to insure delivery to the job site when needed. The contractor found the construction management method advantageous. It relieved him of some of the risk (assumed under a traditional bid method) and shared it with the owner.

The differences in the construction management method – both contractual and operational – are shown in Figure 1.3. The CM/C and architect are on equal terms, having a cooperative relationship.

The contractual relationships between the owner and the contractor and between the owner and the architect are similar to those in the traditional bid method, but with striking differences in the contract language. A comparison of AIA Document A101, *Standard Form of Agreement Between Owner and Contractor, Stipulated Sum* (commonly used for traditional bid work), and AIA contract form A111, *Standard Form of Agreement Between Owner and Contractor, Cost of the Work Plus a Fee* (commonly used for construction management), reveals vast differences in contract language and the relationship and management of the project (see Chapter 10 for an analysis of these two types of contracts). Also apparent is a new element known as *Guaranteed Maximum Price (GMP)*. GMP is used instead of the traditional lump sum as a basis for the cost of the work.

The point in the process when these prices are presented or produced is different. Bid prices are received at the completion of 100% of the design drawings. Construction management GMP pricing is given during the design process, typically around the 60% to 70% completion point. The value of the work is priced at 100% of the anticipated cost by a method known as "scoping". Scoping defines, in words, the remaining work that has not been described by the plans.

There are two approaches to construction management: one involves a professional construction manager (PCM), and the other requires a construction manager/contractor (CM/C). The professional construction manager is a consultant who does no building, but supervises contractors who do the work. The construction manager/contractor is a consultant who also builds.

In the 1960's, the early applications of this process involved contractors who assumed the role of the CM/C. The contractors acted as an estimating service and consulted on construction-related issues in the early stages of the design. Once the GMP was issued, the CM/C assumed the role of the builder. As the process became more sophisticated, especially through the applications of computer scheduling and cost accounting, the construction manager became more of a consultant, more of a professional construction manager. An illustration of this relationship – both contractual and operational – is shown in Figure 1.4.

In professional construction management, the responsibility for the project's success, both contractual and operational, becomes more fragmented. The professional construction manager acts as an advisor. His role is that of central responsible party, yet he is not an agent of the owner, so does not have that authority. Construction management is a team effort; it lowers the traditional barriers that bid work raises between the contractor and the architect. It does not, however, provide for a single entity responsible for the successful completion of the project.

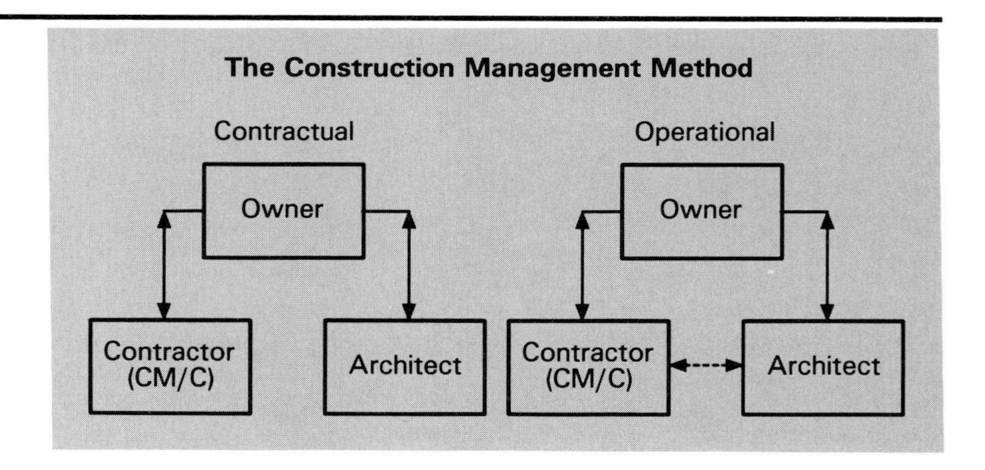

Figure 1.3

The Design/Build Method

The design/build approach reveals a difference not seen in the previous three methods. See Figure 1.5 for illustrations of the design/build process from contractual and operational points of view.

Examination of these diagrams and those illustrating bid construction management and professional construction management reveals that the design/build method furnishes a single point of responsibility for the success of the project. The design/build company is contractually related to the owner, the contractor, and the architect.

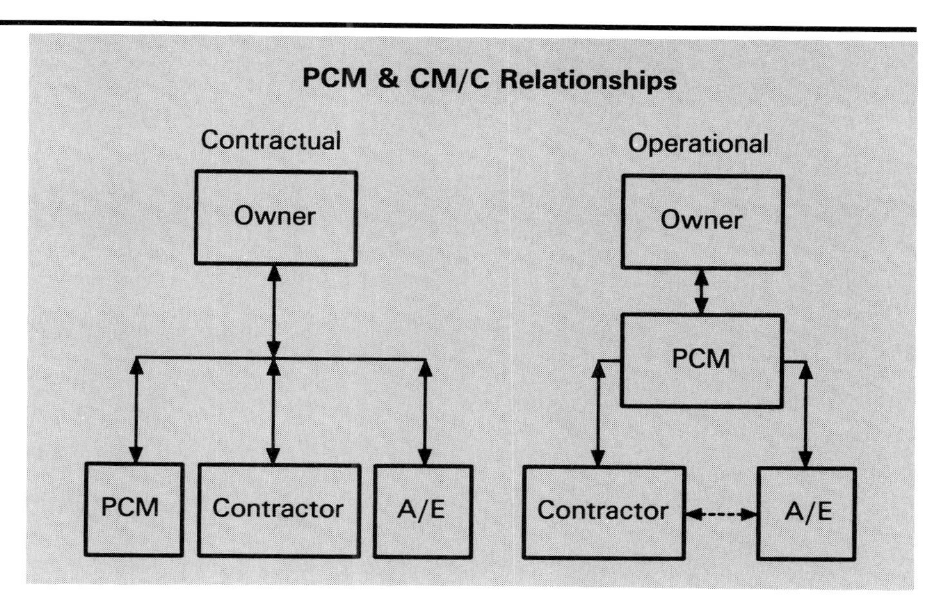

Figure 1.4

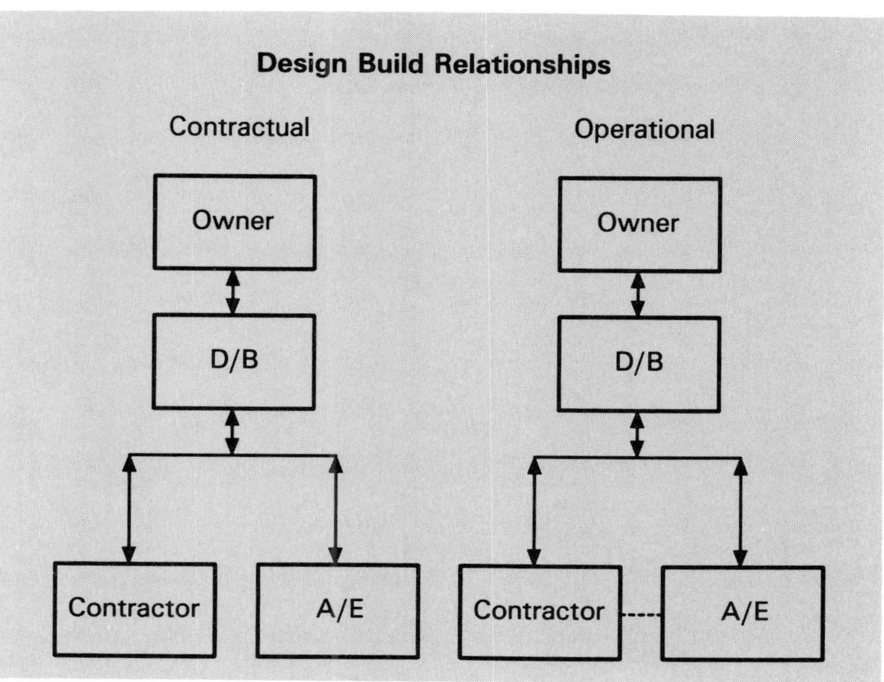

Figure 1.5

Operationally, the relationship remains the same; the cooperation between the contractor and designer can also be found between the design/build contractor and the architect. The design/build company is responsible for the concept, design, and satisfactory completion of the project. Thus, the project is not fragmented nor seemingly left to the devices of separate entities as can often be the case with the other methods. Design/build can be compared to a manufacturing process because in both cases, the product is controlled from inception to completion. This control is what makes design/build work.

The features of the four methods: bid, construction manager/contractor, professional construction manager, and design/build can be compared as shown in Figure 1.6.

The significant difference between design/build and the other methods lies in the level of responsibility for design and construction and at what design level the Guaranteed Maximum Price (GMP) may be given. In design/build, the GMP, or the maximum cost of the project, is provided in the conceptual/schematic design stage with 10-15% of the design complete. In a Construction Management/Contractor (CM/C) project, the GMP is given at the design development/construction stage – 50-90% complete. Professional Construction Management (PCM) requires a 100% completed construction design; even then, the price is never guaranteed. The bid method always requires 100% completed documents.

The design of a building evolves through various levels of completeness until arriving at the construction, or buildable, set of documents. These levels are termed *concept, schematic, design development*, and *construction* (see Figure 1.7). The design/build 10-15% document completion level would include different portions of the project at different levels of completion: concept level to schematic level architectural drawings, schematic to design development structural drawings, and concept level MEPS (mechanical, electrical, plumbing and sprinkler) drawings, depending upon the degree of difficulty of the project. In addition to the plans, an outline specification and a quantitative, assemblies

Comparison of Construction Methods

Items	D/B	CM/C	PCM	BID
Sales/Marketing	Leads/Referrals	Same	Same	Same
Type of Contract	GMP	GMP	Service	LS
Data Gathering	Owner/Code/Other	Same	Same	Same
Scheduling	Fast Track	Same	Traditional	Traditional
Total Responsibility				
For Design	Yes	No	No	Architect
For Construction	Yes	Yes	No	Contractor
Design level at GMP	10-15%	50-90%	100%	100%

Figure 1.6

conceptual estimate complete the definition and documentation of the project. Each of these three components makes a unique contribution to the GMP and the terms of the contract. The drawings define the building graphically through the use of "hard" line; the outline specification defines it narratively; and the conceptual estimate is an empirical description through the use of quantities and unit costs.

The design level required for the construction manager/contractor (CM/C) method differs from that which is required for design/build. It is more detailed than the conceptual drawings, but need not be as extensive as the complete working drawings. The structural drawings for the CM/C method must be done at the construction stage and the MEPS drawings at design development.

Professional construction management (PCM) does not guarantee the cost; it brings the level of design through the construction stage and then the work is put out to bid in the same manner as the traditional bid method. Benefits of the PCM method include the consulting input on costs, building systems, and construction issues, without early contractual commitment for the construction before the expense of detailed drawings and specifications. The PCM contract is similar to that of any other consultant who provides a service.

In summary, the design/build construction method provides:

- Total responsibility for the design, construction, and satisfactory completion of the project.
- Solutions tailored to the owner's problems and concepts that meet the owner's goals.
- Price testing of concepts (before design changes become very expensive).
- Guaranteed maximum price early in the process. This assures that the owner will get the building he wants, as defined, at the agreed set cost.
- Use of the fast track method, which allows the project to be completed more quickly.
- Close association and coordination with all of the parties involved and a quicker response to design or construction problems.

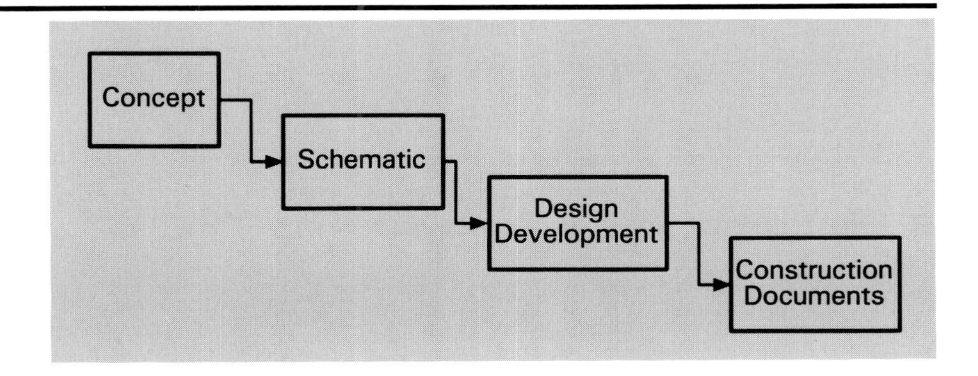

Figure 1.7

Part Two

PREPROPOSAL

Chapter Two

START-UP

A number of decisions need to be made by those who are considering going into the design/build business. The major decision concerns the financial wherewithall needed to establish and remain in the business. This resource is called capitalization. Capitalization provides the start-up money that is needed while the new company seeks clients and jobs. It sustains the payroll, taxes, contract retainage, and other expenses. An established company's need to find sources of capital may not be as important, whereas the new start-up company will find capitalization its biggest single hurdle.

At least one year's start-up time should be expected for a new design/build business to achieve the cash flow necessary to operate. For example, the capital requirements would include a full time salesman working on salary, plus a percentage of captured sales – before income is produced. Figure 2.1 lists some examples of items to consider in starting up a design/build business.

In this example, the total requirements for the first year amount to $315,000.00. Clearly, this amount represents a serious commitment. Initial start-up costs may be reduced by excluding a project manager and salesman. However, reduction can only go so far. Under-capitalization is a prime cause of failure for a start-up company. Large or small, a new company requires dedication if it is to succeed.

Start-up finances are only one of the items that must be considered. Other issues such as determining and analyzing the market and deciding where to locate the main office (in order to be close to that market) must also be resolved.

The location of the office may be a secondary concern in the face of a $100,000.00 to $300,000.00 capitalization. If the market covers a broad geographic area, the office location should not be a problem. However, if the market is confined to a small region, the location of the office may be a strategic consideration and a major factor in the success of the company.

Being able to identify target markets is key to the success of a design/build company. The foremost reason for this approach is the nature of the product being sold. This business cannot be approached by a "shot-gun" method such as mass mailings or media saturation advertising. Mass mailings sell the same product over and over. Design/build sells a different product (but the same service) every time. Since design/build companies are selling a custom building service, face-to-face contact is an important element, setting the stage for follow up contacts.

Start-Up and Operational Expenses

Salaries	Initial Outlay	Monthly
Owner		$ 3,445.00
Salesman		3,800.00
Project Manager		2,947.00
Secretary		1,733.00
Taxes, W.C., FICA, etc. 33%		3,935.00
Company Cars	$ 30,000.00	
Expenses		400.00
Office Rent 1500 S.F.		3,125.00
Utilities		125.00
Taxes/Maintenance		70.00
Telephone		350.00
Hospitalization (4)		840.00
Insurance		150.00
Disability		140.00
Office Expenses		200.00
Dues/Subscriptions		50.00
Accounting		200.00
Legal		150.00
Marketing Expenses		300.00
Incorporation Taxes	500.00	
Incorporation	2,500.00	
Accounting (set-up)	500.00	
Legal	1,000.00	
Furnishings	5,000.00	
Computer	4,000.00	
Rent Security Deposit	3,125.00	
Telephone/Supplies	1,500.00	
Promotional Material	3,000.00	
	$ 51,125.00	$21,960.00
Total Capitalization		
Initial Outlay	$ 51,125.00	
$21,960.00 x 12 =	263,520.00	
	$314,645.00	

Figure 2.1

Selecting a Market

An established company may continue with its existing client base and the type of work already demanded. A client may opt for a design/build arrangement because of the benefits it offers over bidding, and because of the design/build company's past reputation and the type of construction for which it is best known. For a new company, however, finding the right job market may be more a matter of default, at first, than a predetermined selection.

Light Industrial Market

The most popular market for design/build construction is light industrial work, which includes light manufacturing, warehousing, and transient storage facilities. This market is most suitable for both the established and the new design/build company. The reasons are listed below:

Large Market: This area of the business is constantly in demand.

Simplicity: The market is desirable due to its simplicity in design and construction, provided there are no site work complications.

A Forgiving Nature: Mistakes can be more easily rectified because these buildings are usually only one story in height and any mistakes are single occurrences. High rise construction, by contrast, suffers more heavily from mistakes because each represents a multiple occurrence.

Short Duration: The relatively short construction period allows for quicker turn-around of the project and a better cash flow. "Retainage", the money held back to insure performance, is not left unpaid for a long period. This is important, especially if some of the work is performed by your own work force. The retainage may end up as the job's profit. The sooner received, the better.

Suburban Location: Most light industrial buildings are built in a suburban location with large amounts of land available for support. This setting does not require the kind of close coordination of men, material, and equipment involved with an urban project. Consequently, a less experienced (less expensive) superintendent may be able to control the work.

Retail Market

The commercial market includes retail stores, in strip-type buildings on narrow parcels of land, and retail malls on large tracts of land. This market is very similar to that of industrial construction. Commercial projects are generally located in suburban areas; they are one-story, simple, forgiving kinds of structures. The duration for release of retainage may be longer, however, due to the nature of rental-stimulated projects. Commercial projects also involve a great deal of finish work.

Medical Market

Hospitals are a highly specialized and risky venture for the inexperienced contractor. For design/build, this market is limited due to state and federal regulations and specific operational requirements. The selection of the contractor is generally carried out by committee and is by the traditional bid method.

Doctor's office buildings are much better suited to the design/build method because they are essentially small office buildings with particular amenities. The challenge of this market is the design of interior space. Doctor's suites, while sharing certain common traits, generally require custom design. This part of the work proceeds slowly due to the particular and specific needs of the individual clients.

Office Market

Office buildings in a suburban setting represent a good market for design/build. This type of project shares many of the characteristics of industrial work. Offices differ in their multiple story height and longer construction duration. High-rise construction should not be attempted until a company is experienced. The close coordination of men and material needed for elevated work areas, life safety, and control is extremely difficult; a young company often has neither the capability nor the capacity to undertake such projects.

Housing Market

The residential market has many characteristics of the other types of markets, depending on the type of construction and the setting. Mid-rise apartments are like office buildings, completely finished. A condominium type of building is similar, but may be less finished in order to accommodate customizing. Housing in one- and two-story structures tends to resemble the commercial market. Housing, other than single family homes, may be built for a developer or in a "turn-key" arrangement, for example, for a local agency which provides housing for the elderly.

Turn-Key

Turn-key is a form of design/build, except that the design criteria is set. This approach sacrifices a certain amount of the design concept and flexibility that is part of design/build. To work in the turn-key market involves conforming to predetermined design, securing land, financing the entire project, and dealing with zoning issues if necessary. This involvement goes beyond simply building and requires developer capability. A small design/build company could be overtaxed in attempting to perform well in the turn-key market. For turn-key, one must have the financial strength to be able to acquire loans, work with real estate specialists, and be willing to participate on a modified bid basis. This is not a highly specialized market, but one that requires good financial stability and specialization as a company.

Examining the market options available, a new company should select one or two areas. Why would it not be prudent to prospect in all the markets? The reason is three-fold. First, a design/build company must market a product. A broad appeal to all the markets in one advertising copy appears like the "shot-gun" effect. Good marketing strategy instead dictates custom literature to specific clients. The second reason for narrowing the market is the high cost of producing custom literature specific to each market. The third reason is the ability of the new or even the experienced small company to cover sales calls, follow up, and service all of these market areas, activities which are both time consuming and difficult.

Work Program

A work program is a determination of a company's capability. This capability relies on manpower, profit, and minimum earnings necessary to stay in business. The purpose of the work program is to plan the marketing strategy and to set realistic goals. In effect, the plan is developed by asking, "How many jobs can I do in a year with the staff I have, and will that be enough work to make a profit?" Assuming that an initial goal is set at $3,000,000.00 of job volume a year, and the average job size is anticipated to be $250,000.00, then:

$$\frac{\$3{,}000{,}000.00}{250{,}000.00} = 12 \text{ jobs}$$

Can the company handle 12 jobs a year with the current staff, assuming no further hiring at this time? If not, then how many jobs can the company handle? If the answer is six, for example, then:

$$\frac{\$3{,}000{,}000.00}{6} = \$500{,}000.00 \text{ Average job size}$$

This analysis of the work program has determined that the average job cost using the current staff should be $500,000.00. This kind of analysis may not be completely reliable because a major consideration is being overlooked – return on investment. Return on investment is the minimum profit required of a business. Doing work without profit, in general, is doing it for practice. Doing work below the minimum profit is doing it at a loss. If a person must earn 15% on his investments and he realizes less, then it would not be worth investing in that particular endeavor.

Assuming an initial investment of $100,000.00, and an additional amount of $50,000.00 a year to keep it going, the minimum amount needed to be earned in the first year would be the following:

Start-up funds	$100,000.00	
Ongoing costs	50,000.00	
Required return on investments	15,000.00	(15% of $100,000.00)
Reserve for retainage	15,000.00	(10% own forces)
Minimum required earnings	$180,000.00	

If 5% gross profit (not including job related general condition costs) can be realized on average, then the volume needed is:

$$\frac{\$180{,}000.00}{.05} = \$3{,}600{,}000.00$$

Therefore, the amount of work needed to be contracted and performed in a given year is $3,600,000.00 and not the $3,000,000.00 originally envisioned. This analysis does not allow for amortization of the start-up funds were they available to be borrowed.

A towering influence on a design/build company is bonding capacity. By analyzing a company's financial structure, current worth, and other factors the bonding company determines the design/build company's capacity, financial capability, and professional ability to do work and the appropriate dollar volume of their work program. This judgment determines the size of the individual jobs that they will bond and the aggregate dollar amount, or total jobs they will bond at one time. Since the design/build construction process requires risk, the prudent client will require some insurance. That insurance will often be in the form of Performance and Labor and Material Bonds.

Another factor that influences work volume is the margin requirements. *Margin* is a term that refers to a fee or gross profit. In the previous example, margin would be the 5% multiplier of job cost. That 5% may not be fixed or may only apply to jobs of a certain dollar size. To determine the margin of a company, we must first define the other costs.

The conceptual estimate will be composed of so called "hard" and "soft" costs. The hard construction costs are those associated with the labor and material necessary to produce the building. This would include the concrete, structural steel, etc. and the labor to install these items of work. The soft costs are those in support of the building project and consist of General Conditions costs, design costs, and the design/build fee. General Conditions costs cover the salaries of the project manager, superintendent, assistants, field clerical, trailers, telephones, lighting and power, and the like. The design costs should include all expenses for the architect, engineers, inspections of the work, and testing requirements for design and quality assurance. The fee represents the gross profits which include net profit, a portion of main office costs, non-chargeable salaries of corporate officers, and similar kinds of costs (see Figures 2.1 and 2.2). For example:

$$\frac{\$180{,}000.00}{6 \text{ jobs}} = \$30{,}000.00 \text{ average min. margin/job}$$

Assuming a net profit of 3% before taxes (1-1/2% after taxes is the industry norm), then a project having hard costs, General Conditions costs, and design costs of $1,000,000.00 will require a minimum margin of 6%, as follows:

Minimum expense	$ 30,000.00
Net profit (3% of $1,000,000)	30,000.00
	$ 60,000.00

$$\frac{\$\ 60{,}000.00}{\$1{,}000{,}000.00} = 6\% \text{ margin}$$

Conversely, the minimum expense at a rate of $30,000.00 on a $5,000,000.00 project would represent less than 1%. The total margin then would be as follows:

Minimum expense	$ 30,000.00
Net profit (3% of $5,000,000.00)	150,000.00
	$180,000.00

$$\frac{\$\ 180{,}000.00}{\$5{,}000{,}000.00} = 3.6\% \text{ margin}$$

The considerations for margin will vary as the size of the project varies. Each case is approached individually and considered in the negotiations of the contract.

Hiring

When hiring someone to fill a role within a company, one should anticipate a period of adjustment and indoctrination before this person will be working at full potential. During that period, the efficiency of that employee is not 100% and the return on his worth is not 100% realized. If a project manager for the company is required to handle two to three jobs, a new manager will not be able to immediately take on this complete responsibility. That loss is reflected in the company profit structure, and for a small company, it can have a large impact. The decision to hire is one that requires careful consideration and financial analysis.

Having selected a market area or areas and realistically determined the work volume and margin requirements, a company should then observe and study the methods of the competition. In design/build, as in any competitive market, it is the best prepared, most cost-effective company that will be selected and be successful.

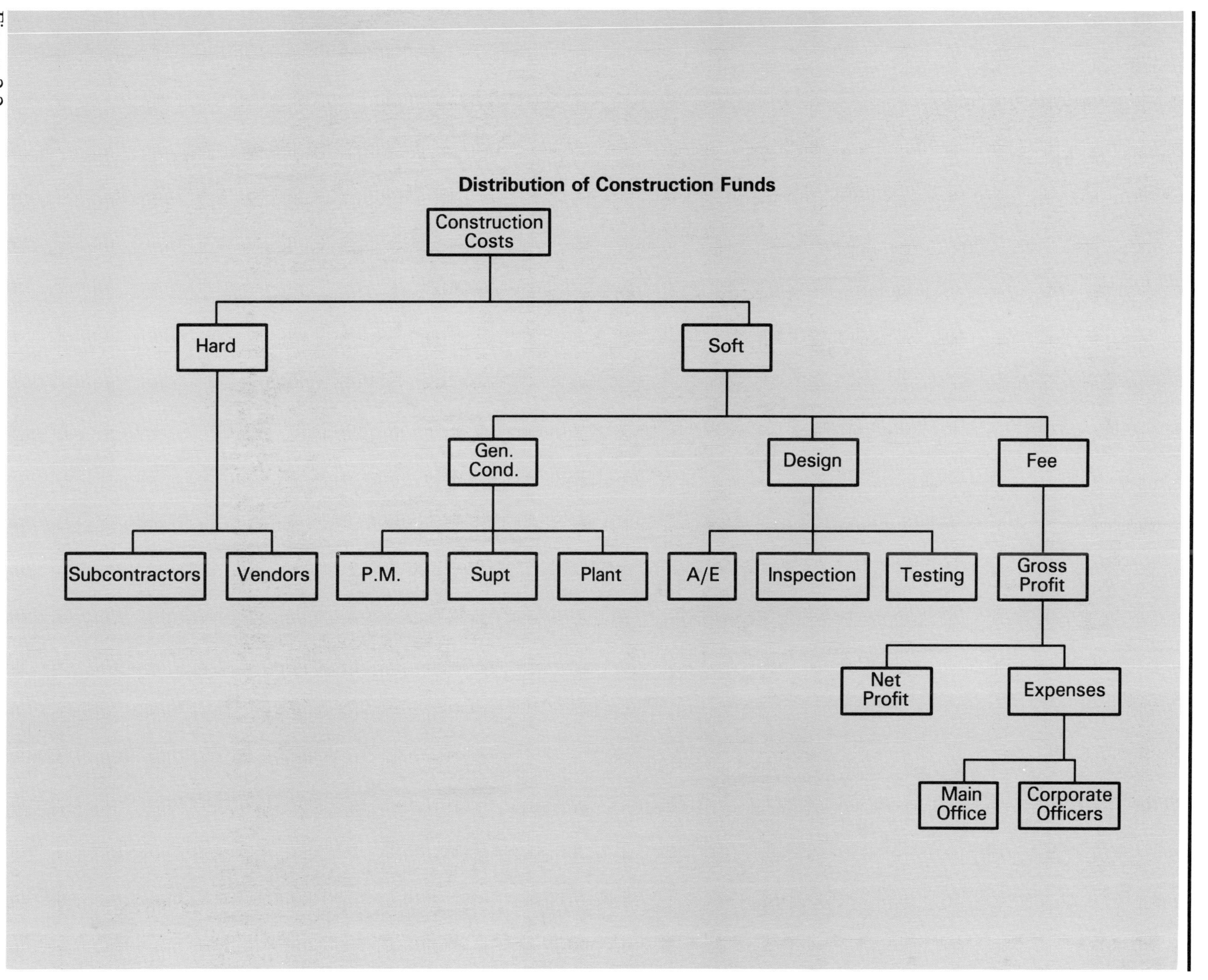
Distribution of Construction Funds
Construction Costs
Hard
Soft
Gen. Cond.
Design
Fee
Subcontractors
Vendors
P.M.
Supt
Plant
A/E
Inspection
Testing
Gross Profit
Net Profit
Expenses
Main Office
Corporate Officers

Figure 2.2

Chapter Three

MARKETING

Chapter 1 compared the features of the bid, design/build, construction management/contractor, and professional construction management methods. While design/build construction requires that the company seek clients, bid work is secured through public notification and other available sources such as advertisement in Dodge Reports, government publications, and by invitation for private work. The only criterion necessary to bid work is to be qualified. Qualification is a process by which a company's financial strength and past performance and experience with a particular type of work are examined and measured against a standard. If found acceptable, that contractor is thus qualified to bid the available work. If successful by virtue of submitting the lowest cost quotation, then that contractor may be awarded the job. None of the steps in a bid process involve a marketing effort or salesmanship. The "leads" are not developed, but are simply a response to advertisements.

Design/build, on the other hand, requires more effort in the "getting" or securing of work. This "getting" leads to the opportunity to negotiate, which may result in being awarded the contract.

Finding the Opportunity

Prospective clients may be referred to as "opportunities". Finding these opportunities may depend on good marketing techniques. Some of the marketing techniques used in design/build are widely known and used in other businesses. Other approaches are unique to design/build construction.

Brochures

Brochures are a form of advertising used to identify and describe a company. The simplest type of brochure is the two-fold letter-size format. A more elaborate version might include coordinated folders with individual job description sheets, and/or bound booklets. In either case, the brochure should sell the features and benefits of a company and establish name recognition.

If the company is newly formed, the brochure's emphasis should be on the company philosophy and the experience of the personnel. Projects completed by the staff members, even while in the employ of another company, may be highlighted. Professional and industry affiliations and other such organizations should also be mentioned. If the company has been in business for awhile, then the company achievements should be emphasized, along with the personnel and other interest-generating items.

The brochure can be composed by the design/build company itself, or by an advertising professional who will write the copy, arrange the artwork, design the logo, recommend the type of paper and color printing, and secure a typesetter and printer. This professional method is the preferred approach.

Direct Mail

Direct mail is geared to specific markets in which the design/build company hopes to find opportunities. The mailing could be derived from the following groups:

1. Existing and prior clients
2. Fortune "500" companies in the designated areas
3. A newspaper clipping service
4. Dun and Bradstreet lists
5. Trade magazine ads and ad promotion items
6. "The Million Dollar" book of corporations with sales in excess of one million dollars
7. Referrals.

1. *Prior satisfied clients* represent a good source because they are already familiar with the work of the design/build company; extensive promotional efforts would not be required to be awarded an assignment. Furthermore, satisfied clients represent a good source of referrals.
2. *Fortune "500" companies* may be thought of as only doing business with large, established design/build and construction management companies. Nevertheless, these companies also have projects of a size geared to smaller construction firms. While this source may classify as a long shot, it is still a viable prospect.
3. *Clipping services* find leads in various newspapers. Such a service is provided based on a particular region and type of information required. For example, a typical request might be as follows: "The New England States, all items concerning industrial work specifically with companies whose sales volumes are in excess of $5,000,000.00 and personnel in excess of 500 people." Further instructions might be to clip those articles involving corporate quarterly reports, personnel promotion announcements, or rental of space. The purpose of using a clipping service is to gather leads regarding companies that appear to be on the verge of facility growth. A clipping service should be closely monitored for effectiveness. The information it produces will only be as good as the instructions given.
4. *The Dun and Bradstreet* lists are derived from the resources of this organization which collects financial data. A subscription to this service provides access to financial information regarding public corporations and many privately held companies. Dun and Bradstreet also gathers data on the companies' work forces, condition of their offices and/or plants, names of company officers and owners, outstanding conditions that affect how they do business, and lists of subsidiaries, branch offices, and plants.

From this data base, a number of reports may be extracted to provide specific information for sales needs. One example is a list of companies in a particular region, doing business of a certain type and volume, and having a specified number of employees. Such a list, coupled with newspaper clippings on quarterly reports, could identify companies in growth industries who might have the need for new facilities. Another service offered is called *trend*. What trend provides is data over a period of years with empirical judgments on growth. For example:

Year	Sales Volume	No. Employees
1980	1,000,000	100
1985	3,000,000	250
growth	40%/yr. average	30%/yr. average
1986	3,400,000	280

Depending on the nature of the growth (either in sales only or in both sales and personnel), and the ability of the existing facilities to absorb this growth, such information could define a potential client in need of a larger or new facility.

5. *Trade magazines* concern themselves with developments in industry. Subscriptions to such magazines, within the target market, could be useful for gathering leads. The area to concentrate on is growth information. Published financial reports may be helpful in this regard. Current data and future projections can provide leads and insights into future business.
6. The *Million Dollar Book* and *Who's Who in Business* can be found in the reference section of the public library. These books list the names, titles, and business addresses of all the officers of the companies as well as the board of directors and owners. The major stockholders are also listed. This book can be useful in determining the proper names, titles, and addresses to which brochures and other promotional information may be sent. Direct mail for design/build should always be in the form of a letter addressed and directed to the person who would be interested and responsible for the services offered.
7. *Referrals* come from a wide range of sources. The best source, as previously explained, is past, satisfied clients. The next best source can be friends, associates, relatives, and fraternal and business organizations. This network of personal connections is the best because there is an existing familiarity and trust – crucial elements which cannot be quickly established. Referrals are a distinct plus and represent a reduction in the sales budget.

Following are some factors that affect the success of direct mail advertising: 45% of all mail is delivered on Monday (or the previous Saturday, but not opened until Monday). Tuesday's mail is about 8%, and the balance of the week's deliveries are distributed among Wednesday, Thursday, and Friday, about equally. Consider the busy executive with just so much time allocated to reading his mail each day. It doesn't take much to realize that Tuesday's mail gets more attention than the mail that arrives on Monday or any other day of the week. Envelopes with windows are generally attended to first. Why? – because checks and bills come in envelopes with windows.

Direct mail may seem like a mass appeal "shot-gun" approach, but the design/build mailing volume is geared only to a number of anticipated leads that the company is capable of following up. The follow-up of the letter is the phone call for the appointment and the resulting visit. Obviously, that reduces the number of contacts that can be made to the limit of time available to pursue the opportunity.

Source of Leads

- *Personal Contact*. Friends, relatives, business associates, members of social clubs, former clients, fraternal associates, political and charitable activities, co-workers, and contacts made at seminars and conventions. Personal contact is the best source because of the built-in confidence factor.
- *Advertising* in trade magazines and in the publications of markets of interest. Advertising in local newspapers and on local radio stations is seldom done, but may, in some cases, be a worthwhile source if good ad copy is created.
- *Industrial Realtors*. This group is a good source because of the association of land and structure. The realtors are interested in selling the land, and welcome the appeal of the design/build concept for use of that land.
- *Industrial planning commissions*. Local and state commissions are interested in attracting developers to build in local areas and in the state as a whole. These groups are charged with promoting development and are usually willing to share inside information with design/build companies who can help realize their common interest.
- *Architects who have participated in the process*. As a rule, most architects see design/build as a competitor and would not be a good source. However, those architects who have participated in the process in the past see a mutual cooperative interest in sharing the lead.
- *Chambers of Commerce*. This source depends upon the location, make-up, and level of activity of a particular Chamber of Commerce. Urban groups generally have an aggressive attitude toward business in their area and fund full time personnel for these associations as well as sponsoring marketing efforts. Rural groups may only function on a social club level and may not be a viable source.
- *Banks can be a secondary source* if they are familiar with your work and are asked for names of companies. The confidential nature of the banking business often precludes furnishing direct information to design/build companies.
- *Utility companies* are sometimes contacted early, but as with banks, are a secondary source.
- *Local zoning and appeal boards* which have been asked to clear land of certain legal encumbrances. Useful information can be gathered by following the proceedings of these committee meetings. This is an area that can be specified to a news clipping service.
- *Major support companies that depend on a particular industry*. The growth of a major industry generates offshoot prosperity to subsidiaries and associated companies. The Dun and Bradstreet service, for example, lists such associated companies as part of the information they provide. Trade journals and publications would also have such lists. While the larger firms may be committed to the larger design/build companies, the fallout of business to the support companies makes them a prime source for the smaller design/build companies.

- *The Wall Street Journal*. This publication is a good source of financial news that may generate leads and is also a source of advertising.
- *The U.S. Department of Commerce*, which also publishes trend information and data on different products.
- *Dodge Reports*. This information is primarily geared to bid work which is not normally a good source for design/build work. Nevertheless, reading these reports may reveal a situation wherein the bids received are over the estimate. If that is the case and re-bidding does not remedy the cost problem, that owner may become a candidate for design/build. However, in order to stay within the budget specified, the existing design will most likely be scrapped, and a new, more economical design and drawings created, using the design/build approach.
- *Large Commercial Developers*. Many large tract buildings are now built by a work force that is not part of the commercial developer's own company. At one time, the large developers did do all of their own work, but this is no longer the case. Thus, they may be a good source of leads.
- *Prefabricated metal building franchises*. The franchise will provide the leads, the marketing material, and estimating services, along with design services. Well known product reliability and reputation provide a pre-existing level of confidence. Many of the established design/build companies have started in this way.
- *Reputation*. Reputation is built not only on past achievements, but also on the impression one leaves with potential clients. Communications should always be courteous and cooperative.

Qualifying the Opportunity

Qualifying the opportunity is the process of identifying the actual intent of the prospective client. It is important to understand how the client comes to the realization that a facility should be built or enlarged. The most common motivation is actual or anticipated growth, along with increased profit. Actual company growth is easily observed. Most often, it is simply the expectation of growth that motivates an owner to expand or build a new facility.

A manufacturer situated in an antiquated, inefficient building may come to recognize the need for more efficient space. An improvement in efficiency would reduce the unit cost to produce the product. The product could then become more competitive, he reasons, and thus increase sales. The gross profit increase would provide the equity and mortgage payments needed for the new facility. The amount to be gained must then be stacked up against costs for land, interest payments, legal costs, accounting fees, taxes, insurance, and other "soft" costs. Added to this figure is the remainder of the work – the "hard" construction costs. The question the client asks himself is "Can I have it designed and built into the facility that is necessary to generate the profits in the first place?" This "back door" approach to the design and construction cost presents a good case for using the seldom understood and often deemed "risky" design/build method instead of the more readily accepted and "safe" bid method, to build the facility.

The first test in qualifying the opportunity is to see if the client has a realistic budget for the concept he envisions. This is done by square foot estimating the costs. This is an estimating technique used in sales, but not used for guaranteed pricing.

Figure 3.1 shows an example of the square foot method based on the case study project (see Appendix A for the complete project documentation).

The next qualifying test is whether or not the client will accept the concept of design/build. What he hears from the design/build company may be something like the following: "We (design/build company) will guarantee to build the project based on the data provided to us by you, required in accordance with the information we have furnished." This information should, at the very least, consist of a rudimentary site plan, a building "footprint", some elevation drawings, and a building cross section or two, but with no details. Specifications are in outline form and are accompanied by a conceptual estimate. To anyone more accustomed to detailed drawings and completed specifications, this concept should sound quite foreign. It may seem like buying "a pig in the poke". Of course that is not the case, but if the client cannot understand and accept the design/build concept and approach, then there is little possibility for a sale.

To avoid wasting your time or that of your client, an understanding must be reached at the outset of the relationship regarding these two issues: 1.) the client's realistic budget for the project he envisions, 2.) the client's acceptance of the design/build approach. If both of these points can be resolved, then there is a basis for continuing the relationship and proceeding with the next phases of the design/build process.

Salesmanship

No one truly knows what sells. The abundance of advertising in infinite forms and content is proof of this statement. The rapid cloning of successful ad copy shows a lack of confidence in new, and therefore risky, methods.

What is known is that selling involves name association, acceptance, and an ability to close the deal. Name association may get the salesman in the door. Acceptance is earned through the process of confidence

Case Study
Square Foot Analysis

Office	9,600 S.F. @ $50/S.F. =	480,000
Warehouse/Staging	39,060 S.F. @ $35/S.F. =	1,367,100
Assembly Area	24,590 S.F. @ $40/S.F. =	983,600
		2,830,700
	Design Fee @ 4%	113,200
	Sitework 5-1/2 Acres @ $50,000/Acre =	275,000
		3,218,900
	Say $3,220,000.00	

Figure 3.1

building. The client must have faith in the contact person and believe in his ability to do what he says he can do: design and build the building. Closing the deal is the culmination and consequence of this effort.

Sales techniques are the means by which the above goals are carried out. A market plan and source of leads should be established. The process of direct mail can also be used to introduce the company and its representatives. Next is the follow-up phone call and subsequent appointment with the appropriate person. It pays to be conservative in dress and manner regardless of the client's demeanor. Business persons do not necessarily take to those who "do their own thing". Most important is the company's reputation, which involves building confidence and trust. Demonstrate that trust through the use of references and letters of recommendation that your company has received for past good performance. Communicate your versatility, problem solving ability, training, and experience.

Finding the opportunity, qualifying the opportunity, and using good salesmanship to close the deal – these activities can make or break a design/build company. Becoming proficient in these areas allows a design/build company to prove itself as an economical, practical building construction alternative.

Time Utilization

The time spent on the tasks necessary to maintain a design/build business may depend upon the support available. If the company structure allows for a full-time salesperson, a full-time conceptual estimator, a full-time project manager, and various other support personnel, then the owner of the company is better able to concentrate on running the business. However, the opposite is generally the case with a small company. In small companies, the owner is responsible for generating the leads, making the sales calls, and closing the deals. If, on top of these tasks, the company owner also must do the estimating and the project management, then he is busy indeed. Along with these responsibilities are the everyday decisions of accounting, payroll, and office administration. All of these factors make clear the need to use time most efficiently.

No one person can perform all of the tasks mentioned above. Obtaining the work may be the owner's primary responsibility. The total time available assuming a 10-hour day, 6 days a week comes to 60 hours. Many businessmen may work longer hours, but may not be able to maintain this pace. Sixty hours appears to be a reasonable schedule.

To apportion the time properly, the 60 hours should be dispersed as efficiently as possible. Figure 3.2 demonstrates the need for support personnel and for the rapid qualifying of an opportunity. This figure shows that the weekly hours available to the owner of a design/build company cannot satisfy all of the functions. If the owner's strength lies in sales, then his schedule might be somewhat like the following: sales calls (24.5 hours), proposals (20 hours), negotiating contracts (4 hours), and office administration (15 hours). This adds up to a total of 63.5 hours, more than enough to fill his time. The operation activities of 46.5 hours average per week certainly require support personnel. In time, even the owner's sales effort would have to be cut back in favor of spending more time running the company.

Time Apportionment			
Sales:			
Assume 10 Phone Calls = 1 Sales Call			
		Est. Hrs/Week/Average	Total Hrs
Mailing — Compose/Process		2	
Phone Calls		12 1/2	
Appointments		5	
Travel		5	
		24 1/2	24 1/2
Proposals:			
Assume 1 Per Week			
Site Investigation		2	
Travel		1	
Design Activity		3	
Estimating		8	
Presentation		4	
Closing		2	
		20	20
Operations:			
Assume 12 Contracts/Yr.			
Contract Negotiation:	Review	1	
	Draft	2	
	Revise	1	
		4	4
Design Review/Coord.		9	
Procurement		32	
Project Administration		5 1/2	
		46 1/2	46 1/2
Office Administration			15
		Total Per Week	110 Hrs

Figure 3.2

Chapter Four

DATA GATHERING

Data Gathering is the process of collecting information needed to formulate the conceptual design of the project. It involves researching the client's goals and needs. The data is collected from a number of varied sources.

Owner

Organizing the many forms of gathered data can be difficult. Using preprinted forms may be helpful. Such a form can serve two purposes. First, it allows the information to be recorded in one place in an organized manner. Secondly, the form serves as a checklist, and can be used to ensure that the inquiries are complete. If the owner does not have the time to complete the form, a few meetings with the client and a tour of the current facilities by the design/build representative should answer most of these questions. Such a tour would also allow for further observations, leading to new questions and, if necessary, more details. The client and his current facilities will be the biggest source of data.

Existing Facility

The data gathering process is illustrated in the following example. The client is a switch manufacturer, producing switches that range in size from infinitely small microswitches to huge two hundred pound disconnect switches. This piece of information has the following meaning for the design/build company.

Bench Work

A certain amount of assembly or bench work is required in this type of business. The bench work might require the following: task lighting at the work stations, electric power outlets for testing, soldering irons and small tools, process air outlets, and natural gas outlets.

Extruders

Does the business use extruding machines to form plastic or metal encasement elements? Extruders may require structural support or unusual load consideration of the slab. Extra electrical power and mechanical ventilation as well as space for this equipment would also require design input.

Ovens

The use of baking ovens would require fuel, such as natural gas, along with mechanical ventilation, electrical power, air conditioning, and fire protection. In addition, explosion-proof fixtures may be required.

Paint Booths

Paint or spray booths used in this business involve design considerations similar to those for the ovens.

Structural Supports for Cranes or Crane-way Rails

The size of some of the switches suggests the need for cranes (and appropriate access) in order to move the product from point to point.

Material Handling

This category includes raw stock receiving, handling, and storage as well as finished goods handling, packaging, moving, storage and shipping. These activities would generate design considerations involving loads on the structure and floors, handling equipment, maintenance stations for repair, fire protection and heat, humidity and cooling control.

Red Label Areas

Highly or potentially explosive material must be kept in special confinement called "red label" rooms. These rooms require special treatment for lighting, ventilation, and wall construction. All of the lighting fixtures must be explosion-proof. One wall to the outside should be "weak" to direct the blast in the event of an explosion.

All of the above considerations represent typical design conditions for this type of business and would be considered a potential part of the completed structure.

Gathering this information can be accomplished by observation or through the use of broad questions. In some cases, the client will be very helpful. Many questions have been answered by preliminary studies that were made in considering the need for the project. These studies might take the form of flow charts of the client's manufacturing process.

One of the important goals of design/build is to custom design the building to a client's needs and not to "shoe horn" him into a preconceived concept. A major concern is the number of people who will be working at the new facility. How many people on each shift? How many in production and how many in clerical and support? The data given will affect requirements such as toilet facilities (which is determined by formula in the building code), parking spaces (required by the local zoning by-laws), and space allocation of work areas.

Some of this information is directly useable, such as the people count and developing of parking spaces. Other data requires conversion to quantity through the use of predesign factors. Still other information is used in an indirect way. For example, the client may state that the raw material and the finished product are to be stored on storage racks. The racks are to be supplied and installed under contract between him and the supplier. This piece of information brings up several considerations. First is the matter of structural floor load. Secondly, the client states that the rack storage will be four levels high. For these shelves to be efficient and useful, they should be unobstructed.

Lighting fixtures and sprinkler heads in the aisle ways should be high enough so as not to interfere with the loading of the racks. Furthermore, for the racks to be fully accessible for storage, they must have adequate vertical space. This vertical space is termed the "clear height". In Figure 4.1, the clear height is shown as dimension "C". This illustration represents a cross sectional view of an industrial building having a structure composed of beams and bar joists, metal roof decking, roofing, insulation, built-up roofing and a sloping roof. Vertical dimensions are developed from this design as follows:

4 high racking (4 x 4′-6″)	18′-0″ (clear C)
Joist depth	2′-0″
Metal roof deck	0′-4″
Roof insulation	0′-3″
Roofing	0′-0″
Sloping of steel joists for drainage	0′-8″
Overlap of foundation wall	0′-6″
Total vertical height of the exterior wall	**21′-9″ ±**

The height of the structure can now be calculated, as well as the area of the exterior wall.

Code Review

The building code should be reviewed as it applies to the conceptual design of the structure. Design items needed to conform to code should be catalogued. If the code requirements are a hindrance to the use of the building, a variance or relief from certain stipulations may need to be filed. Such procedures are, of course, taken with the consent of the client. The building code review will not only provide parameters for the design, but will also have cost implications. Types of code ratings from one commonly used code are as follows: Type 1 – Fireproof; Type 2A & 2B – Protected; Type 2C – Unprotected; Type 3A – Mill (Timbers); Type 3B – Masonry, Ordinary Protected; Type 3C – Masonry, Ordinary Unprotected; Type 4 – Wood Frame. Each type specifies certain building characteristics that will affect the cost of the work.

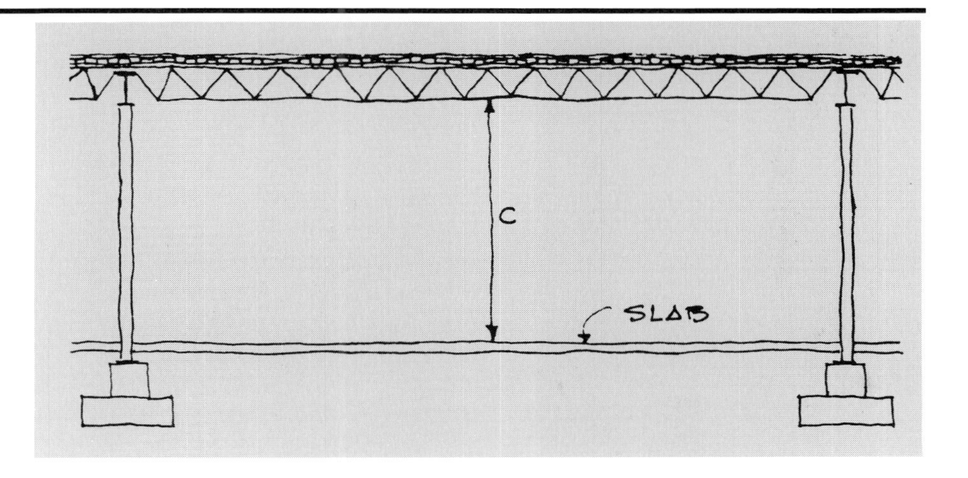

Figure 4.1

Since the design/build method includes design, as the name implies, the obligation to comply with the building code belongs to the design/build company. A code-obliged work requirement that is not estimated does not constitute a contract change.

The code review should include every article in order to determine which standards are applicable to the project. These standards set the limits and restrictions for the concept design. The code review is done either before the project concept is developed in order to establish the design parameters, or after the concept design, but before the presentation, in order to verify that design.

The code review for the case study project appears, for demonstrative purposes, along with the rest of the project documentation, in Appendix A. While an actual code review would be concerned only with the articles that pertain to that particular project, the case study code review includes comments on all of the articles, as further explanation.

The code source used in this book is BOCA (Building Officials and Code Administrators), 1986. Many states utilize BOCA verbatim. Others use BOCA as a standard from which sections are modified to suit that state's particular requirements. Regions may use the same approach, as seen in the Southern Building Code.

Zoning By-Laws

Each local community may have an independent set of codes relating to the design requirements for different zoning or land use designations. These items, like the basic building code, can affect the design and cost of the work. One example is a local code requirement for parking, depending upon the use of the land. The local code might stipulate a formula based on a ratio of the building's area (or number of people using the building) to parking spaces. The local code may also require that parking be provided on the basis of the maximum use of the structure. If it is to be used as a warehouse facility, for example, having only eight men constantly working in the building, the code may nevertheless oblige 160 parking spaces. The highest use of the building could be determined as manufacturing, with a work force that requires that number of parking spaces. This kind of information naturally affects the design and the cost.

Site Visit

A site visit will reveal the apparent characteristics of the land and give a hint as to the underlying soil conditions. A wooded or littered lot presents some visual impairments, but basic characteristics can still be noted, and certain cost implications can be quantified. Obvious signs of ledge outcroppings, for example, lead to the suspicion of ledge below the surface. Standing water or an earthy odor might indicate the presence of clay, loamy soil or peat. This condition leads to a suspicion of poor drainage and poor structural supporting soil.

The design of a structure starts with the understanding of what is to hold it up. This "carrying" or supporting is within the soil capacity. The weight of the structure and its contents constitutes the load. The load is in two categories: one is the dead load and the other is the live load. The dead load is defined by the constant weight or static load such as the structural steel, concrete, all fixed equipment, and items of that nature. The live load is defined by the variable weight, such as amount of snow on the roof, and number of people and equipment that come and go. The sum of the known dead load and the anticipated and allowed for live load represents the total load of the building. That total load is carried or "transmitted" to the soil via the columns to the foundation and then to the footings. The size of the footings is related to both the load characteristics coming down from the building and the soil load capacity of the ground upon which this footing rests. See Figure 4.2.

Some common soil capacities are shown in Figure 4.3 from *Means Assemblies Cost Data*, 1987.

Although visual observation can give a clue to the nature and characteristics of the soil, a more precise knowledge is gained through the application of test pits and/or soil borings. Generally, these investigations are not available in the preliminary stages and even to the point of presenting the proposal. In this case, the soil bearing capacity is estimated and an allowance is developed which is the best reasonable assumption at that time. As better information is obtained through subsurface investigation, the allowance is adjusted.

Subsurface investigation by test pit is the digging of pits to observe, catalogue, and sample the various strata exposed. Analysis of the samples and the visual identification of the strata thus determines the bearing capacity. Test borings are a more elaborate process involving a drilling machine and a method of extracting soil samples at various depths. The samples, the stress necessary to gather them (as measured by blows of a calibrated weight), and the amount recovered all are analyzed; the bearing capacity at the required depth is then determined. The reader is referred to a text on the subject of core drilling for a more detailed explanation.

The site visits also provide data on the available utilities in the area and their approximate location. For example, a manhole in the street should be labeled (embossed into the cast cover) according to its use. Electrical manholes, gas, storm water, and so on should each be identifiable. Investigation of the plans at the utility company and Department of Public Works (DPW) office provides further data on size and capacity.

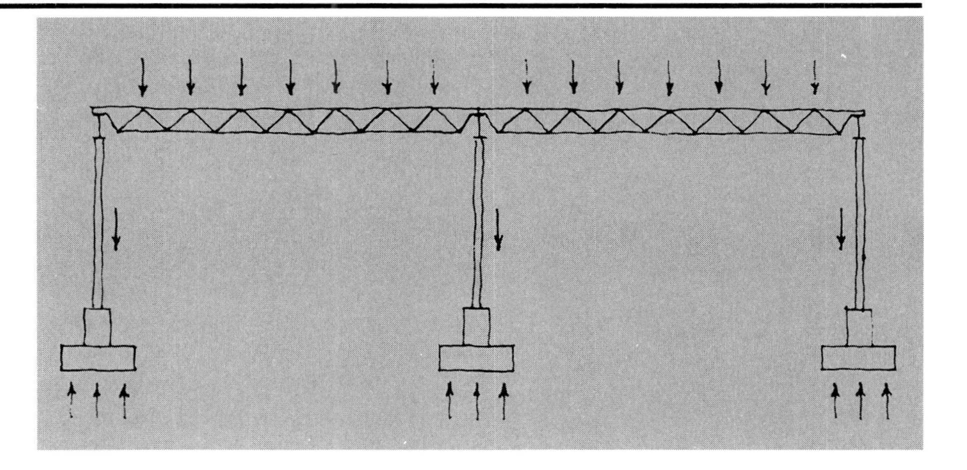

Figure 4.2

FOUNDATIONS | **C1.1-120** | **Spread Footings**

General: A spread footing is used to convert a concentrated load (from one superstructure column, or substructure grade beams) into an allowable area load on supporting soil.

Because of punching action from the column load, a spread footing is usually thicker than strip footings which support wall loads. One or two story commercial or residential buildings should have no less than 1' thick spread footings. Heavier loads require no less than 2' thick. Spread footings may be square, rectangular or octagonal in plan.

Spread footings tend to minimize excavation and foundation materials, as well as labor and equipment. Another advantage is that footings and soil conditions can be readily examined. They are the most widely used type of footing, especially in mild climates and for buildings of four stories or under. This is because they are usually more economical than other types, if suitable soil and site conditions exist.

They are used when suitable supporting soil is located within several feet of the surface or line of subsurface excavation. Suitable soil types include sands and gravels, gravels with a small amount of clay or silt, hardpan, chalk, and rock. Pedestals may be used to bring the column base load down to the top of footing. Alternately, undesirable soil between underside of footing and top of bearing level can be removed and replaced with lean concrete mix or compacted granular material.

Depth of footing should be below topsoil, uncompacted fill, muck, etc. It must be lower than frost penetration (see local code or **Figure 15.3-502** in Section 15) but should be above the water table. It must not be at the ground surface because of potential surface erosion. If the ground slopes, approximately three horizontal feet of edge protection must remain. Differential footing elevations may overlap soil stresses or cause excavation problems if clear spacing between footings is less than the difference in depth.

Other footing types are usually used for the following reasons:

a. Bearing capacity of soil is low.
b. Very large footings are required, at a cost disadvantage.
c. Soil under footing (shallow or deep) is very compressible, with probability of causing excessive or differential settlement.
d. Good bearing soil is deep.
e. Potential for scour action exists.
f. Varying subsoil conditions within building perimeter.

Cost of spread footings for a building is determined by:

1. The soil bearing capacity.
2. Typical bay size.
3. Total load (live plus dead) per S.F. for roof and elevated floor levels.
4. The size and shape of the building.
5. Footing configuration. Does the building utilize outer spread footings or are there continuous perimeter footings only or a combination of spread footings plus continuous footings?

COST DETERMINATION

1. Determine Soil Bearing Capacity by a known value or using Table 1.1-121 as a guide.

Table 1.1-121 Soil Bearing Capacity in Kips per S.F.

Bearing Material	Typical Allowable Bearing Capacity
Hard sound rock	120 KSF
Medium hard rock	80
Hardpan overlaying rock	24
Compact gravel and boulder-gravel; very compact sandy gravel	20
Soft rock	16
Loose gravel; sandy gravel; compact sand; very compact sand-inorganic silt	12
Hard dry consolidated clay	10
Loose coarse to medium sand; medium compact fine sand	8
Compact sand-clay	6
Loose fine sand; medium compact sand-inorganic silts	4
Firm or stiff clay	3
Loose saturated sand-clay; medium soft clay	2

Figure 4.3

The street poles in the area are another source of information. Street poles have a caste system of sorts, and certain services are usually placed in the same relative position. The electric utility's distribution feed (high voltage) is topmost on the pole (note that this feed may not be present on all poles). Three insulated wires on the top cross members indicate that three phase power is available. [The word "phase" in electrical circuitry refers to time relationships of 60 cycle power. The standard of measure is degrees, where 360 degrees is equal to a duration of one cycle (i.e., 1/60 of a second). Three-phase power has three voltages which peak 120 degrees apart. Single phase has only one or two voltages which are 180 degrees apart.]

Nominal distribution voltages could be 208V, 480V, 4160V, 13,800V (13.8KV), or 26,000V (26.0KV) and are almost always three-phase circuits. The next level down is the electric secondary distribution feed (line voltage), which goes to the buildings. Two insulated wires (with a bare neutral wire) on the top generally indicates that only single phase (120/240V) power is available. The lowest levels on poles are for cable T.V. and telephone.

Residential buildings and very small commercial projects can be powered with single phase 120/240V power. As the project size (or load) increases, three-phase power at higher voltages quickly becomes an economic necessity. For office buildings and modest commercial facilities, 480V power is very convenient. Larger commercial and most industrial facilities can best be served by 4160V services. The largest industrial and manufacturing plants may require feeders at 13.8KV or 26.0KV.

Utility Companies

Contact with local electric and gas companies is a necessary part of the planning process. Conversations with the commercial representatives for the area where the site is located provide information on the quantity and quality of service available. Such representatives may also be able to furnish approximate costs, if any, to provide the service to the proposed building. If service is not available in the immediate vicinity, the utility company representatives should be able to advise on the company's position regarding furnishing of service in remote areas. They may also supply approximate "ball park" costs. For example, if the project is to be located on a tract of land sufficiently remote from main streets and other improved properties, then the cost of installing the service to the site may be unusual and costly. Natural gas as fuel for the heating system may have to be abandoned if the estimate to install a new system becomes too costly for the project. Electric heating or the use of propane may then become the design criteria.

Contact with the local water department and Department of Public Works (DPW) is another source of needed information. The water department can furnish data on the availability of potable water, as well as the size of the main pipes, their age, the quantity of flow, and the pressure contained within the system closest to the site. The estimate of costs, if any, to install a continuation of the water mains from more remote termination points can then be developed. The DPW will furnish information on the location of storm and sanitary sewers in the area. This data will have a bearing on cost, depending upon the location of the mains and whether the work will be performed by the DPW. The age and current usage of the system are other relevant facts that affect performance and pressure.

Building Officials

The building codes for many states follow and subscribe to the standard set by the Building Officials and Code Administrators (BOCA). Modifications to this standard are made state by state and region by region as required for the unique prevailing conditions. Another code, the Southern Building Code, for example, developed by the Southern Building Code Congress, is used throughout the south. Provisions within a particular code allow the local official charged with enforcement the latitude to make decisions involving the interpretation of specific design elements. This latitude is needed due to the custom nature of building construction. The procedure is to make a code review with the specific building in mind and then meet with the official and discuss the findings. Issues regarding the type of construction and life safety can be reviewed and agreed upon. Design can then proceed with an assurance that there will not be any major changes. Such changes could be costly in redesign and in construction. Since the design/build company is responsible for code compliance and furnishes an early guaranteed price, a code review and discussion with the building official is imperative. This official can also furnish the rates and requirements for building permits as well as the specifics for receipt of a Certificate of Occupancy (C of O). This last document allows occupancy of the building (see Chapter 15 on permits).

In most communities, unpaid or nominally paid local citizens serve on the planning, zoning, conservation, and other boards and committees that approve local building construction projects. The day to day operations of these boards are generally left to a secretary whose position is full time. This person is usually very knowledgeable in the permit issuance and approval process in that community and can be a fountain of knowledge on such matters. If the property includes or borders wetlands, for example, obtaining permits could be complicated and time-consuming. The design/build company should be familiar with the standard application and approval procedures and time frames. In this way, realistic plans can be made. Applying for, presenting, and securing approvals is an owner's responsibility, but the design/build company needs to know the probability of success in these matters as well as the time frame. These factors influence not only the cost escalation factor, but also the feasibility of the project.

The *Health Department* input is more directly related to the design/build company, rather than the owner. A major Health Department issue is the need for on-site disposal of sewage. Septic systems are needed in a majority of building sites away from urban centers. Regulations pertaining to percolation, design capacity, and accepted layout are all found either with the Health Department or with the Community Engineering Department. *Percolation of the soil* is a process by which the seepage of a measured amount of water is observed and timed. The rate of penetration (percolation) into the soil is then formularized to provide design criteria for the septic or cesspool systems. The size of the solids holding tank is determined by the projected use (i.e., number of people and their theoretical use of the system) per day. The percolation test must meet certain minimum standards. Its actual percolation, and hence capacity, dictates the size of the dispersing field. The dispersing field is a series of porous pipes placed under the ground in a prescribed geometric pattern. The dispersing field requirements become part of the design information for the site. The precise location of the septic system on the site (in relation to other systems, well water, streams – both above and underground – and abutting property) is prescribed and documented in the state or local statutes.

Fire Marshal

In most areas, the local Fire Marshal or Fire Safety Officer will need to be consulted after review of the building code. This discussion should result in an agreement on fire safety requirements for the proposed building. This official may also have certain discretionary powers beyond those specifically listed in the code.

The Fire Marshal would be interested in the fire alarm system, use designations for sprinkler design and deluge density, egress and access for fire personnel and fire apparatus, and other concerns regarding fire safety peculiar to that region or city. Most states and communities within the states have individual requirements.

Local Surveyor

The local surveyor is a major source of information, partially due to his experience in the community. Through the years, the surveyor may have investigated numerous parcels of land pertaining to survey lines, wetlands boundaries, and limits of conservation land. In the course of property line layout and building layout, he has gained a good knowledge of the local terrain. Through observation of excavations, the surveyor is also familiar with the substrata prevalent in the various parts of the community.

A local surveyor may be helpful in additional, less obvious ways. For example, information on the utility services available in a given area may be more specifically known by the local surveyor than by the community official responsible for that service. Furthermore, a local surveying company should be familiar with the officials of the community and should understand the channels through which documents must be processed.

Local Excavator

The local excavator is another good source of information in a community. An excavator could be knowledgeable about the geological characteristics, ground water levels, and conditions that affect the structural stability of the soils. His knowledge of the underlying soil strata would also include indications of unsuitable materials, and he might have suggestions for the disposal of such material including rock or ledge. This information is not only useful design input, but also may affect the siting of the building, use of the excess material (spoils) on site, and costing of the work.

Good information on excavation characteristics not only helps in planning the work, but also contributes to a more accurate estimate. The less information available, the greater the allowance must be for the work. As the excavation data accumulates, the allowance portion may be reduced or even eliminated. Conceptualizing underground work will most always be an allowance, as part of the Guaranteed Price, to be adjusted when better information is collected.

An understanding of the ground water conditions in a given region is also an obvious necessity from a design and costing point of view. The location of underground streams is part of this needed information.

In mining states, the potential of ground collapse into mine shafts is another consideration. The local excavator would be a fine source for this information and probability of occurrence.

Use of Forms

As previously stated, forms can be valuable tools in the process of data gathering. The most important use of forms for data collection or any other aspect of the design/build business is for self checking. Design/build involves the application of conceptualized data. As such, a second party can only check the arithmetic, since it is difficult to check and verify a concept.

Fire Underwriters

Fire Underwriters are insurance companies that write Highly Protected Risk (H.P.R.) property insurance. These companies employ engineering firms that specialize in loss prevention. The engineers are concerned with these two areas of loss prevention: human conservation and property conservation.

The engineering groups work for different types of insurance companies. Mutual insurance companies generally use Factory Mutual (F.M.) as the reviewing engineering group. Stock or stock owner insurance companies use Industrial Risk Insurers (IRI) as the engineering group. Some insurers such as Kemper Insurance Companies and The Chubb Group of Insurance Companies have their own engineering groups.

To obtain insurance for the property, the design/build company, (on behalf of the owner) must submit complete working drawings to the Fire Underwriters for their approval. At the data gathering, conceptual estimating stage, there are no working drawings, but the design/build company can still incorporate the Fire Underwriters' requirements into the plan by consulting published data issued by the appropriate engineering group. In the conceptual estimate, it should be noted that the budget is based on the best judgment of the design/build estimator, and that costs could change based on the Fire Underwriter's official review of final plans.

When the design is reasonably defined, it is submitted to the reviewing engineering group as required by the insuring company. The engineers evaluate the design and make recommendations which have a bearing on the insurance premium. In addition to the building design, the sprinkler layout and calculations are to be submitted for approval.

The published data used in this evaluation process suggests certain treatments for various aspects of the construction. Examples are roof construction, the need for smoke and heat vents, sizes of roof drains as defined by storm intensity and by formula, the need for smoke curtains, and sprinkler design.

Case Study

Following is an excerpt from the case study conceptual estimate (see Appendix A for the complete case study). It shows items which are affected by the requirements of the Fire Underwriters. In this case, smoke and heat vents are carried as part of the roofing work. Perimeter fastening of the insulation, a normal requirement, is also carried. The sprinkler system is designed based on the available knowledge of the Fire Underwriters' and N.F.P.A. (National Fire Prevention Agency) requirements. The outline specification under Interior Fire Protection includes this statement:

> "The above fire protection specificity is our best estimate of the requirements of the Fire Underwriters. The actual final requirements will be determined by negotiation between the owner and his insurance company. Any additions or deletions from the above requirements inclusive of final building materials

acceptance will most likely result in a price adjustment which will be fully documented and substantiated."

The purpose of this statement is to alert the client to a potential change which the design/build company has taken care to minimize.

The data gathered represents the essential elements necessary for both the design of the project and the quantitative conceptual estimate. As a precursor to the actual design documents, data gathering should be extensive, as it allows the design/build company to formulate the data essential to guaranteeing the price at such an early stage.

Chapter Five

PREDESIGN FACTORS

Predesign factors serve as approximate conversion factors. They are used to translate the information developed in the data gathering process into quantifiable items. Predesign factors can be developed for a specific project or type of project. They also can be used in a more general way for a range of projects. These factors may be developed independently, solicited from subcontractors and vendors, or gathered from available literature and manuals. Means publications, *Square Foot Costs* and *Means Assemblies Cost Data*, are good sources. (The 1987 editions are used in the case study, shown in Appendix A.)

The predesign factors are needed so that a conceptual quantity survey may be made of the building. This survey is then used, with appropriate assemblies prices, to calculate the cost of the work. Since the design is in the very early stages, there is no other way to do a takeoff of the elements that will comprise the building.

It is not always necessary to do a complete quantity survey including the MEPS (mechanical, electrical, plumbing and sprinklers) work. In the process of design/build, there are a number of different ways to generate the cost of work. For example, if you team up with mechanical, electrical, plumbing, and sprinkler subcontractors and commit them to the project (i.e. "If I get the job, so do you"), then that part of the conceptual pricing can be assured. A design/build subcontractor not only has to do a conceptual estimate, but must also write the outline specifications and do the design drawings for the appropriate portion of the work. These requirements place a heavy burden on the subcontractors, who must be experienced in the design/build process and not just in pricing. The same approach can be used for other trades required for the project with the same provision that they have experience in the process and are supportive in obtaining as well as satisfactorily completing the project.

There are potential problems in this approach. For example, if there is an abundance of construction work available, the subcontractors can be choosy, and may not find design/build work to be as lucrative as other work requiring less risk. A newly formed design/build company would not have long term subcontractor associations and might find it difficult to convince subcontractors to "spec" (at no cost) their efforts in anticipation of a job. Even if the design/build company is well established, there are still uncertainties in this approach. For example, a subcontractor with more definite and immediate prospects may put off doing an estimate and proposal for a design/build project. As a result, presentation commitments may be missed. In other cases, the design/build company may not want to commit the subcontractors early in the process, preferring to complete the design and then bid out all of the work. The challenges of subcontracting support the conclusion that a successful design/build company should be knowledgeable of all the trades. Predesign factors are needed to generate quantities and unit prices are used to develop the cost.

Predesign Factors as a Sales Tool

Knowledge of predesign factors is of primary importance as an estimating tool, but it can also serve as a sales tool. There are admittedly two schools of thought on the subject of salesmanship. Some advocate the semi-informed approach, while others feel that a show of knowledge gains important credibility. With the semi-informed approach, the salesman knows just enough, but not enough to fully explain. The design/build salesman is not so effective when appearing semi-informed. A salesman or contact person in this business should be familiar with predesign factors, as this knowledge builds confidence with the client. Knowing how to convert area into tons of air conditioning and R values into inches of insulation will help establish a salesman's professionalism. If, for example, the prospect is a plant manager, he will undoubtedly be using quantities in his discussion and may be put off if the design/build representative is not responsive. Other prospects may not have specific knowledge of design features, but may require some pricing and quantities. The predesign factors aid in this effort.

Salesmen for other kinds of products have an advantage that design/build salesmen do not have – a product that can be shown and literature to explain it in detail. The design/build product is a concept and the client's concerns cannot be alleviated by descriptive literature. Lacking these resources, the design/build salesman must go about the process of building confidence. Being knowledgeable is the best way to accomplish this feat.

Developing Factors

Some predesign factors can be illustrated using the example of an industrial building. This type of building is usually built in suburban communities with a large staging area on the site. The foundations for such structures are generally one of the following two types: one is a simple foundation just four to five feet into the ground, and the other is deeper to allow for truck unloading. Figure 5.1 shows the two types.

Costs for these examples can be found in *Means Assemblies Cost Data*; see examples in Figures 5.2a and 5.2b.

The information in Figures 5.2a and 5.2b is predicated on soil capacity in pounds per square foot; this data should be determined and confirmed for each project. This assembly provides information on the volume of concrete and the area of formwork needed, in addition to cost.

To develop information on the structural steel, it may be necessary to engage a structural engineer to design prototypes. Once the design(s) are made, a quantity analysis can be done. That analysis provides a factor to be used in adaptations for future projects. The key is to be sure that the assumptions for this project remain valid for future projects, and that the final design concept is the same regardless of the design engineer.

Assume the following design concept:

- Structural steel columns and spandrel beams
- Open web joists spanning the long side bay
- No floor loads
- Roof load to be 40 lbs./S.F. live load and 15 lbs./S.F. dead load. (Live load is the term for loads that come and go, such as snow. Dead load is the term for constant loads such as the weight of the roof and roof decking and the steel.)
- Total load @ 55 lbs./S.F.
- Soil capacity @ 3000 lbs./S.F.
- Dead weight of the footing is 8% of the column load.
- Steel to be A 36 (36,000 psi allowable stress).
- Clear height of the building is 20 feet.

(Clear height means the inside height of the building to the first or lowest obstruction, such as the bottom of the joists. The overall height of the building usually is about 2′ to 2′-6″ higher.) The resulting design factors can be displayed as shown in Figure 5.3.

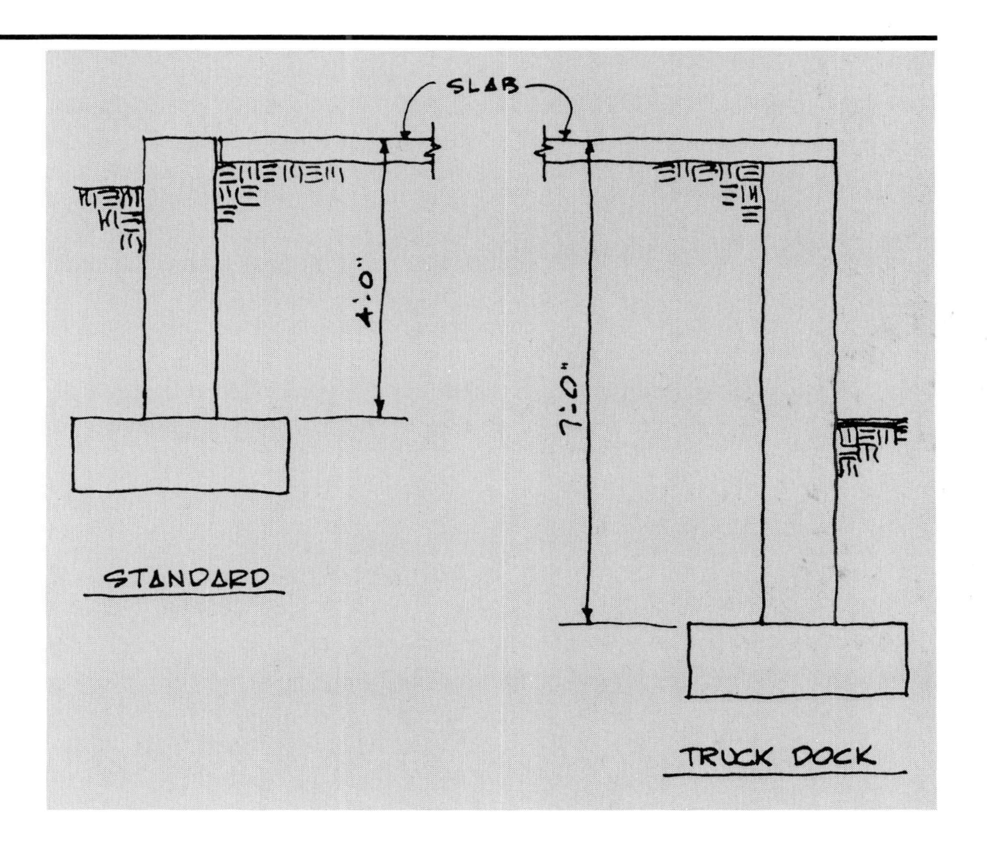

Figure 5.1

FOOTINGS & WALLS | B1.1-140 | Strip Footings

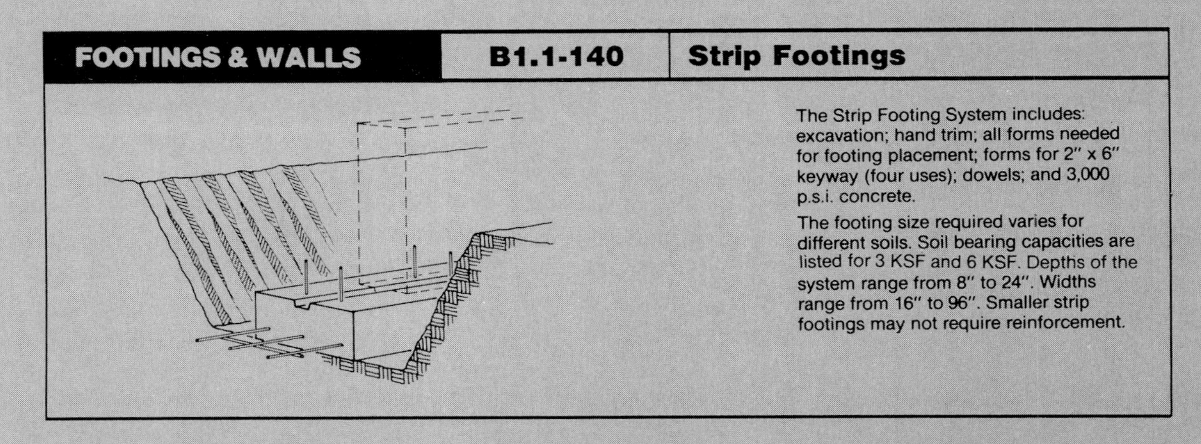

The Strip Footing System includes: excavation; hand trim; all forms needed for footing placement; forms for 2" x 6" keyway (four uses); dowels; and 3,000 p.s.i. concrete.

The footing size required varies for different soils. Soil bearing capacities are listed for 3 KSF and 6 KSF. Depths of the system range from 8" to 24". Widths range from 16" to 96". Smaller strip footings may not require reinforcement.

System Components	QUANTITY	UNIT	COST PER L.F.		
			MAT.	INST.	TOTAL
SYSTEM 01.1-140-2500					
STRIP FOOTING, LOAD 5.1KLF, SOIL CAP. 3 KSF, 24"WIDE X12"DEEP, REINF.					
Trench excavation	.148	C.Y.		.60	.60
Hand trim	2.000	S.F.		.78	.78
Compacted backfill	.074	C.Y.		.12	.12
Formwork, 4 uses	2.000	S.F.	.60	3.90	4.50
Keyway form, 4 uses	1.000	L.F.	.10	.48	.58
Reinforcing, fy = 60000 psi	3.000	Lb.	.84	.75	1.59
Dowels	2.000	Ea.	1.98	8.82	10.80
Concrete, f'c = 3000 psi	.074	C.Y.	4.14		4.14
Place concrete, direct chute	.074	C.Y.		.77	.77
Screed finish	2.000	S.F.		.70	.70
TOTAL			7.66	16.92	24.58

1.1-140	Strip Footings	COST PER L.F.		
		MAT.	INST.	TOTAL
2100	Strip footing, load 2.6KLF,soil capacity 3KSF, 16"wide x 8"deep plain	3.62	9.55	13.17
2300	Load 3.9 KLF soil capacity, 3 KSF, 24"wide x 8"deep, plain	4.57	10.25	14.82
2500	Load 5.1KLF, soil capacity 3 KSF, 24"wide x 12"deep, reinf.	7.65	16.90	24.55
2700	Load 11.1KLF, soil capacity 6 KSF, 24"wide x 12"deep, reinf.	7.65	16.90	24.55
2900	Load 6.8 KLF, soil capacity 3 KSF, 32"wide x 12"deep, reinf.	9.40	17.90	27.30
3100	Load 14.8 KLF, soil capacity 6 KSF, 32"wide x 12"deep, reinf.	9.40	17.90	27.30
3300	Load 9.3 KLF, soil capacity 3 KSF, 40"wide x 12"deep, reinf.	11.05	18.90	29.95
3500	Load 18.4 KLF, soil capacity 6 KSF, 40"wide x 12"deep, reinf.	11.15	19	30.15
3700	Load 10.1KLF, soil capacity 3 KSF, 48"wide x 12"deep, reinf.	12.35	19.95	32.30
3900	Load 22.1KLF, soil capacity 6 KSF, 48"wide x 12"deep, reinf.	13.20	21	34.20
4100	Load 11.8KLF, soil capacity 3 KSF, 56"wide x 12"deep, reinf.	14.50	21	35.50
4300	Load 25.8KLF, soil capacity 6 KSF, 56"wide x 12"deep, reinf.	15.70	22	37.70
4500	Load 10KLF, soil capacity 3 KSF, 48"wide x 16"deep, reinf.	15.30	22	37.30
4700	Load 22KLF, soil capacity 6 KSF, 48"wide, 16"deep, reinf.	15.65	22	37.65
4900	Load 11.6KLF, soil capacity 3 KSF, 56"wide x 16"deep, reinf.	17.50	29	46.50
5100	Load 25.6KLF, soil capacity 6 KSF, 56"wide x 16"deep, reinf.	18.45	30	48.45
5300	Load 13.3KLF, soil capacity 3 KSF, 64"wide x 16"deep, reinf.	20	25	45
5500	Load 29.3KLF, soil capacity 6 KSF, 64"wide x 16"deep, reinf.	22	26	48
5700	Load 15KLF, soil capacity 3 KSF, 72"wide x 20"deep, reinf.	27	29	56
5900	Load 33KLF, soil capacity 6 KSF, 72"wide x 20"deep, reinf.	28	30	58
6100	Load 18.3KLF, soil capacity 3 KSF, 88"wide x 24"deep, reinf.	38	35	73
6300	Load 40.3KLF, soil capacity 6 KSF, 88"wide x 24"deep, reinf.	41	38	79
6500	Load 20KLF, soil capacity 3 KSF, 96"wide x 24"deep, reinf.	41	36	77
6700	Load 44 KLF, soil capacity 6 KSF, 96" wide x 24" deep, reinf.	43	39	82

267

Figure 5.2a

FOOTINGS & WALLS	B1.1-210	Walls, Cast In Place

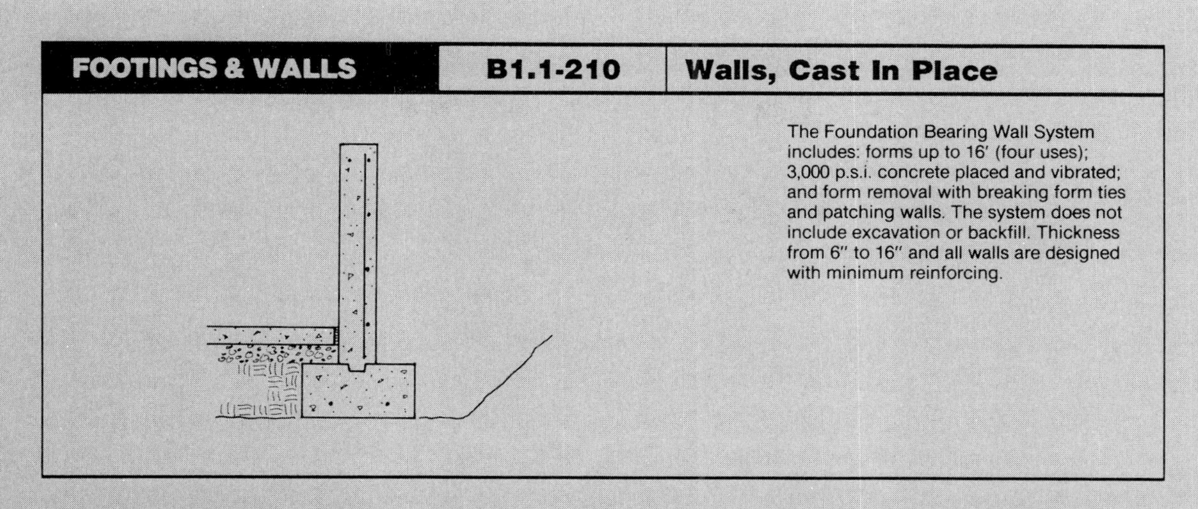

The Foundation Bearing Wall System includes: forms up to 16′ (four uses); 3,000 p.s.i. concrete placed and vibrated; and form removal with breaking form ties and patching walls. The system does not include excavation or backfill. Thickness from 6″ to 16″ and all walls are designed with minimum reinforcing.

System Components	QUANTITY	UNIT	COST PER L.F. MAT.	INST.	TOTAL
SYSTEM 01.1-210-1500					
FOUNDATION WALL, CAST IN PLACE, DIRECT CHUTE, 4′ HIGH, 6″ THICK					
Formwork	8.000	SFCA	2.96	17.60	20.56
Reinforcing	3.300	Lb.	.94	.65	1.59
Concrete, 3,000 psi	.074	C.Y.	4.14		4.14
Place concrete, direct chute	.074	C.Y.		1.03	1.03
Finish walls, break ties and patch voids, one side	4.000	S.F.	.04	1.68	1.72
TOTAL			8.08	20.96	29.04

1.1-210 Walls, Cast In Place

	WALL HEIGHT (FT.)	PLACING METHOD	CONCRETE (C.Y./L.F.)	REINFORCING (LBS./L.F.)	WALL THICKNESS (IN.)	COST PER L.F. MAT.	INST.	TOTAL
1500	4′	direct chute	.074	3.3	6	8.10	21	29.10
1520			.099	4.8	8	9.90	22	31.90
1540			.123	6.0	10	11.60	22	33.60
1560			.148	7.2	12	13.35	23	36.35
1580			.173	8.1	14	15	23	38
1600			.197	9.44	16	16.75	23	39.75
1700	4′	pumped	.074	3.3	6	8.10	22	30.10
1720			.099	4.8	8	9.90	23	32.90
1740			.123	6.0	10	11.60	23	34.60
1760			.148	7.2	12	13.35	24	37.35
1780			.173	8.1	14	15	25	40
1800			.197	9.44	16	16.75	25	41.75
3000	6′	direct chute	.111	4.95	6	12.15	31	43.15
3020			.149	7.20	8	14.90	32	46.90
3040			.184	9.00	10	17.35	33	50.35
3060			.222	10.8	12	20	34	54
3080			.260	12.15	14	23	34	57
3100			.300	14.39	16	25	35	60
3200	6′	pumped	.111	4.95	6	12.15	33	45.15
3220			.149	7.20	8	14.90	34	48.90
3240			.184	9.00	10	17.35	35	52.35
3260			.222	10.8	12	20	36	56
3280			.260	12.15	14	23	37	60
3300			.300	14.39	16	25	38	63

268

Figure 5.2b

Given a building type with the characteristics stated, it is a simple matter to locate the proper factor and multiply it by the area of the building. The result is a number of pounds; this figure is factored by dividing by 2000 lbs./ton. The same can be done for the open web joists.

The case study building has 30′ x 40′ bays. Using the chart (Figure 5.3), the roof steel weight may be found as follows:

Steel 63,650 S.F. x 2.36 ÷ 2000 = 80 Tons
Joists 63,650 S.F. x 2.88 ÷ 2000 = 92 Tons

The roof structural system when designed will look like Figure 5.4.

To carry the factors further, the same engineer could design the footings of the building. Calculations for the footings are shown in Figure 5.5. They would be displayed in a fashion similar to the formation used in Figure 5.3.

"Spread" footings are used under the individual columns. The figure allows quick quantification of footings for similar buildings.

As the type of building changes, or similar building concepts begin to differ more radically, then new factors have to be developed. If there is enough time prior to a presentation, a custom concept design could be developed and a conceptual takeoff priced. This is not always possible, nor is the investment in professional design services always warranted. A number of predesign factors need to be developed to anticipate different conditions. Adjustments are then made as may be needed. The list of factors can be virtually endless.

Structural Predesign Factors

Bay Size (L.F. x L.F.)	lbs./S.F. Steel	lbs./S.F. Joists	lbs./S.F. Total
S x J			
25 x 35	2.00	2.33	4.33
30 x 40	2.36	2.88	5.24
40 x 40	2.40(1)	2.88	5.28
50 x 25	3.68(2)	1.62	5.30
50 x 40	2.76(3)	2.88	5.64

Notes: S = Structural Steel Span J = Bar Joists Span
(1) = 55% V 50 Steel Included (50,000 Allowable Stress)
(2) = 64% V 50 Steel Included
(3) = 61% V 50 Steel Included
Structural steel is measured as net tonnage.... add 5% for miscellaneous items.
Open web joists are measured as net tonnage.... add 10% for bridging and miscellaneous items.

Figure 5.3

Predesign factors may be obtained from subcontractors' information or based on standard designs as described above. They may also be developed from information found in manufacturers' literature (such as roofing, drywall, and insulation). Other factors involve the need for special features requiring conformance to an agency, authority, or building code. The sprinkler systems, for example, must adhere to NFPA (National Fire Protection Agency) standards. These standards require a certain sprinkler coverage over a specified area for a given storage use or for a particular class of construction. The NFPA manuals provide the information needed to determine the number of sprinkler heads for a given space or a given building.

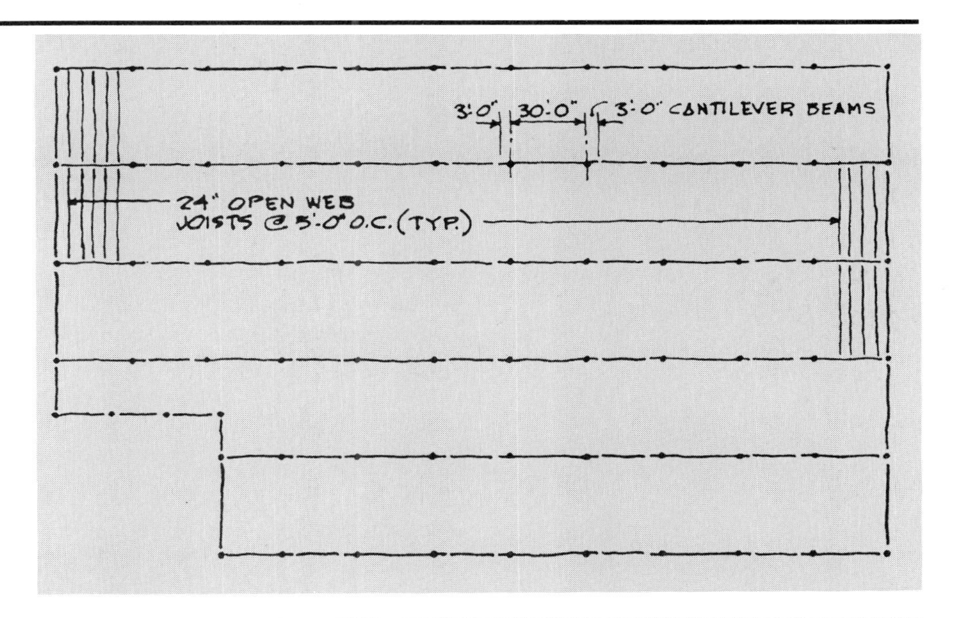

Figure 5.4

Foundation Predesign Factors

Bay	Col. Load (in pounds)	Calc. Area (S.F.)	Actual Footing Size (LxWxD)	Area (S.F.)	B.Y.	Barb (Size)
25 x 35	52,000	17.4	4x4x1	20.3	0.79	5-#6ew
30 x 40	72,000	24.0	5x5x1	25.0	1.06	6-#6ew
40 x 40	95,000	31.7	6x6x1	36.0	2.00	6-#7ew
50 x 25	75,000	25.0	5x5x1	25.0	1.06	6-#6ew
50 x 40	119,000	39.7	6x6x1	42.3	2.35	7-#7ew

Figure 5.5

Cautions

Predesign factors, when properly applied, are an asset to the design/build method of construction. They furnish the quantities necessary to perform the conceptual estimate. However, such a system can also be a trap for the unwary and unthinking. The estimate developed is conceptual, and should be considered in the order of the design. The factors should be applied with reason, and tailored to the specific project. They should not be used in "cookbook" fashion and the quantities applied verbatim. Predesign factors must be tested and reasoned out. For example, the formula for calculating the number of roof drains provided by FM* (Factory Mutual Insurance) is as follows:

$$N = \frac{A \times i}{100\ Q}$$

WHERE:

N = Total number of drains needed
A = Total roof area (S.F.)
i = Rainfall intensity (in./hr.) from rainfall intensity over 50-year period, 60 minute duration (Boston: 2.4)
Q = Drain capacity
4″ = 180 gal./min.; 6″ = 540 gal./min.
8″ = 1170 gal./min.

Assume a building (in Boston) dimensioned as 270′ x 400′ = 108,000 S.F. How many 4″ drains are needed? Using the formula:

$$N = \frac{108{,}000 \times 2.4}{100 \times 180} = 14$$

*Factory Mutual Data Sheet 1-54.

Taking this quantity as derived verbatim, or "cookbook" style, the estimate would be one drain short. Roof drains are installed to service the entire roof and therefore are modular. The number of roof drains actually needed as determined by the final design would be 15 each (see Figure 5.6). No matter what the formula or factor or other standard may determine, reason and design dictates a more "precise" answer. This type of mistake happens frequently and can be avoided through a careful analytical approach and application of common sense.

Using the case study roof area of 63,650 S.F. and the Factory Mutual formula, we find the following:

$$N = \frac{A\ i}{100\ Q} = \frac{63{,}650\ (2.4)}{100\ (180)} = 8$$

Conclusion: Use 8 4″-diameter roof drains.

From the configuration of the roof (see Figure 5.7), eight roof drains (two rows of three and one row of two) are adequate to drain this roof. See Figure 5.7 for the roof plan.

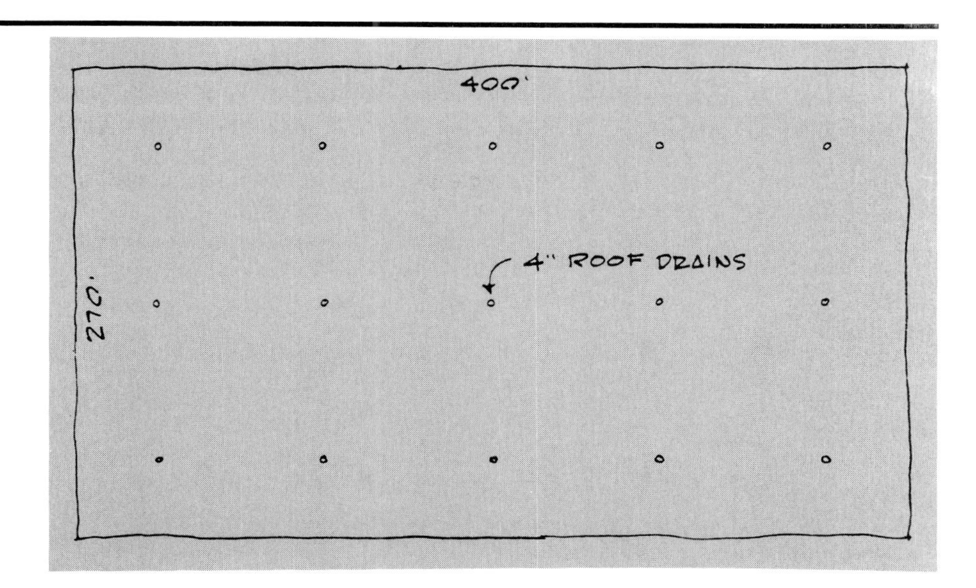

Figure 5.6

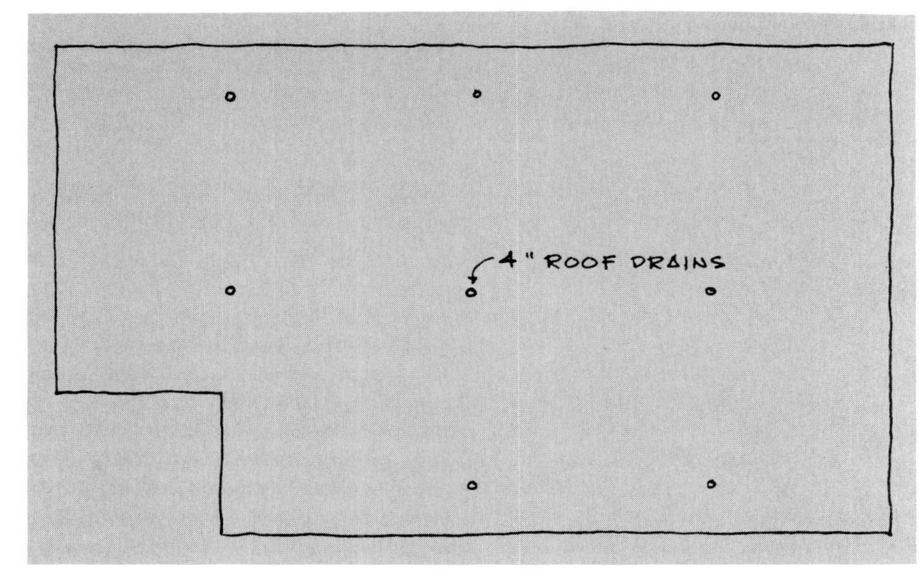

Figure 5.7

Means Annual Construction Cost Data

In addition to sources named above, R.S. Means Company, Inc. publishes annual construction cost reference books to help in the conceptual estimating of a project. Two books in particular are used for the case study; these are the 1987 editions of *Means Square Foot Costs* and *Means Assemblies Cost Data*.

For example, based on the assembly area of 24,590 S.F., we can find the approximate number of fluorescent fixtures required. Tables and charts in *Means Assemblies Cost Data* book (see Figure 5.8) indicate 100 foot candles as appropriate for medium bench and machine work.

Also from *Means Assemblies Cost Data* is a table (Figure 5.9) showing 100 foot candles as roughly equivalent to 3.5 watts per S.F. (using Line A).

The number of two-tube (40 watt/tube), 4 foot long fixtures is:

$$\text{No. Fixtures} = \frac{\text{Area x Watts/S.F.}}{\text{Watts/Fixture}}$$

$$= \frac{\text{24,590 S.F. x 3.5 Watts/L.F.}}{\text{80 Watts/Fixure}}$$

$$= \text{1076 Fixtures}$$

This number of fixtures can be used if the area is open, without obstruction, and the fixtures are to be spaced evenly throughout the area.

Examples

To further illustrate the determination of predesign factors, we will use four examples from the case study, as follows:

1. Determine the number of fluorescent fixtures required for the case study building.

   The lighting for the warehouse area may be calculated as follows:

   Figure 5.8 indicates 20 foot candles for material loading. Figure 5.9, line A, shows that 20 foot candles are roughly equal to .8 watts/S.F. using fluorescent lighting.

   The number of one-tube (80 watts/tube) 8-foot long fixtures is:

   $$\text{No. Fixtures} = \frac{\text{Area x Watts/S.F.}}{\text{Watts/Fixture}}$$

   $$= \frac{\text{39,060 x .8}}{80}$$

   $$= \text{391 Fixtures}$$

2. Find the number of roof ventilation fans for the assembly and warehouse portion of the building (assuming two air changes per hour are desired).

   A typical roof fan would be selected; say 26″ square, propeller type, with an 8750 CFM (cubic feet per minute) rating. See Figure 5.10 (from *Means Mechanical Cost Data*, 1987).

   If the clear height is to be 23′-2″, the considered volume of the space is 23′-2″ x 63,650 S.F. = 1,474,770 C.F. The number of fans needed is:

   $$\text{No. Fans} = \frac{\text{Volume x Changes/Hour}}{\text{Fan Capacity x 60 Min./Hr.}}$$

   $$= \frac{\text{1,474,770 C.F. x 2/Hr.}}{\text{8750 C.F./Min./Fan x 60 Min./Hr.}}$$

   $$= \text{5.6 (use 5) Fans}$$

ELECTRICAL	B9.2-200	General

General: The cost of the lighting portion of the electrical costs is dependent upon:

1. The footcandle requirement of the proposed building.
2. The type of fixtures required.
3. The ceiling heights of the building.
4. Reflectance value of ceilings, walls and floors.
5. Fixture efficiencies and spacing vs. mounting ratios.

Footcandle Requirements: See Table B9.2-204 for Footcandle and Watts per S.F. determination.

TABLE 9.2-201 I.E.S. Recommended Illumination Levels In Footcandles

Commercial Buildings			Industrial Buildings		
Type	Description	Foot-Candles	Type	Description	Foot-Candles
Bank	Lobby	50	Assembly Areas	Rough bench & machine work	50
	Customer Areas	70		Medium bench & machine work	100
	Teller Stations	150		Fine bench & machine work	500
	Accounting Areas	150	Inspection Areas	Ordinary	50
Offices	Routine Work	100		Difficult	100
	Accounting	150		Highly Difficult	200
	Drafting	200	Material Handling	Loading	20
	Corridors, Halls, Washrooms	30		Stock Picking	30
Schools	Reading or Writing	70		Packing, Wrapping	50
	Drafting, Labs, Shops	100	Stairways	Service Areas	20
	Libraries	70	Washrooms	Service Areas	20
	Auditoriums, Assembly	15	Storage Areas	Inactive	5
	Auditoriums, Exhibition	30		Active, Rough, Bulky	10
Stores	Circulation Areas	30		Active, Medium	20
	Stock Rooms	30		Active, Fine	50
	Merchandise Areas, Service	100	Garages	Active Traffic Areas	20
	Self-Service Areas	200		Service & Repair	100

9

Figure 5.8

Table 9.2-204 Procedure for Calculating Footcandles and Watts Per Square Foot

1. Initial footcandles = No. of fixtures x lamps per fixture x lumens per lamp x coefficient of utilization ÷ square feet
2. Maintained footcandles = initial footcandles x maintenance factor
3. Watts per square foot = No. of fixtures x lamps x (lamp watts + ballast watts) ÷ square feet

Example — To find footcandles and watts per S.F. for an office 20′ x 20′ with 11 fluorescent fixtures each having 4-40 watt C.W. lamps.
Based on good reflectance and clean conditions:
Lumens per lamp = 40 watt cool white at 3150 lumens per lamp
Coefficient of utilization = .42 (varies from .62 for light colored areas to .27 for dark)
Maintenance factor = .75 (varies from .80 for clean areas with good maintenance to .50 for poor)
Ballast loss = 8 watts per lamp.

1. Initial footcandles:

$$\frac{11 \times 4 \times 3150 \times .42}{400} = \frac{58212}{400} = 145 \text{ footcandles}$$

2. Maintained footcandles:
145 x .75 = 109 footcandles

3. Watts per S.F.

$$\frac{11 \times 4 \times (40 + 8)}{400} = \frac{2112}{400} = 5.3 \text{ watts per S.F.}$$

Table 9.2-205 Approximate Watts Per Square Foot for Popular Fixture Types

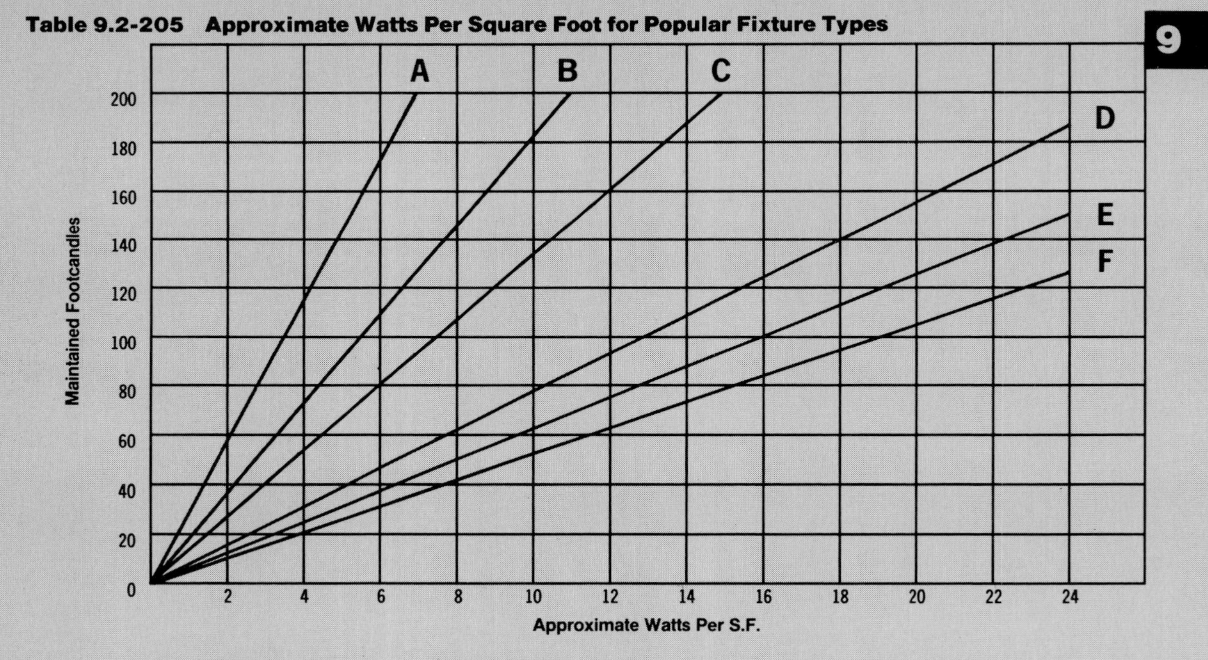

Due to the many variables involved, use for preliminary estimating only:
A. Fluorescent — industrial System B9.2-212
B. Fluorescent — lens unit System B9.2-213 Fixture types B & C
C. Fluorescent — louvered unit
D. Incandescent — open reflector System B9.2-222
E. Incandescent — lens unit System B9.2-223
F. Incandescent — down light System B9.2-222

Figure 5.9

8

Table 8.4-004 Recommended Ventilation Air Changes
Range of Time in Minutes Per Change for Various Types of Facilities

Facility	Minutes	Facility	Minutes	Facility	Minutes
Assembly Halls	2-10	Dance Halls	2-10	Laundries	1-3
Auditoriums	2-10	Dining Rooms	3-10	Markets	2-10
Bakeries	2-3	Dry Cleaners	1-5	Offices	2-10
Banks	3-10	Factories	2-5	Pool Rooms	2-5
Bars	2-5	Garages	2-10	Recreation Rooms	2-10
Beauty Parlors	2-5	Generator Rooms	2-5	Sales Rooms	2-10
Boiler Rooms	1-5	Gymnasiums	2-10	Theaters	2-8
Bowling Alleys	2-10	Kitchens-Hospitals	2-5	Toilets	2-5
Churches	5-10	Kitchens-Restaurant	1-3	Transformer Rooms	1-5

CFM air required for changes = Volume of room in cubic feet ÷ Minutes per change.

Table 8.4-005 Ductwork (Duct Weight in Pounds per L.F. Straight Runs)

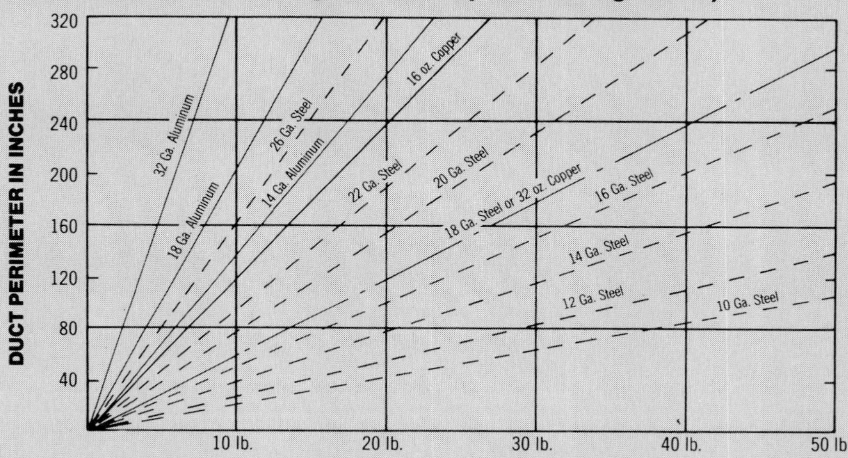

Add to the above for fittings: 90° elbow is 3 L.F.; 45° elbow is 2.5 L.F.; offset is 4 L.F.; transition offset is 6 L.F.; square-to-round transition is 4 L.F.; 90° reducing elbow is 5 L.F. For bracing and waste, add 20% to aluminum and copper, 15% to steel.

Table 8.4-006 Diffuser Evaluation

CFM = V × An × K where V = Outlet velocity in feet per minute. An = Neck area in square feet and K = Diffuser delivery factor. An undersized diffuser for a desired CFM will produce a high velocity and noise level. When air moves past people at a velocity in excess of 25 FPM, an annoying draft is felt. An oversized diffuser will result in low velocity with poor mixing. Consideration must be given to avoid vertical stratification or horizontal areas of stagnation.

400

Figure 5.10

The down rounding is due to the warehouse space occupied by stored material and partially devoid of the total volume of air.

3. Determine the number of unit heaters needed to maintain an inside temperature of 70 degrees F. in the warehouse and assembly areas.

   First, an ambient design temperature is selected for the location. Typically, 0 degrees F. is standard for many parts of the U.S.

   Next, a heat loss factor must be determined. The chart (from *Means Assemblies Cost Data*, 1987) in Figure 5.11 gives typical loss factors for a variety of buildings and uses. This building type is rated at 6.9 BTUH of heat loss per C.F.

   BTU/Hr. = 1,474,770 C.F. x 6.9 BTU/Hr./C.F.

   = 10,175,900 BTU/Hr. or

   10,200 MBH

4. Choose a type of unit space heater. In this case, large, gas-fired, roof-mounted units with a 1200 MBH rating each are selected.

$$\text{No. of Heaters} = \frac{\text{Total Heat Loss/Hr.}}{\text{BTU/Hr. Per Heater}}$$

$$= \frac{\text{10,200 MBH}}{\text{1,200 MBH/Heater}}$$

$$= \text{8.5 (use 9) Heaters}$$

In the search for quantities for a conceptual estimate, predesign factors play a major role. Without them, conceptual estimating is square foot costing guesswork. With them, as shown in the examples, conceptual estimating deals with the precise quantities, the value of which can be accurately determined.

Table 8.3-011 Factor for Determining Heat Loss for Various Types of Buildings

General: While the most accurate estimates of heating requirements would naturally be based on detailed information about the building being considered, it is possible to arrive at a reasonable approximation using the following procedure:

1. Calculate the cubic volume of the room or building.
2. Select the appropriate factor from Table 8.3-011. Note that the factors apply only to inside temperatures listed in the first column and to 0° F outside temperature.
3. If the building has bad north and west exposures, multiply the heat loss factor by 1.1
4. If the outside design temperature is other than 0° F, multiply the factor from Table 8.3-011 by the factor from Table 8.3-012
5. Multiply the cubic volume by the factor selected from Table 8.3-011. This will give the estimated BTUH heat loss which must be made up to maintain inside temperature.

Building Type	Conditions	Qualifications	Loss Factor*
Factories & Industrial Plants General Office Areas 70°F	One Story	Skylight in Roof No Skylight in Roof	6.2 5.7
	Multiple Story	Two Story Three Story Four Story Five Story Six Story	4.6 4.3 4.1 3.9 3.6
	All Walls Exposed	Flat Roof Heated Space Above	6.9 5.2
	One Long Warm Common Wall	Flat Roof Heated Space Above	6.3 4.7
	Warm Common Walls on Both Long Sides	Flat Roof Heated Space Above	5.8 4.1
Warehouses 60°F	All Walls Exposed	Skylights in Roof No Skylights in Roof Heated Space Above	5.5 5.1 4.0
	One Long Warm Common Wall	Skylight in Roof No Skylight in Roof Heated Space Above	5.0 4.9 3.4
	Warm Commom Walls on Both Long Sides	Skylight in Roof No Skylight in Roof Heated Space Above	4.7 4.4 3.0

***Note:** This table tends to be conservative particularly for new buildings designed for minimum energy consumption.

Table 8.3-012 Outside Design Temperature Correction Factor (for Degrees Fahrenheit)

Outside Design Temperature	50	40	30	20	10	0	-10	-20	-30
Correction Factor	0.29	0.43	0.57	0.72	0.86	1.00	1.14	1.28	1.43

Figures 8.3-013 and 8.3-014 provide a way to calculate heat transmission of various construction materials from their U values and the TD (Temperature Difference).

1. From the Exterior Enclosure Division or elsewhere, determine U values for the construction desired.
2. Determine the coldest design temperature. The difference between this temperature and the desired interior temperature is the TD (temperature difference).
3. Enter Figure 8.3-013 or 8.3-014 at correct U value. Cross horizontally to the intersection with appropriate TD. Read transmission per square foot from bottom of figure.
4. Multiply this value of BTU per hour transmission per square foot of area by the total surface area of that type of construction.

8

Figure 5.11

Chapter Six

UNIT COSTS

A unit cost is the detailed price or value of work to be performed. The labor portion of a unit cost is based on productivity, which represents the time it takes for a task to be performed. Establishing productivity standards provides a method for determining the efficiency of labor. In construction, the greater the efficiency, assuming all other things being equal, the more competitive a price can be.

Productivity starts with the observance of a given task being performed. The length of time required to perform the task is then multiplied by the cost of the performer (worker or equipment). This figure is then divided by the quantity of work performed. The result is a cost per unit quantity, which can be defined as follows:

$$\frac{\text{Time x Cost}}{\text{Quantity of Work}} = \text{Cost/Unit Quantity}$$

Further observations may reveal other factors such as:

- the effects of repetitiveness
- time saving methods
- introduction of time saving aids such as certain tools and machinery.

Labor is the most variable part of the unit costs because it is influenced by so many factors. Since labor requires human response, productivity is apt to vary. This variation may be due to time of day, health of the worker, difficulty of the task, and the individual's experience and abilities, as well as other factors. Applying unit costs for labor requires judgment based on experience.

Material unit costs are easier to estimate because they represent a finite cost for a given item at a certain point in time. Allowing a certain amount for waste, the material unit cost should only vary due to the cost of the individual element or a change in its use.

Equipment unit cost relates to productivity in the same manner as do labor unit costs. Although the equipment is not subject to variation, except within its own mechanical efficiency, it is controlled by an operator who is subject to variable performance.

To determine a unit cost, examine the following excavation problem. Find the unit cost to excavate a trench having the following cross section. See Figure 6.1.

For the purpose of this example, we can assume that a worker's time working with the excavation machine is the same as the machine time, even though the labor only involves hand grading the last six inches at the bottom of the trench.

The cross section illustration of the trench shows the following dimensions:

$$\frac{9' + 4'}{2} \times 5' = 32.5 \text{ C.F./L.F. of trench}$$

Assuming 1000 L.F. of trench:

$$32.5 \frac{\text{C.F.}}{\text{L.F.}} \times 1000 \text{ L.F.} = \frac{32{,}500 \text{ C.F.}}{27 \text{ C.F./C.Y.}} = 1{,}204 \text{ C.Y.}$$

If the cost of a backhoe with operator (based on one week rental) is $722 per day and the backhoe is observed to excavate, on average, 600 C.Y. of this type of trench in a day, then the unit cost is:

$$\frac{1204 \text{ C.Y.}}{600 \text{ C.Y./day}} = 2 \text{ days}$$

$$\frac{\text{Time x Cost}}{\text{Quantity of Work}} = \text{Cost/Unit Quantity}$$

$$\frac{2 \text{ days x \$722.00/day}}{1204 \text{ C.Y.}} = \$1.20\text{/C.Y.}$$

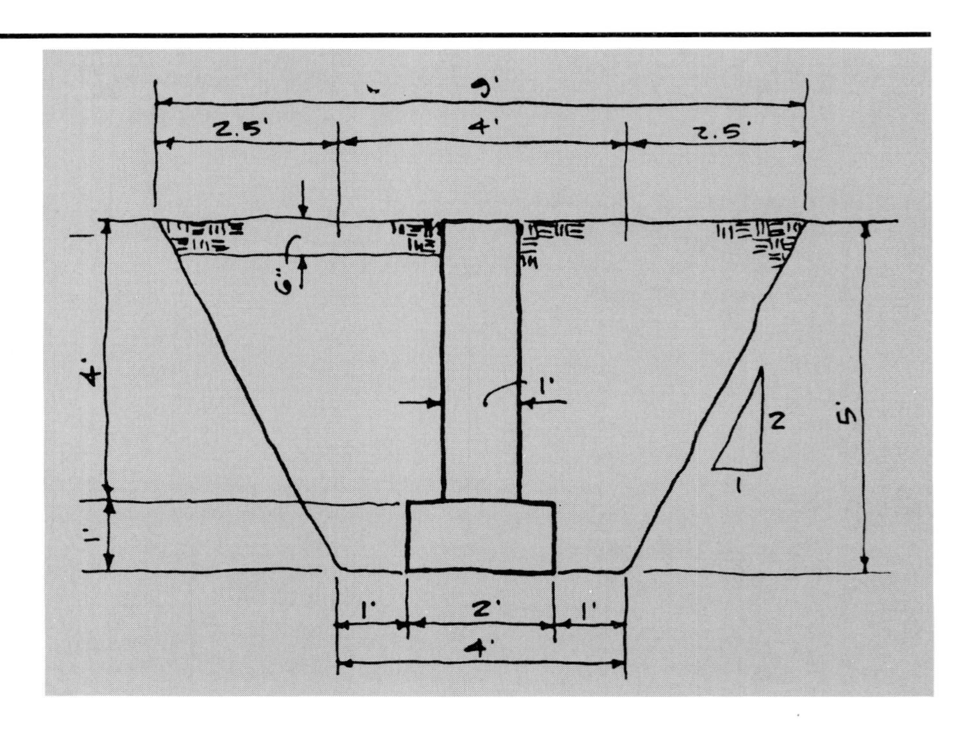

Figure 6.1

To similarly calculate the labor unit cost for the same work, both the cost of labor and the productivity rate are needed. In this case, the productivity for the labor is the same as that for the equipment.

Certain specific items might also be associated with this work, such as: shoring the sides of the trench, pumping out ground water seepage from the trench, placing stone or gravel base material, compacting the bottom of the trench, and backfill. Unit costs for these items can be calculated in a similar manner.

Unit costs are used in the detailed estimation of work. They are applied to quantities of work items determined from the quantity survey or takeoff to arrive at the value for the work.

Quantity Takeoff

The quantity takeoff for unit cost estimates is a survey, or inventory, of work items. This detailed inventory is usually based on a bid-quality set of drawings and requires meticulous scrutiny of each sheet to locate all of the items. Each construction item is counted individually. Quantities of similar items are summarized and appropriate unit costs are applied to arrive at a value for the work. This time consuming unit cost method is usually performed for competitive bidding and may not be done unless there is a reasonable chance of receiving the contract.

For estimates at the conceptual stage, there is not enough time or detail of design to perform unit cost estimates. Design/build construction frequently requires estimated costs performed quickly during the concept or design development stage. The method used by the design/build companies is known as *conceptual estimating* and uses a format called *composite costs*.

Conceptual Composite Costs

The conceptual estimate looks like the quantitative bid estimate except that being conceptual, it has no detail to quantify, and utilizes an assemblies cost. Predesign factors can be used to generate many of the quantities of a conceptual estimate in the same manner as for a quantity survey. The big difference is that the conceptual survey is composed of collective, or assembled quantities. Certain work operations are combined in a convenient system or assembly. To estimate, the system cost must also be developed in a compatible manner. For example, if the conceptual estimator sees the need for a hollow metal door, he would quantify this door as one each furnished and installed, including typical associated items (e.g., frame and hardware). To enable the conceptual estimator to calculate the cost of such work, he would need a complete assembly cost. This cost is developed by totalling all of the separate tasks, as follows:

Furnish hollow metal frame	$ 67.00
Furnish hollow metal door	120.00
Furnish hardware	200.00
F & I wood blocking	50.00
Install hollow metal frame	26.00
Install hollow metal door	23.00
Install hardware	28.00
Paint finish door and frame	30.00
Total cost per door assembly	$544.00

The unit cost method estimates each item separately, whereas the conceptual method estimates each system or group of items (in this case, each door assembly at $544.00 per door).

Using the example of the trench excavation (Figure 6.1), the individual unit cost for labor and equipment for excavation and backfill may be a composite as follows:

Equipment excavation	$1.20/C.Y.
Labor excavation	0.66
Equipment backfill	0.95
Labor backfill	0.41
	$3.22/C.Y.

This summary reflects the unit cost for the entire operation of work in cubic yards. This unit could also have been developed on the basis of lineal feet of the trench described. The only criteria is that the takeoff quantity unit be matched with the comparable unit price unit. For this exercise, the cubic yard unit is preferred because it is more flexible and can be used with trenches in a specific range (say from 3 to 6 feet deep) and with the approximate same slope of the excavation. The lineal foot unit would be limited to trenches of that specification only.

A wide number of other assemblies costs can be developed in advance of their use and listed for quick reference. Continuing with the same example, an assembly can be developed to find a unit cost for the construction of the concrete footing. The procedure involves establishing a theoretical wall length with a theoretical cross-sectional configuration such as shown in Figure 6.1. The takeoff and individual quantification of all the items of work can be represented as follows:

Item	Unit	Mat	Labor	Equip	Total
Form footing	S.F.C.A.	x	x	x	x
Install reinforcement	lb.	x	x		x
Place concrete	C.Y.	x	x	x	x
Finish exposed top surface	S.F.		x		x
Set keyway	L.F.	x	x		x
Cure/protect	S.F.	x	x		x
Strip forms	S.F.C.A.		x		x
Strip keyway	L.F.		x		x
Clean, oil, move forms	S.F.C.A.	x	x	x	x
Total cost					x

This total cost can then be converted to any convenient unit cost. If the assemblies unit needs to be C.Y. (cubic yards), then the total cost is divided by the quantity of cubic yards. If the unit cost is stated in S.F.C.A. (square foot contact area), then the total cost is divided by the area of the form work.

Chapter Seven

CONCEPTUAL ESTIMATING

Levels of Estimating

There are four basic estimating approaches used to determine the cost of a structure. In order from the least to the most accurate, they are: Order of Magnitude, Square Foot, Composite or Assemblies, and Unit Price.

Order of Magnitude uses large scale items in the estimate. Examples are: cost per unit apartment, cost per house, and cost per unit length of roadway. These terms present quick, overall cost identifications. They provide assurance that the overall costs are in order. The same comparison can be made based on the systems used in the construction of a building. Some examples are as follows: total cost for a mechanical variable air volume (VAV) system, exterior curtain wall, EDPM roofing, structural steel, bar joists and metal decking. This level of estimating allows for the selection of design elements to meet budget constraints.

Square Foot estimating bases its costs on building area. The salesman uses this method to qualify an opportunity. Square Foot estimating can also be used to establish gross budgets.

Composite, or Assemblies, estimating is the level used in Conceptual Estimating. This method gathers individual unit costs to produce a cost per assembly (e.g., walls, floors, roofs).

Unit Price estimating is based on the costs of individual work units. This form of estimating is used to arrive at an estimate for the Lump Sum Bid method of construction.

Conceptual Estimating

Conceptual estimating is a technique of pricing work before complete plans have been drawn or comprehensive details are available. The items of work are still a concept, visualized by the estimator. Since design/build construction guarantees the cost of the project early in the design effort, it relies on this estimating technique exclusively. Conceptual estimating can include Order of Magnitude or Square Foot estimating, at early stages in the design/build process, but a guaranteed price is never prepared using these estimating techniques.

Square Foot Estimating

Annually updated square foot costs can be found in books such as *Means Square Foot Costs* for residential, commercial, industrial, and institutional buildings. In this book, major building types and systems are priced by square foot. An example page from the 1987 edition is shown in Figure 7.1.

The information in this example page gives the price for the type of building by the overall cost, which could be per square foot of ground or floor area, or per complete building. These models are not very detailed and rely on a total cost for each system to define its value in the estimate. Individual units that make up the total cost are not given and are not needed in this type of survey. This survey is only for comparative analysis of similar structures and cannot be used as a fixed price, but only as a budget. The accuracy of a developed project cost estimated in this fashion is problematic. The trouble lies in the nature of buildings and building designers. No two buildings having the same basic elements are ever exactly alike. Likewise, no two sites are alike. Conceptual estimating is not a detailed quantity survey produced from a 100% completed set of working drawings and specifications – as in a bid situation. A complete set of drawings and specifications is not available in the early stages of the design/build process; it comes later *after* the price and contract are agreed to and executed.

Quantity Survey

A quantity survey (in a lump sum bid situation) from 100% completed drawings and specifications usually involves the listing of every item of work, which are then summarized under categories, or divisions. The 16 MASTERFORMAT divisions, created by the Construction Specifications Institute (CSI), are used in most cases, by the architect as well as the contractor. Such use may be a condition of the bid. In Massachusetts, for example, there is a law called the *File Sub-bid Law* which requires certain trade work to be bid independently of the base bid. The "file sub bids" are then made available to the general contractor who includes the low bid in each segment as part of his cost. Gathering of costs within those divisions would then be a bid requirement. Lacking such a statute, a general contractor is still likely to organize the costs of the work into the MASTERFORMAT divisions, which are the known and accepted standard in the industry. There are 16 CSI divisions used to organize items of work and their costs. These divisions are as follows:

Division 1 General Requirements
Division 2 Site Work
Division 3 Concrete
Division 4 Masonry
Division 5 Metals
Division 6 Wood and Plastics
Division 7 Thermal and Moisture Protection
Division 8 Doors and Windows
Division 9 Finishes
Division 10 Specialties
Division 11 Equipment
Division 12 Furnishings

COMMERCIAL/INDUSTRIAL INSTITUTIONAL | 2.360 | Hotel, 8-24 Story

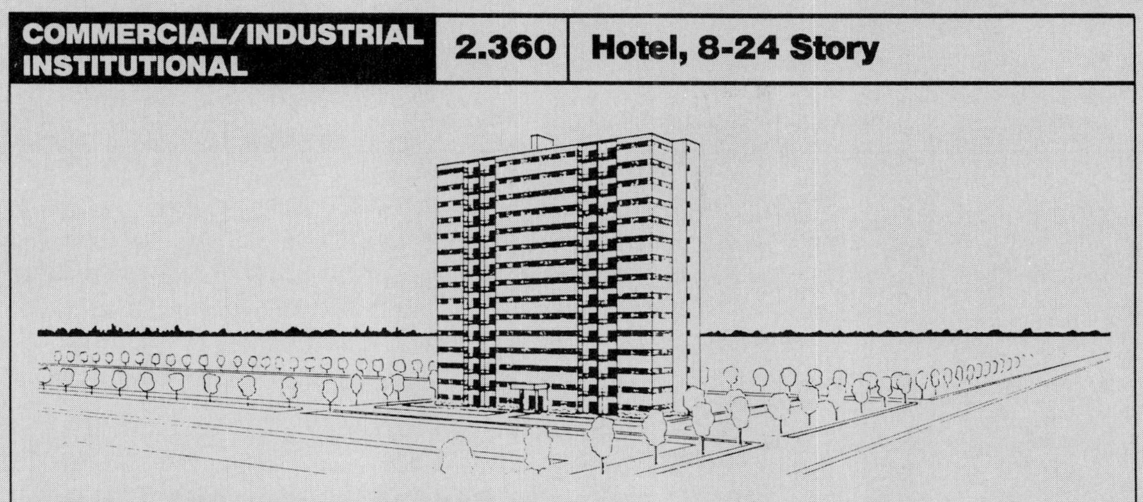

BUILDING TYPES

COSTS PER SQUARE FOOT OF FLOOR AREA

EXTERIOR WALL	S.F. Area	140000	243000	346000	450000	552000	655000	760000	860000	965000
	L.F. Perimeter	403	587	672	800	936	1073	1213	1195	1312
Face Brick with Concrete Block Back-up	Steel Frame	66.15	63.20	61.05	60.15	59.70	59.40	59.15	58.40	58.25
	R/Conc. Frame	68.25	65.35	63.20	62.25	61.80	61.50	61.30	60.50	60.35
Face Brick Veneer On Steel Studs	Steel Frame	64.90	62.15	60.20	59.40	59.00	58.70	58.45	57.80	57.65
	R/Conc. Frame	67.30	64.60	62.60	61.80	61.35	61.10	60.85	60.20	60.05
Glass and Metal Curtain Walls	Steel Frame	72.30	68.40	66.05	65.00	64.45	64.00	63.70	63.00	62.85
	R/Conc. Frame	74.70	70.85	68.45	67.35	66.80	66.40	66.10	65.40	65.25
Perimeter Adj. Add or Deduct	Per 100 L.F.	2.40	1.40	1.00	.80	.65	.50	.45	.40	.35
Story Hgt. Adj. Add or Deduct	Per 1 Ft.	.95	.80	.65	.60	.55	.55	.55	.45	.45
FOR BASEMENT, add $15.60 per square foot of basement area										

The above costs were calculated using the basic specifications shown on the facing page. These costs should be adjusted where necessary for design alternatives and owner's requirements. Reported completed project costs, for this type of structure, range from $43.00 to $93.00 per S.F.

COMMON ADDITIVES

Description	Unit	Cost
BAR, Front Bar	L.F.	$200
Back bar	L.F.	155
BOOTH, Upholstered, custom, straight	L.F.	93-150
1/4 Circle, "L" or "U" shaped	L.F.	105-165
CLOSED CIRCUIT SURVEILLANCE, One station		
Camera and monitor	Each	985
For additional camera stations, add	Each	525
DIRECTORY BOARDS, Plastic, glass covered		
30" x 20"	Each	370
36" x 48"	Each	545
Aluminum, 24" x 18"	Each	320
48" x 32"	Each	545
48" x 60"	Each	1025
ELEVATORS, Electric passenger, 8 stops		
3000# capacity	Each	79,300
5000# capacity	Each	83,500
Additional stop, add or deduct	Each	6875
EMERGENCY LIGHTING, 25 watt battery operated		
Lead battery	Each	280
Nickel cadmium	Each	430

Description	Unit	Cost
LAUNDRY EQUIPMENT		
Folders, blankets & sheets, king size	Each	$34,500
Ironers, 110" single roll	Each	12,300
Combination washer & extractor 50#	Each	8500
125#	Each	23,400
SAUNA, Prefabricated, complete		
6' x 4'	Each	3225
6' x 6'	Each	3975
6' x 9'	Each	5125
8' x 8'	Each	5075
10' x 12'	Each	7250
SMOKE DETECTORS		
Ceiling type	Each	115
Duct type	Each	285
SOUND SYSTEM		
Amplifier, 250 watts	Each	1175
Speaker, office	Each	92
Industrial	Each	175
TV ANTENNA, Master system, 12 outlet	Outlet	130
30 Outlet	Outlet	100
100 Outlet	Outlet	92

Use LOCATION FACTORS on pages 389 to 393

142

Figure 7.1

Division 13 Special Construction

Division 14 Conveying Systems

Division 15 Mechanical

Division 16 Electrical

Each division is further broken down into subheadings to specialize the estimate into more specific categories of work. For example, Division 10 – Specialties contains various types of partitions:

toilet partitions
wire mesh partitions
demountable partitions
accordion partitions
sliding partitions

All of these items deal with special partitions. On the other hand, drywall partitions and prefinished panels are sub-listed in Division 9 – Finishes. Division 7 includes certain exterior partitions. Masonry partitions are in Division 4, and glass partitions and window walls are in Division 8. Separating similar types of work into different divisions is not complicated when all of the pieces are readily identified on the plans and defined in the specifications. The estimator simply "takes off" or quantifies each item of work and gathers similar components together in the division and subdivision that have been designated for that work. The price extension for each work item is a matter of multiplying the quantity by the respective itemized unit cost to arrive at a price.

Quantity of Units x Unit Cost = Total Price

The quantities are systematically gathered and totalled in each CSI MASTERFORMAT subdivision until all of the work is taken into account. The resulting compilation is the estimate of costs. Design/build construction may also require that the working estimate be in the CSI MASTERFORMAT. The working estimate is needed in this form for the procurement of subcontractor trades and vendors for material items of work. The conceptual estimating, however, cannot be performed in this manner because there are no drawings or detailed work to "take off". Conceptual estimating serves other purposes, and is later converted to a working estimate, by another process.

Conceptual Survey

The methods used to survey or gather work items for a conceptual estimate are different from those used for a working estimate. The details of a building shown on a completed drawing are not available to a conceptual estimator. He needs to develop a quantity survey, just as if he had a completed drawing. This survey is based on the type of cost units the estimator is able to formulate. These cost units are then applied to the quantity to produce a price.

Quantity x Cost per "conceptual unit" = Price

This is the same fundamental formula as used for a lump sum bid, but the cost and quantity are derived differently.

Predesign factors, based on a variety of information sources, are used to quantify work. Once quantified, these factors, or requirements, should be gathered into categories in preparation for pricing. The unit cost and the quantity need to be in compatible "units". Let us take, for example, a hollow metal door from *Means Assemblies Cost Data*, 1987, as shown in Figure 7.2.

DOORS	B6.4-220	Interior Doors & Frames

The Metal Door/Metal Frame Systems are defined as follows: door type, design and size, frame type and depth. Included in the components for each system is painting the door and frame. No hardware has been included in the systems.

Steel Door, Half Glass
Steel Frame

Steel Door, Flush
Steel Frame

Wood Door, Flush
Wood Frame

System Components	QUANTITY	UNIT	COST EACH MAT.	COST EACH INST.	COST EACH TOTAL
SYSTEM 06.4-220-1200					
STEEL DOOR, HOLLOW, 20 GA., HALF GLASS, 2'-8"X6'-8", D.W.FRAME, 4-7/8" DP					
Steel door, flush, hollow core, 1-3/4" thick,half glass, 20 ga.,2'-8"x6'-8"	1.000	Ea.	170.50	24.50	195
Steel frame, KD, 16 ga., drywall, 4-7/8" deep, 2'-8" x 6'-8", single	1.000	Ea.	77	33	110
Float glass, 3/16" thick, clear, tempered	5.000	S.F.	20.65	18.35	39
Paint exterior door & frame one side 2'-8"x6'-8", primer & 2 coats	1.000	Ea.	2.59	25.41	28
TOTAL			270.74	101.26	372

6.4-220 Metal Door/Metal Frame

	TYPE	DESIGN	SIZE	FRAME	DEPTH	COST EACH MAT.	COST EACH INST.	COST EACH TOTAL
1000	Flush-hollow	20 ga. full panel	2'-8" x 6'-8"	drywall K.D.	4-7/8"	210	86	296
1020				butt welded	8-3/4"	245	91	336
1160			6'-0" x 7'-0"	drywall K.D.	4-7/8"	390	140	530
1180				butt welded	8-3/4"	425	155	580
1200		20 ga. half glass	2'-8" x 6'-8"	drywall K.D.	4-7/8"	270	100	370
1220				butt welded	8-3/4"	305	105	410
1360			6'-0" x 7'-0"	drywall K.D.	4-7/8"	520	170	690
1380				butt welded	8-3/4"	555	180	735
1800		18 gal. full panel	2'-8" x 6'-8"	drywall K.D.	4-7/8"	235	89	324
1820				butt welded	8-3/4"	265	94	359
1960			6'-0" x 7'-0"	drywall K.D.	4-7/8"	445	145	590
1980				butt welded	8-3/4"	480	160	640
2000		18 ga. half glass	2'-8" x 6'-8"	drywall K.D.	4-7/8"	305	105	410
2020				butt welded	8-3/4"	335	110	445
2160			6'-0" x 7'-0"	drywall K.D.	4-7/8"	615	170	785
2180				butt welded	8-3/4"	650	185	835

222

For expanded coverage of these items see *Means Interior Cost Data 1987*

Figure 7.2

Doors

This assembly of work items is quantified and organized to include components from the following categories:

- Metal doors
- Metal frames
- Glazing
- Painting

The following items are added as required:

- Locksets
- Hinges
- Closers
- Exit devices

Instead of the eight segments with eight different unit costs, the conceptual estimate assembly unit quantifies this work activity as follows:

> F & I (furnish and install) a painted hollow metal door and frame, with glass.

All of the activities would be included as part of that assembly. The assemblies cost, composed of all work activities, must be established. Since the quantity is taken as one complete door, the cost reflects one complete door. The assemblies cost gathers all of the activities associated with the door installation and totals them for an amount per door. Since the quantities are estimated and not necessarily shown, this form of quantity survey and pricing is simplified for the conceptual estimate.

Another example is:

Gypsum Plaster on Concrete Block Partition

This activity of work can include: unit masonry (block), dampproofing, plaster, and painting.

See Figure 7.3 for an illustrated concrete block partition (also from *Means Assemblies Cost Data*, 1987).

Instead of thinking in division headings or in subdivision groupings, a conceptual estimator thinks in terms of building systems. It is a process of mentally constructing the building as it would be built in reality. To begin, the conceptual estimator must think about clearing the land and excavating for the building and the foundations. He then proceeds through the installation of foundations, erecting the structure, applying the roof and siding, and through each element and system to completion.

It should now be apparent that a CSI MASTERFORMAT, unit price approach to conceptual estimating would be too time consuming and a difficult task. In the process of trying to keep tidy each particular category and individual item of work, the potential for missing items becomes very great. Those missed items can be very costly since the conceptual estimate is intended to contain all of the elements necessary for a complete building, within the agreed guidelines. This complete building is what is guaranteed by contract.

The origins of the conceptual estimate quantity survey and the assemblies costs method are developed from the same source – the unit price cost survey. Detailed cost and quantity information gathered in the construction of buildings is recorded and used to assist in the estimating of future work.

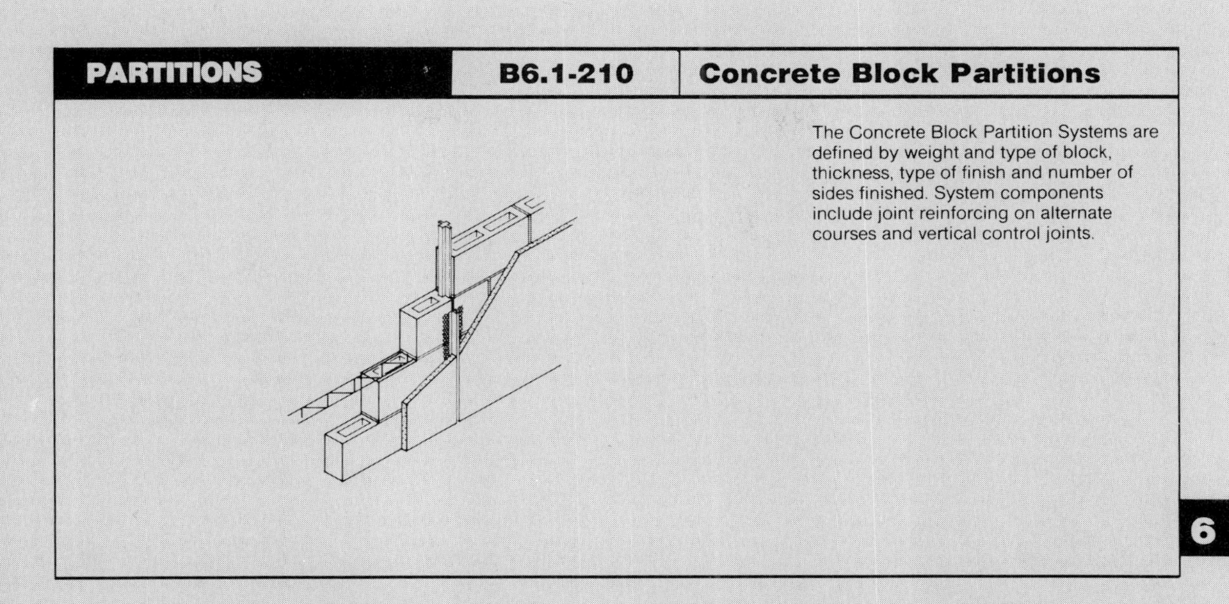

PARTITIONS | B6.1-210 | Concrete Block Partitions

The Concrete Block Partition Systems are defined by weight and type of block, thickness, type of finish and number of sides finished. System components include joint reinforcing on alternate courses and vertical control joints.

6

System Components	QUANTITY	UNIT	COST PER S.F. MAT.	COST PER S.F. INST.	COST PER S.F. TOTAL
SYSTEM 06.1-210-1020					
CONC. BLOCK PARTITION, 8" X 16", 4" TK., 2 CT. GYP. PLASTER 2 SIDES					
Conc. block partition, 4" thick	1.000	S.F.	.86	2.51	3.37
Control joint	.050	L.F.	.08	.06	.14
Horizontal joint reinforcing	.800	L.F.	.08	.02	.10
Gypsum plaster, 2 coat, on masonry	2.000	S.F.	.49	2.50	2.99
TOTAL			1.51	5.09	6.60

6.1-210 Concrete Block Partitions - Regular Weight

	TYPE	THICKNESS (IN.)	TYPE FINISH	SIDES FINISHED		COST PER S.F. MAT.	COST PER S.F. INST.	COST PER S.F. TOTAL
1000	Hollow	4	none	0		1.02	2.59	3.61
1010			gyp. plaster 2 coat	1		1.27	3.84	5.11
1020				2		1.51	5.10	6.61
1050			fiber plaster - 2 coat	1		1.28	4.33	5.61
1100			lime plaster - 2 coat	1		1.15	3.84	4.99
1150			lime portland - 2 coat	1		1.14	3.84	4.98
1200			portland - 3 coat	1		1.08	4.03	5.11
1400			⅝" drywall	1		1.54	3.54	5.08
1410				2		1.86	3.82	5.68
1500		6	none	0		1.09	2.78	3.87
1510			gyp. plaster 2 coat	1		1.34	4.03	5.37
1520				2		1.58	5.30	6.88
1550			fiber plaster - 2 coat	1		1.35	4.52	5.87
1600			lime plaster - 2 coat	1		1.22	4.03	5.25
1650			lime portland - 2 coat	1		1.21	4.03	5.24
1700			portland - 3 coat	1		1.15	4.22	5.37
1900			⅝" drywall	1		1.61	3.73	5.34
1910				2		2.13	4.68	6.81
2000		8	none	0		1.30	2.97	4.27
2010			gyp. plaster 2 coat	1		1.55	4.22	5.77
2020			gyp. plaster 2 coat	2		1.79	5.45	7.24
2050			fiber plaster - 2 coat	1		1.56	4.71	6.27

For expanded coverage of these items see *Means Interior Cost Data 1987*

205

Figure 7.3

Pyramid Theory

Figure 7.4 shows the relationship of the various levels of cost information. The base level, unit price information, is the most detailed. Individual items of work and their respective unit costs are provided at this level. This information comes from field work progress reports and cost accounting reports.

The gathering of the individual unit cost items of work into assemblies costs is shown as the next level and is developed through the selective gathering of the "base" cost units. The assemblies quantities and costs should not be used during the construction stage. When the project goes into construction, the assemblies quantity and respective costs are redistributed back to the base units and gathered into the CSI MASTERFORMAT divisions for job use and record keeping.

Information received from field reports concerning productivity, changes in labor rates, and new material costs is used to update the assemblies costs – both for future use and for checking the original conceptual estimate.

A conceptual estimate cannot be checked for accuracy of quantities (before construction) because it is conceived. Only the math can be checked. Therefore, the techniques used for self checking become very important. One of the methods is to compare the square foot cost of the estimated building system to one previously built. Thus, the need is clear for the development of square foot costs per system from known and current data.

The next level in the pyramid is the cost of building assemblies or systems (usually per square foot). These costs are developed by combining the costs that make up each system of a building – for example, all costs comprising the exterior skin of the building or the roofing. The purposes of square foot estimating are strictly self-checking and quick budgeting. Even though the degree of accuracy can be reasonably good when the source of data is accurate, these types of costs should not be used when a more detailed, unit price estimate is required.

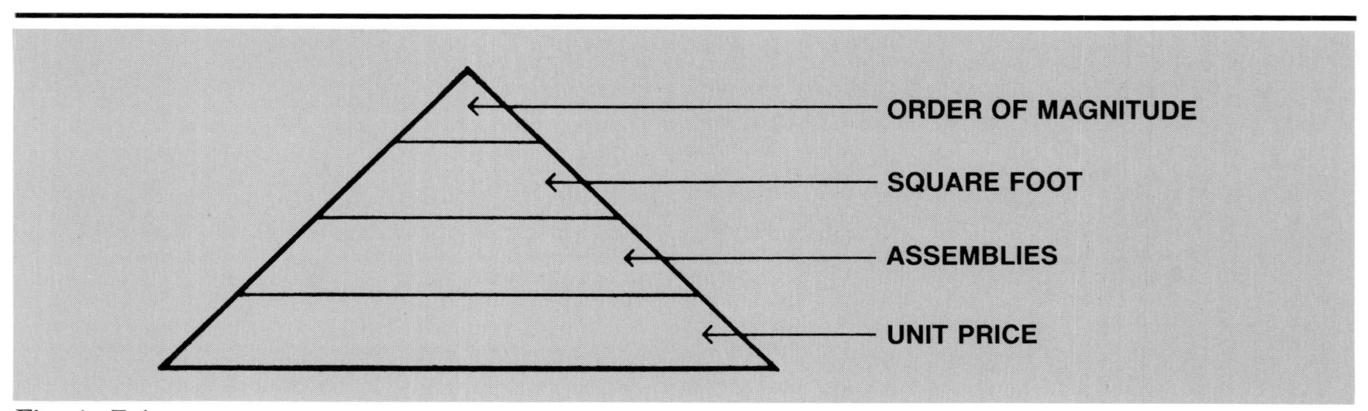

Figure 7.4

The next level up the pyramid is the gathering of preceding costs to generate costs per square foot of complete buildings. This information can also be used for self checking, but is used primarily by the salesmen to qualify a sales opportunity.

Finally, the pinnacle of the pyramid is the *order of magnitude* estimate. This approach is used to obtain a quick, rough budget cost.

All of this costing information can be interrelated. Using a computer, a series of codes identifies cost items which can be gathered from any level to generate data at any higher level.

List of Systems

Conceptual estimating deals with building segments. All work associated with these segments is estimated – regardless of the CSI category. In this manner, the conceptual estimator can focus his attention and experience on the appropriate segment and include all of the work. A suggested *Uniformat* summary is as follows:

1.0	Foundations
2.0	Substructure
3.0	Superstructure
4.0	Exterior Closure
5.0	Roofing
6.0	Interior Wall Construction
	Interior Finishes
7.0	Conveying
8.0	Mechanical
	Plumbing
	Sprinkler
9.0	Electrical
10.0	General Conditions
11.0	Special Construction
12.0	Site Work
	Architectural Fees

The estimate segment for foundations would include formwork for footings and walls, furnishing and placing concrete, furnishing and installing reinforcing steel and mesh, furnishing and installing anchor bolts, dampproofing, and expansion and construction joints. All of the above could be separated according to the CSI MASTERFORMAT as follows:

	CSI Division
Forming of the footings and walls	3
Concrete materials	3
Placing concrete	3
Reinforcing steel	3
Anchor bolts	5
Waterproofing	7
Expansion joints	3
Testing	1

The exterior closure segment can be conceived in the same manner to include all of the elements with regard to CSI MASTERFORMAT, and would include, for example:

	CSI Division
Brick masonry	4
Block & Tile masonry	4
Miscellaneous metals	5
Scaffolding	1
Caulking	7
Light gauge metal framing	9
Dampproofing	7
Precast panels	3
Insulation	7
Glass and glazing	8
Entrances	8
Doors and frames	8
Painting and finishing	9
Specialties	10 & 11

In both of the previous examples above, the foundations and exterior closure, each would be regarded (and estimated) as one complete building system or assembly.

To conceive of all of the building elements in this manner requires knowledge of construction, rather than knowledge of form. The General Conditions items, fee and design costs are usually determined based upon the type of job, the client, and other factors not empirically derived.

Developing the Concept

While conceptual estimating does not confine itself to a specified cost format, it does require the cultivation of a certain mental process. After careful study of the information at hand, images begin to form concerning particular systems and components of systems. The estimator must examine a proposed building as if in the presence of the actual, completed building. This is done by recalling completed projects. This is an effective approach despite the fact that each conceived building is not likely to be exactly the same as the previous buildings.

Some design information is available at this stage. All of the exterior skin, glass, and glazing, as well as the type of structure, roofing system, foundation system, slabs, and even (in most cases) the interiors have been determined as part of the design concept. The mechanical system, the need for a sprinkler system, and the type and the degree of plumbing work have been decided. The survey will determine the quantity of each item.

Where does conceptualization come in? It is necessary when all of the support and peripheral work items need to be quantified. That is not to say that the given design concept is readily readable (on the drawings) or available. In fact, these items are seldom visible. One has to know that an item is installed in a certain place on the building and how a particular element goes around corners or terminates at the roof or the ground. Using this basic knowledge, a quantity can be adequately determined.

The support and peripheral items involve such unseen elements as steel channel frames around and supporting overhead doors and other miscellaneous metals items. Blocking is another "hidden" item, whether for stair railing, toilet accessories and fixtures, roofing, or other applications. Caulking, waterproofing, and dampproofing must also be considered. The roof structures, such as hatches, smoke and heat vents, flashings, pitch pockets, roof drains, curbs, and pavers do not show, but must be included. Decisions must be made as to the supervision and support. Other General Conditions requirements must also be calculated: temporary utilities, job trailer, etc. Being aware that an element of work will need additional support, when neither the element nor the support are shown on the drawings, is part of the conceptualization. Familiarity with buildings similar to the type being estimated allows for this process (which also requires prior knowledge of building construction, the design concept, and state and local code requirements).

Producing a detailed conceptual estimate is not necessary for all potential clients. Not all clients are ready to proceed to the construction phase, but simply need to have an approximation of the costs to use in planning for the future. In those instances, one of the other, quicker estimating methods may be used.

Different Forms of Pricing

The simplest method of pricing a building is based on the cost per square foot of floor area. When a client wants to know approximately what a certain type building would cost, the square foot method can provide a quick "ballpark" figure. As previously mentioned, the square foot method is commonly used in sales to find out if the client has a realistic budget.

The next method that can be used is the costs by overall building system. This method requires a little "takeoff" work prior to the cost applications. Such a takeoff would involve determining the area of siding, roof, and flooring, for example. Costs for siding would be expressed in square feet of surface area. Costs for finishes, structure, and the mechanical, electrical, plumbing, and sprinkler work would be given as cost per square foot of the floor area. Roofing given as a cost per square foot of roof area would include all of the miscellaneous items commonly found on roofs. Mechanical work can also be given as cost per ton of air conditioning if there is a large amount required. Although this method is not as quickly developed as the square foot of buildings estimate, accuracy is greater.

Another approach to pricing is *segmental estimating*. See Figure 7.5 for an illustration. Segmental estimating is somewhat like estimating the building square foot cost, but it also incorporates a systems factor. A "slice" of the building is developed, summarizing all elements cut through. For example, a segment of an industrial building would include, from top to bottom, cost factors for: the roof, elements on the roof, structure, slabs, foundations, mechanical, electrical, plumbing, and sprinkler. This method is used to "mix and match". If pricing is being done to "zero in" on a building design concept, then the segmental estimates of different buildings can be applied exclusive of the exterior envelope. When an acceptable segmental concept is found, it is wrapped in the envelope (of the desired exterior skin) to form the total building design concept. One drawback is the number of segmental sections needed on file due to the unique nature and variety of the information.

The detailed assemblies estimating format is an option. This type of assemblies conceptual estimate can take up to several days to a week to prepare. The unit price estimate can take several weeks to prepare.

In all cases, these estimates may involve a degree of unavoidable unknowns. These unknowns cannot be excluded, but must be treated in a particular manner as part of the estimate.

Contingency

A contingency is an amount of money carried as part of an estimate when there is an uncertainty about the work – a "safety factor". A contingency is for items of work that are generally included or that have the potential for problems. Rehabilitation or remodelling work, for example, carries a contingency. One never knows what kind of work will be needed once a wall is exposed after demolition. In new construction, contingencies are less likely to be carried, but may be included, depending upon the degree of difficulty of the project. Hospitals are an example of work that would require a contingency due to the complexity of the project.

There are two ways in which to develop and include a contingency amount of money. The first method for developing a contingency is the application of a flat percentage (often approximately five to ten percent) to the total cost of construction.

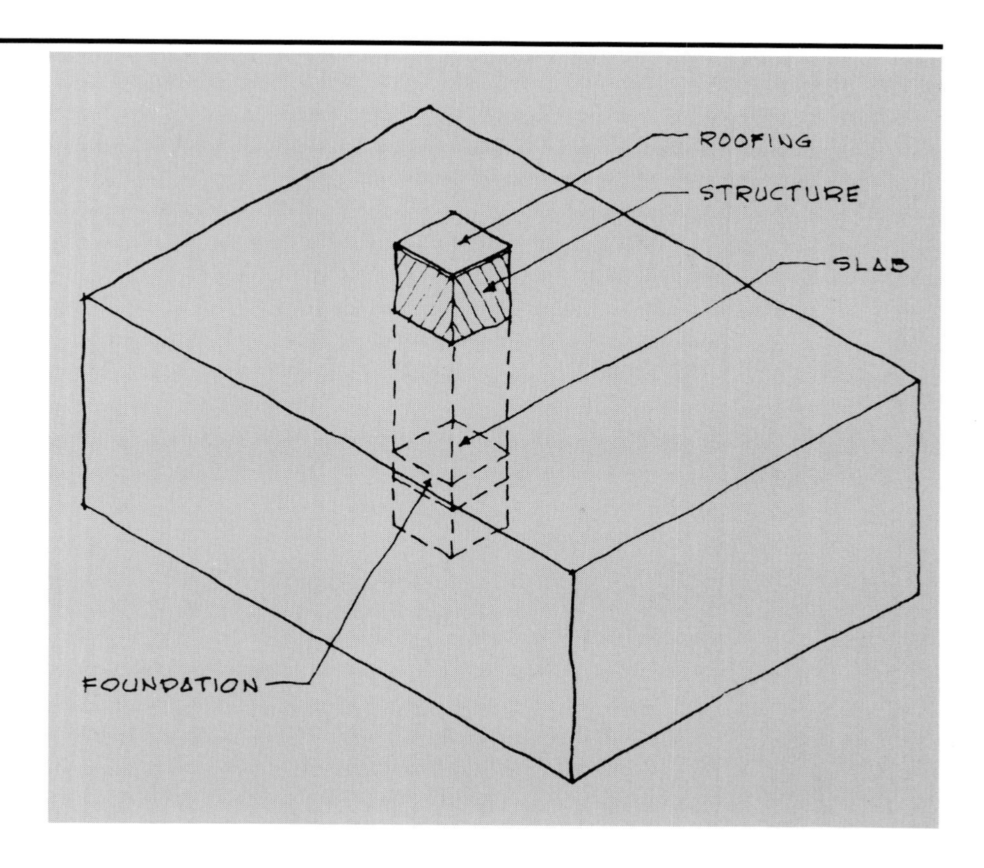

Figure 7.5

The second way to include a contingency is with items that are identified "unknowns". These items (such as miscellaneous metals; waterproofing, dampproofing, and caulking; the mechanical trades; and site work) would each have a separate contingency. Contingencies would not be applied to items that are completely understood. This method demonstrates the skill of the estimator, but carries a higher risk factor than the blanket percentage contingency.

Design/build seldom uses contingencies except in the case of very difficult projects. In the conceptual estimate, allowances are carried when necessary due to uncertainties of the work. Rock excavation, hardware, and landscaping are traditional examples. An allowance would also be carried for work that might affect a fire underwiter's review. For example, Factory Mutual (FM) has criteria for the construction of different types of buildings. These standards must be adhered to if the client is to receive a favorable insurance rate. The specifics, however, are not totally known until after the design is complete and reviewed by FM. For design/build, this is too late for pricing. The conceptual estimator would then carry, as part of the cost, his best guess of FM requirements as an allowance, pending their review.

Self Checking

As previously noted, it is important that the conceptual estimator have methods to check his work. The nature of conceptual estimating, with much of the information in the mind of the estimator, does not allow for the common methods of checking. Apart from having someone else check the arithmetic, the conceptual estimator must be responsible for the numbers in the estimate. The risk of error can be reduced using the following methods:

The first (described earlier in this chapter) is comparing historical system costs per square foot against a compilation of those costs in the current estimate.

Another approach is to use an estimating form as a checklist to prevent the omission of items of work. The summary of composite costs (gathered in the same system categories to be estimated) provides another kind of checklist. Keeping all of the takeoff figures opposite the items of work and on the main estimating sheet (rather than on scratch sheets) also helps in the checking.

Examination of ratios is another way to check the figures. The ratio of "soft", or indirect, costs to "hard", or direct, costs can be useful in the analysis of a particular project. The hard costs are those costs directly involved with the elements of the building such as excavation, foundations, slabs, etc. The soft costs are those costs involving such items as fee, General Conditions costs, and design and testing costs. The ratio of soft-to-hard costs for an industrial building may be in the range of 15%.

$$\frac{\text{Soft Costs}}{\text{Hard Costs}} = .15 \text{ or } 15\%$$

The same ratio for a hospital could be about 10%.

Another method of checking is to review the cost reports of other recent or current jobs. The purpose of this review is to compare the costs from a similar job based on items required for both projects. The conceptual estimator does well to use several of these techniques which can help him to minimize the risk of error.

Annual Cost Data Books

Finally, a good reference for double-checking estimate costs is annually updated construction cost data books such as those published by R.S. Means Company, Inc. Users of these books are chiefly interested in obtaining quick, reasonable, average prices for construction estimates. In addition to unit price, assemblies, and square foot costs, Means annual books also provide productivity data and national labor rates. Tables and charts for location and time adjustments are included to help the estimator tailor the nationally averaged prices to a specific project. Means' costs are obtained through contact with manufacturers, suppliers, dealers, distributors, and contractors across the country. There are many versions of Means cost data books which cover all types of construction.

Chapter Eight

OUTLINE SPECIFICATIONS

Construction Specifications

Plans and Specs are the familiar terms for construction documentation. The *specs*, or construction specifications, establish the quality level of the items that will be used to make up the project. They complement the drawings, or *plans*, which illustrate the arrangement of components to be used in the project. The plans are separate and distinct from the specs; each has a specific role.

The specifications are usually organized using a format that repeats in each section. The following list is an example of a typical architectural specification section. The format (numbering system) used here represents one individual approach; there are many other alternatives. The choice may depend on the preference of the architect.

1 - General

- 1.01 General Provisions
- 1.02 Work to be performed
- 1.03 Related work
- 1.04 Description of work
- 1.05 Workmanship quality standards
- 1.06 Material quality standards
- 1.07 Submittals
- 1.08 Applicable publications
- 1.09 Definitions
- 1.10 Levels, location, and responsibility
- 1.11 Warranties, Guarantees

2 - Products

- 2.01 Materials
- 2.02 Performance criteria
- 2.03 Specific performance
  - A. Insulation and protection
  - B. Sheet flashings
  - C. Metal flashings
  - D. Plastics
  - E. Etc.

3 - Execution

3.01 Pre-installation conditions
3.02 Cutting, removal and preparation work on existing
3.03 Fabrication
3.04 Fabrication and installation
3.05 Infilling and modification of existing
3.06 Hanging and fitting
3.07 Installation of specialties
3.08 Storage and protection
3.09 Measurements
3.10 Cleaning
3.11 Testing and adjustments
3.12 Final inspection
3.13 Maintenance recommendations

Clearly, all of the categories of this format relate to quality standards, performance, and assurance. The mechanical trades specifications could be separated into sections, each organized in the format shown above. Plumbing might be separated as follows:

Section 15B1 Pipes and Pipe Fittings
Section 15B2 Valves
Section 15B3 Piping Specialties
Section 15B4 Supports and Anchors
Section 15B5 Plumbing Fixtures
Section 15B6 Insulation
Section 15B7 Potable Water Systems
Section 15B8 Soil and Waste Systems
Section 15B9 Storm Water Systems
Section 15B10 Pumps
Section 15B11 Natural Gas Systems
Section 15B12 Water heaters

Each of the mechanical sections is then developed to include the following categories:

1. General
2. Products
3. Execution

The format is thereby the same and the content includes: quality standards, performance, and assurance.

Outline Specification

Like lump sum bid construction, the design/build method must also have "plans" drawn and "specs" written in order to carry out the construction. Such specifications must conform to the same statements on quality standards, performance, and assurance. In the proposal stage of the design/build process, however, a document called an *outline specification* is written. Whereas the construction specification deals primarily with quality, the outline specification has an additional role. It also states quantity, location, and function. The section format for an outline specification follows the estimating format. If the conceptual estimate is developed based on a building segment analysis, then the outline specification is written in the same way. An outline specification may not use the same category format as a construction performance specification, and may not be a regimented or detailed form. The outline specifications for the interior walls and plumbing of the case study are shown in Figures 8.1 and 8.2. (Complete outline specs for the case study project can be found in Appendix A.)

6 INTERIOR WALLS

Interior walls shall be constructed of the following materials:

- The walls in the Manufacturing Area and Warehouse Area shall be constructed of eight-inch (8") concrete block installed to ten feet (10') high.
- The common Warehouse/Office wall shall be constructed of eight inch (8") concrete block installed to the underside of deck.
- The Assembly/Warehouse demising wall shall be constructed of eight-inch (8") concrete block installed to the underside of the deck.
- The common mezzanine wall with the Assembly Area shall be constructed of 3 5/8" metal stud with 2 layers of 5/8" F.C. each side to the underside of deck and include sound attention insulation.
- All walls in the Office Area shall be constructed of three and five-eighths inch (3-5/8") metal stud and one-half inch (1/2") sheetrock, each side, nine feet (9') high through the finished acoustical ceiling.
- The interior surface of the exterior and interior masonry walls at the Office Area shall be furred and receive sheetrock.
- All remaining metal stud and sheetrock walls shall be painted.
- Walls in Warehouse toilets shall receive Tru-Glaze. Walls in Assembly and Office toilets shall be full height ceramic tile.
- Metal sash with one-quarter inch (1/4") tempered glass shall be installed at the Shipping Office and Guard Station.
- Four (4) 8' x 8' overhead doors in the demising wall.
- All pass doors within the masonry walls shall be 18-gauge hollow metal set in 16-gauge metal frames.
- All doors within the metal stud and sheetrock walls shall be particle-filled, solid-core birch set in 16-gauge metal frames.
- All interior hardware shall be heavy duty.
- All doors and frames shall be painted.
- Ceiling and walls of the Assembly Area to be painted.

Figure 8.1

8 PLUMBING

The Plumbing System shall be designed to meet governing codes and shall consist of the following:

- An interior roof drainage system consisting of roof drains and piping to ten feet outside the foundation wall.
- An interior waste and vent system from plumbing fixtures and floor drains to ten feet outside the foundation wall.
- An interior domestic water system beginning ten feet outside the foundation wall, including a water meter, with piping distribution to serve plumbing fixtures.
- An interior domestic hot water system, including hot water heaters, with piping to serve plumbing fixtures.
- A gas piping system will be provided from the building side of the meter with piping distribution to serve mechanical equipment.
- The plumbing fixtures shall consist of the following, and be American Standard, Kohler, or Crane.
- Eleven wall-hung vitreous china water closets with flushometers.
- Five wall-hung vitreous china urinals with flushometers.
- Four wall-hung vitreous china lavatories with centerset fittings.
- Six vitreous china lavatories to be set in a countertop.
- One vitreous china lavatory wall-hung in shipping toilet.
- Two enameled cast-iron service sinks.
- Four electric water coolers.
- A stainless steel sink will be provided in kitchen.
- Three exterior hose bibbs will be provided for exterior watering.
- Color coding of piping and hook up of owner's equipment is not included.

Figure 8.2

Compared to the outline specifications, the construction performance specification provides more detail about the quality of materials, and less information about their location or use. For interior walls, the following statements might be included in a construction performance specification:

- Hangers: 8 gauge galvanized steel wire
- Screw studs: 25 gauge USG galvanized steel
- Runners for screw studs: 25 gauge galvanized steel
- Internal reinforcement: 20 gauge galvanized steel
- Wall board: Standard USG sheetrock brand, 1/2" thick
- Fasteners: USG type "S" bugle head screws
- Laminated adhesive: Durabond joint compound
- Taping: USG Brand standard manufacturer treatment

Note that the construction specification does not stipulate painting, as this item appears in another section. No quantity or location was stated, because this information (type of wall, quantity, and placement) is shown on the drawings. The outline specifications, on the other hand, do need to include that level of detail since it does not show on the concept drawings.

The style of writing used for outline and construction performance specifications would not be considered good literature, but the brief "shorthand" style is sufficient to convey the requirements. "Bullet statements" are used in order to be direct and concise. Note the example of a construction performance specification below.

> "Joint and internal corner finish treatment: USG joint treatment or equal. Use compound for treatment of all non-caulking joints, screw heads, and other depressions."

Purpose of the Outline Specification

As previously stated, the outline specification complements the other documents. It is a part of the presentation, and defines the building. The basic documents package also includes the following:

- The concept plans, which describe the project graphically.
- The outline specifications, which describe the project verbally.
- The conceptual estimate (if furnished to the client), which describes the project in terms of quantities and costs.

These documents, along with any other information that is important to the building's definition, constitute the presentation material and subsequent parts of the contract documents. Alone, each of these elements is incomplete, but together, they form the basic body of information for the project.

Included Items

The included items are part of the project and part of the pricing. An example was given in Figure 8.1 for the interior walls. In addition, statements concerning standards, code compliance, and performance criteria are also part of the outline specification. Statements in the outline specifications pertaining to codes for electrical work, for example, are included in Figure 8.3.

9 ELECTRICAL

GENERAL:

The scope of work consists of the installation of all materials to be furnished under this Section, and without limiting the generality thereof, includes all equipment, labor, and services required for furnishing, delivering, and installing the principal items of work hereinafter listed.

STANDARDS:

All materials and equipment shall be new and comply with the applicable standards of the following authorities:

Massachusetts Electrical Code
Underwriters Laboratories, Inc.
National Electrical Manufacturers' Association
Institute of Electrical & Electronic Engineers
American Society for Testing Materials
United States American Standards Institute
National Board of Fire Underwriters
Insulated Power Cable Engineers Association
Applicable Insurance Underwriter
National Electrical Safety Code
Occupational Safety and Health Act

TEMPORARY LIGHT AND POWER:

A 200A, 208/120V 3-phase, 4-wire temporary service will be provided by the electrical subcontractor from a source provided by the utility company for electric light and power requirements while the building is under construction and until the permanent feeders have been installed and are in operation. Energy charges will be borne by the Design/Build Contractor.

Figure 8.3

It is the design/build company's obligation to conform to code requirements. The specification sections should list the applicable codes, for clarity. Performance standards usually apply to elevators, sprinkler systems, and similar items that rely on manufacturer standards. Included are items that are provided completely assembled, such as pumps and heating and cooling equipment. The mechanical work is usually given as a performance specification. Figures 8.4a and 8.4b are from the case study and outline the HVAC design criteria:

This form of specification defines the limits of performance of the yet-to-be designed heating and cooling systems. This performance statement does not relieve the design/build company of its obligation to design within the applicable codes; it simply stipulates the parameters of the design. The cost of this level of performance is in line with the criteria provided. Change the criteria and one changes the cost and the price of the project. At this level of design development, it is difficult to be more specific. The performance criteria does say that the mechanical system will be designed in conformance with code standards, and that the system designed will maintain an inside temperature of 75 degrees Fahrenheit in summer [when the outside temperature is at 91 degrees Fahrenheit measured on a dry bulb (FDB) thermometer], at a 50% relative humidity.

Similarly, the inside heating of 70 degrees can be maintained when the temperature outside goes to 0 degrees F. This method of defining the work is clear and should be fully explained at the presentation.

Excluded Items

It is almost as important to stipulate the excluded items as it is to list the included items. Since there is so little visual information, a statement of exclusions helps to avoid misunderstandings concerning the scope of the work. A client may expect to have a particular feature incorporated into the building, yet nowhere in the documents is it stated that this feature is included. Items of work that have the potential to create confusion should be specifically addressed as *in* or *out*. If traditional work is not to be carried or needs qualification, then this too, must be said. An example of such an item (not included) is ledge removal. The statement listed under the title, *Work Not Included* is as follows:

> "Ledge removal is not anticipated in the excavation of the building and site and has not been carried in the cost or as part of the work. If rock is encountered, its definition, for removal, shall be as follows: A single piece larger than 1 cubic yard and/or a continuous strata hard enough to require blasting to break up. Any ledge removable by equipment (not blasted) will not be considered as ledge excavation. All required ledge excavation is to be considered as extra cost to the project."

If this work had been estimated as an allowance, there would have been a statement to that effect, as well as a method to reconcile the allowance developed as part of the outline specification. Another area to consider for the "items not included" category involves the installation of the client's own equipment. If the owner's equipment is to be installed by his own employees, but coordination of this activity is required of the design/build company to insure a smooth transition, say so.

8 HEATING, VENTILATING & AIR CONDITIONING

I. DESIGN CONDITIONS:

A. The heating, ventilating and air conditioning systems shall be designed within the limits of the following conditions:

OFFICE AREA

Summer: Outside - 91° FDB 78° FWB

Inside - 75° FDB 50% RH at the above outside conditions

Winter: Outside - 0° FDB

Inside - 70° FDB

Ventilation: Ventilation shall conform to the applicable codes, and as required to offset the heat gains and losses of spaces being served. Summer ventilation shall be provided at a rate of air change to limit the space temperature rise to 20°F above the outside summer ambient temperature indicated above.

WAREHOUSE/ASSEMBLY AREA

Winter - 0° - 70°FDB

B. The following codes, organizations, jurisdictional bodies, and insurance organizations shall be considered applicable:

1) F.M./F.I.A.
2) OSHA
3) NFPA
4) BOCA (Mechanical Code Appendix)
5) ASHRAE
6) SMACNA

Figure 8.4a

8 HEATING, VENTILATING & AIR CONDITIONING (cont'd.)

II. DESCRIPTION OF SYSTEM:

A. HEATING: Heating for the Office Area shall be accomplished by means of gas-fired rooftop air unit(s) equipped with fan, burner, filters, and air intake plenum with damper. A supply duct distribution system shall be furnished to provide optimum air distribution. Terminal air diffusers shall provide air diffusion at the specific area or room, and the ceiling void shall be used as a return air plenum.

Gas-fired unit heaters shall serve the heating needs of all other spaces including the Shipping & Receiving Area.

B. COOLING: Cooling shall be accomplished by means of a package rooftop air conditioning unit(s) equipped with fans, compressors (electric), condenser, filters, and outside air intake plenum with damper. A supply duct distribution system shall be furnished to provide optimum air distribution. Terminal air diffusers shall provide air diffusion at the specific area or room, and the ceiling void shall be used as a return plenum.

A total nominal 30 tons of cooling for the Office Area and 5 ton supplemental cooling for Conference Room.

C. Ventilation: Ventilation shall be accomplished by means of the appropriate air handling device applicable to the specific area. Ventilation shall be accomplished according to the following schedule, unless specified otherwise hereinafter:

Space	Number of Air Changes
Offices - General -	4 - 6 Max.
Assembly -	2 (Summer) - 1 (Winter)
Warehouse -	2 (Summer) - 1 (Winter)
Toilets -	10 air changes or 40CFM per fixture (exhaust)
Janitor's Closets -	10 air changes or 40CFM per fixture (exhaust)

D. Miscellaneous:

1. The Conference Room shall be equipped with a supplementary 5 ton cooling coil and rooftop condensing unit to provide optimum environmental control.

2. The kitchen will be fitted with exhaust ductwork and an up-blast type exhaust fan.

Figure 8.4b

Particular attention needs to be paid to the mechanical trades when it comes to items not included. Sprinkler work, for example, involves performance criteria. Therefore, statements concerning this item need to be particular and specific regarding work not included and allowance items. In addition to the local building official's review of the sprinkler design (prepared by the sprinkler subcontractor), the fire underwriter's engineers also require a review and approval. An example outline specification for interior fire protection for the case study is shown in Figure 8.5.

The insurance companies rely on reviewing engineers to establish design parameters for different types and uses of buildings. Not all of the criteria is the same for each. In addition to the insurance company engineers, the local fire marshal has the authority to influence the design in the interest of public safety. In the process of gathering data, much of this information is made available to the design/build company and is included in the specifications. The problem is that the fire marshal and the insurance company engineers reserve the right to change their minds upon review of the plans. To cover a contingency for added work, certain work may either be excluded or made part of an allowance. This strategy is not intended to circumvent designing by code, but rather to avoid an obligation to include added items of work not normally required.

Like mechanical items, electrical work must also be closely examined for items commonly excluded. Some examples are listed below.

- Utility charges for primary service
- Bus duct required for owner's equipment
- Wiring or connection to owner's equipment
- Telephone work in wiring of service
- Building security systems
- Television systems
- Data processing hook-up
- Public address systems
- Music systems

If any of this work is included, then its clear scope would be part of the included items. (See the complete electrical outline specifications for the case study in Appendix A.)

Simply put, it is necessary to have a complete document. This document not only needs to explain, in words, what constitutes the bits and pieces of the project, but should also be equally clear on what is not included. A huge shopping list of items not included is not advocated; only those items that might reasonably cause confusion should be listed. For example, there is no point in stating that brick siding is not included when it has been previously stated in the "included" section that the siding is metal panel.

8 INTERIOR FIRE PROTECTION

- A wet ordinary hazard sprinkler system shall be installed in all areas of the building.

- Two (2) eight-inch (8") sprinkler entries, complete with wall indicator valves, shall be provided.

- The Manufacturing and Warehousing Areas shall be protected by means of brass upright heads at a ratio of one (1) head per each one hundred (100) square feet of floor area.

- The Office Area and Warehouse Office Areas where suspended acoustical ceilings shall be installed, shall be protected by means of chrome pendant heads at a ratio of one (1) head per each one hundred and thirty (130) square feet of floor area.

- Six (6) hose stations complete with one hundred feet (100') of one and one-half inch of woven fabric hose shall be installed at convenient locations in the Manufacturing and Warehousing Areas.

- All tees of the sprinkler piping shall be painted red in compliance with OSHA specifications.

- Sixteen (16) four foot wide by eight foot long (4' x 8') smoke and heat vents for Warehouse/Assembly Areas are included as required by Local Code. Design criteria low heat release content, 1 s.f. per 150 s.f. of floor space.

ITEMS NOT INCLUDED

- A.D.T., booster pumps, rack sprinklers, draft curtains, hose carts, fire extinguishers, firewalls, fireproofing, field painting, or any such items.

- The above fire protection specificity is our best estimate of the requirements of the fire underwriters. The actual final requirements will be determined by negotiation between the owner and his insurance company. Any additions or deletions from the above requirements, inclusive of final building materials acceptance, will result in a price adjustment which will be fully documented and substantiated.

Figure 8.5

Items for Consideration

Items for consideration are additional and deductive items of work. They are included as part of the outline specification. This category is designed for two purposes. First, it gives the client an opportunity to either customize the proposed building with added features or to reduce the cost by selective deducts. Secondly, it allows the sales contact an opportunity to follow up with the client. Following are some examples of items for consideration:

Substitute metal panel siding for brick: Deduct: $200,000.

Furnish a VAV (variable air volume) HVAC system in lieu of the heat pump system specified:

Add: $350,000.

Like many other aspects of the design/build method, preparation of the outline specifications must be accomplished with care. The contract for the work depends heavily on this document for definition. The outline specs are not to be considered as "fluff" to fill the presentation book, but rather as one of the three principal documents that defines the project – a project which the design/build company is pledging to build, if accepted, for the amount stated.

Chapter Nine

The Presentation

The presentation is the meeting that introduces the client to the project. It is at this time that the concept drawings, the outline specifications, and the conceptual estimate, among other documents, are presented and the building explained.

The Players

Participants in the presentation generally include: the salesman, the conceptual estimator, the conceptual architect (if necessary), and the client and his staff. The roles of the participants should be clearly established so that there is no overlapping or duplication of information. These roles are discussed, generally just prior to the meeting. Typically, the participants take on the following roles:

The **Sales Contact**, is responsible for the introductions and acts as moderator throughout the meeting. The sales contact may at this time also introduce the company to the client through the use of brochures and financial statements. The sales contact is the person who closes the meeting as well. The importance of this assignment is tied to the nature of the job. The sales contact will try to conclude the sale at the end of the presentation. Failing to achieve this, he will try to arrange a follow-up meeting to close the deal at a future date.

The **Conceptual Architect**, if included, is responsible for introducing the rendering (usually furnished for "full-dress" presentations), the plan, pertinent elevations, sections and details, if any. The conceptual architect explains his concept of the building and how this concept meets the needs of the client. Many times, this member of the team is not included as part of the presentation, and his information is relayed by the conceptual estimator.

The **Conceptual Estimator** is responsible for explaining the estimated elements of the project. This explanation involves a review of the outline specification and then the conceptual estimate (if the latter is to be given as part of the information). Not all design/build companies will furnish the conceptual estimate due to a certain feeling of proprietary costing knowledge that they are not willing to share. The outline specifications are read sheet-by-sheet, and explained through the use of the concept plans (which are usually mounted on a board and placed on an easel for ease in presentation). After reviewing the outline specifications, the conceptual estimator proceeds through the conceptual estimate using the same sheet-by-sheet, item-by-item approach.

At any point in the presentation meeting, as questions are asked, the player who is responsible for that section answers. For example, the following questions are matched with the individual who should appropriately answer them:

Flow of Products – Conceptual Architect

Allowance Items – Conceptual Estimator

Design/Build Financial Status – Sales Contact

Contract Terms – Authorized Company Spokesman

Fee and Saving Provisions – Authorized Company Spokesman

Prior Experience – Sales Contact

The authorized company spokesman could be any one of the three (conceptual architect, conceptual estimator, or sales contact). If not, then the person who is empowered to enter into the agreement must also attend the presentation. This is an important point. Many times, the absence of an authorized person and/or the time delay in getting him to the meeting, could lose the deal.

More specific information concerning cost accounting need not be demonstrated at this time and could be saved for a future meeting. If the client insists upon seeing that information, then cost reports should be furnished and the company person (among those present) who is most qualified should speak. Bringing in an accountant to explain the cost reports is not advised, as this level of information may be in excess of what a client wants at this stage of the process. If the presentation takes place in the design/build company office, then appropriate support personnel can be introduced as needed. If the client's interest is centered around the cost accrual and reporting aspects of the project, then by all means, someone completely knowledgeable in this area should attend. This is, however, an unlikely scenario. The thing to avoid is the "cast of thousands" approach. Bringing in all participants works well with certain types of construction projects such as hospital construction which is usually controlled by a committee interested in meeting everyone involved – from the field superintendent to the president of the company. This is, however, not the best approach for design/build, which trades on expertise and confidence building. The project requires support from a variety of personnel, which is understood, but it is not necessary to parade everyone through as part of this meeting. Fewer representatives are more effective at this stage.

The **Client** contact who attends the meeting will (hopefully) bring the decision maker. The initial contact is usually a different person. Throughout the sales process, the decision maker may be met, but usually only in passing and not long enough to establish a contact or make a point. In the actual presentation, it is important for the decision maker to be present. Otherwise, the presentation or some portion of it may have to be repeated. This is not to say that it cannot be repeated, but the impact and "moment" is lost in repetition. Nevertheless, the decision maker may not be ready to make the decision and thus, a repeat may be unavoidable. As long as there is dialogue, the prospects continue.

The Formats

A sample agenda is listed below, with the appropriate person matched to each part of the presentation.

1. Introduction – Sales
2. Introduction of the Design/Build Personnel – Sales
3. Introduction of the Project Concept – Architect
4. Introduction of the Outline Specifications – Estimator
5. Introduction of the Conceptual Estimate – Estimator
6. Questions – All
7. Terms and Form of Contract – Authorized Representative
8. Concluding – Sales

The agenda is broad and need not be developed more specifically unless the sale warrants it.

The documentation is presented in a bound book with graphics created for this particular project. The documents presented may include the following:

- Rendering – colored or black and white perspective of the proposed building
- Mounted concept plans
- Bound outline specifications and conceptual estimate
- Plot plan, if available
- Site plan, if available
- Flow diagrams
- Equipment schedules
- Boring or test pit logs
- Code review report
- Life safety analysis
- Letters from officials on elements of the design
- Any other documentation pertinent to the definition of the project and clarification of the contract agreement

This "full dress" procedure is not performed every time. Only those "hot" prospects that appear serious about building get this treatment. Creating and assembling all of this documentation and information is costly and the decision to proceed with this course should be made with care. This is not to say that less advanced prospects are not given some form of presentation.

Short Form Presentations

In many instances, a "prospect" may, in fact, only be interested in obtaining a budget. Such situations may involve plant facility managers who are putting together long-range improvement budgets for management. Or, small-scale manufacturers may be "testing the waters" in anticipation of future expansion. Another possibility may be a client planning to do the project by lump sum bid, and looking for a free estimating service. This type of false lead should be recognized early in order to avoid wasting time or resources. Others among potential clients may not have stated their intentions in advance for fear of not receiving the estimating service. That is not to say that the design/build company would not provide such a service, but that it would do so in a briefer way. The design/build company is interested in the contact and the goodwill achieved by this service. It is simply a question of recognizing the difference between the need for a short approach and the full treatment. The sales contact is responsible for making such a decision.

The short form of presentation may only require the preparation of a letter containing brief estimate information. Short forms of conceptual estimating are available and the concept design may only require a "footprint", or plan view of the project. The outline specification need be only a few paragraphs. These brief elements would constitute the documentation. Investigation of the site might be limited to simply viewing it. Investigation of building codes or local zoning particulars is not necessary at this time. A brief version of an outline specification is shown as part of an example short-form letter in Figure 9.1.

The sales contact must maintain communications with the client until the deal is made. The method used is called follow-up and involves periodic phone calls and meetings when appropriate.

Peaking

"Peaking" is the term used when a sales contact runs out of reasons to recontact a potential client. One can always call to check in and say hello, but the communication is more effective if there is a concrete reason to get back into the client's office to conclude some form of business.

A good salesman should never peak. The way to avoid it is to build into each encounter the potential for another meeting. The period prior to the presentation provides many opportunities for meetings. Not only are preparations being made for the presentation to come, but the data gathering process also requires contact with the client. Once the presentation has been made, but the deal has not yet been closed, then it becomes critical to re-establish contact. The client may need help at this stage. If he had contacted more than one design/build company, the client is now faced with a dilemma. Each of the concepts and the pricing of those concepts is unique and there may be no apparent basis of comparison. As a rule, the sales contact who can retain a legitimate reason to make contact in the client's office will have the best shot at closing the deal. One way of doing this is to introduce, but not discuss (at the presentation), the *items for consideration*. A follow-up call to discuss them in the client's office could lead to further services required by the client; these needs generate the occasion for further contact, and so on.

Another issue that the client may want to discuss is the repricing of certain items based on the schemes of other design/build companies. Or, he may want further explanations which the sales contact may put off in order to avoid "peaking".

All of these subtleties are important to the success of the design/build company. It is a business based on trust – which has to be cultivated. The design/build concept can be difficult to accept, but the more positive the contacts, the greater the potential for a contract.

(Date)

(Client)
(Address)
(City, State, Zip Code)

Re: Proposed Manufacturing/Warehouse & Office Facility

(Dear):

In accordance with the criteria outlined in our meeting of __September 15,__ we have established a conceptual budget for the above referenced project in the amount of __$3,300,000.00.__

The above costs include the necessary clearing and preparation of the site (assuming a balanced site), and installation of all utilities and hook-ups to the city system. The building, containing foundations (five-inch slab on grade) and truck dock, has a metal skin and masonry (at the office facade). The interior of the office is finished with drywall partitions (200 linear feet included), acoustic lay-in ceiling, solid core doors, and medium grade hardware, and is painted throughout. The structural steel and bar joists are shop-painted, with touch-up in the field included. The toilets have ceramic floor tile and base, and the walls are ceramic tiled to five feet above the floor. The office floor is both carpeted and vinyl composite tiled (VCT). The remaining floors are sealed exposed finish.

Sprinklers and Plumbing have been included as required by code and in conformance to your requirements and people-count furnished. Electrical lighting and power will conform to normal office and warehouse requirements. Mechanical work will include a roof-top heating/cooling unit for the office, and gas-fired unit heaters for the warehouse.

Necessary Supervision, Design Services and General Conditions Items of Work are included.

This budget anticipates the start of construction in six months.

We appreciate the opportunity to provide this budget service, and look forward to working with you on this project.

Sincerely,

Figure 9.1

Closing

No one really knows exactly what sells; if they did, there would be no need for advertising agencies. This statement, which has been made previously in the text, also applies to the subject of closing. The actual closing of the deal requires certain actions and the right approach, but it is also the culmination of a successful sales process. While there is no simple formula, there is good technique. Other than avoiding peaking and allowing for follow-up, there are other approaches that a sales contact can use in order to win over a potential client. Some of the basic strategies are outlined below.

In the initial contact at the client's office, never offer a business card when first meeting, but rather leave the card at the conclusion of the meeting. The reason for this approach is that it forces the client to remember your name. Be pleasant with the receptionist as well as the potential client; ask and remember this person's name. He or she may have a great, if subtle, influence on the response you get.

If you find that the meeting is to take place in a conference room, it is helpful if you are sitting on the same side of the table with the client, preferably with your competition on the other side if it is a combined meeting. This arrangement reinforces the impression that you are on the same side as the client, figuratively as well as literally. At the presentation, the sales contact should sit next to the client for the same reason.

Work through your fears in advance so that you can be calm throughout the sales call. This calm is based on a thorough knowledge of your subject, and the confidence that familiarity brings. Be a winner. There is nothing more infectious than a person speaking positively and enthusiastically about his company, a construction technique that can be used on the client's building, or most importantly, about the client's project in total. Joining in the client's excitement is quite effective.

Do not disable yourself by trading on past history. The past accomplishments of a design/build company provide an entrance through the client's door. From that point on, it is what you do with the present and future that makes the deal. Do not think that you can be just an order taker. The winner must work for the contract, which means good salesmanship and doing your homework regarding the client and the project.

Sizing up a prospect in terms of business interests and priorities is another tactic that might be used to close a deal. For example, the theory states: If a client is neat, well dressed, has a tidy desk, and appears conservative, he or she tends to like the technical approach. If the client dresses or decorates his office with a lot of color, has lots of mementos, and is warm and friendly, the sales presentation should feature the aesthetics of the project. If the client appears urgent or aggressive, the sell should be fast with the emphasis on results and the bottom line. If the client is opinionated, uncompromising, and somewhat erratic, the sell should emphasize the company's innovative approach, problem solving abilities, and goal-oriented direction.

The question, of course, is: Does any of this work? Who knows? These are, however, tried and often successful techniques which, along with your personality, friendliness, sincerity, natural talent, and perseverance . . . are usually effective.

Chapter Ten

CONTRACTS

The standard form of contract between a design/build company and a client is much broader in content than the type of contract used for a lump sum bid. The relationship between parties is different for each of these two types of construction methods. Lump Sum places all of the construction responsibility on the contractor once the single price offer for the work has been accepted, in conformance with the complete plans and specifications. All of the risk for this element of the project is with the contractor. The responsibility for the design is with the architect and the design engineers. The owner's relationship with the contractor is passive, providing the project is completely designed and there are no changes in the work. The legal arrangement between owner and contractor is simply that for the consideration offered, the owner accepts the contractor to construct the building.

Design/build contracts are more involved and require the client to be a more active participant. Since the design and construction are carried out by one entity, the design/build company, there is less opportunity for the kinds of complications and delays that can occur in bid construction. The responsibility for the successful completion of the project is with the design/build company, but the owner also shares some of the risk. The lump sum bid is based on a 100% completed set of plans and specifications, while the design/build Guaranteed Maximum Price (GMP) is not. The contractor for a bid keeps all of the profits from his ability to do the work more efficiently and his ability to negotiate subcontracts. The design/build company shares any such profit. However, both the lump sum contractor and the design/build company bear all of the risk if the cost of the work exceeds the money available to build it. The lump sum bidder must pay the price when the cost exceeds the bid; the design/build company must pay when the cost exceeds the GMP. The big difference between these methods is the *control of the design*. The lump sum bidder does not need the same control because the plans (on which the bid and the construction are based) can be seen. The design/build company, on the other hand, guarantees and *then* designs. With the risk factor involved, why then would a client opt for design/build instead of lump sum bid? The reasons are as follows:

- Faster completion of the project through fast track.
- A Guaranteed Maximum Price (GMP) to use for getting project financing.
- A wider disbursement of funds, thereby reducing loan costs.
- Control of the elements of the project to generate savings.
- Avoiding the possibility of redesign costs, which can occur if the design and the budget are not compatible. With traditional redesign also comes a delay for rebid and increased carrying charges for money spent on the original design.

Contract Comparisons

The standard form of contract frequently used for design/build contracts is the American Institute of Architects (AIA) Document A111, *Standard Form of Agreement between Owner and Contractor, Cost of the Work Plus a Fee*. The standard form contract commonly used for lump sum bid is AIA Document A101, *Standard Form of Agreement between Owner and Contractor, Stipulated Sum*. See Appendix C for copies of these documents. A comparison of these contracts will illustrate the basic differences in the relationship.

Note: Although relatively new, AIA Document A191, *Standard Form of Agreements Between Owner and Design/Builder*, is becoming widely used. This document also appears in Appendix C.

AIA A111 Cost of the Work Plus a Fee Article		AIA A101 Stipulated Sum Article	
1	Contract Documents	1	Contract Documents
2	The Work (Scope)	2	Work-Scope
3	Covenants – Good Faith		(not needed)
4	Commencement/ Completion Date		Commencement/ Completion Date
5	Cost of The Work (Contract Sum)	4	Cost of Work (Contract Sum)
6	Changes in Work		(In General Conditions)
7	Cost to be Reimbursed		(not needed)
8	Cost Not To Be Reimbursed		(not needed)
9	Discounts		(all to contractor)
10	Subcontract Agreements		(not needed)
11	Accounting Records		(not needed)
12	Progress Payments	5	Progress Payments
13	Final Payment	6	Final Payment
14	Miscellaneous Provisions	7	Miscellaneous Provisions
15	Termination of Contract	8	Termination or Suspension
16	Documents Signature	9	Documents Signature

From this comparison, it can be seen that the more involved the owner is in the process, the more terms and conditions are required to define the agreement. Some of the articles are essentially the same – especially in the beginning with the Agreement, Job-Project, and Articles 1 & 2. Examination – particularly of the following articles from AIA Document A111 – illustrates the owner's involvement:

Article 3. Covenants

This is the good faith article. It makes statements such as " . . . accepts the relationship of trust . . . . . (he) covenants with the Owner to utilize (his) best skill, efforts, and judgment . . . . . ." The Stipulated Sum Contract does not include this clause. The contractor does not have to have a relationship with the owner, nor does he need to do anything special in his relationship with the Architect. He does not have to be a "nice guy".

Article 5. Cost of the Work

This article concerns itself with the cost of the work. It is needed because the cost is separate from the fee. Savings provisions, a contingency, design costs, and profit are accounted for separately. The Stipulated Sum, or Lump Sum contract is all cost. Profit included in such costs is not identified and does not need to be accounted for separately.

Article 7. Costs to be Reimbursed

This article defines those items that are accepted as cost of the work. It stipulates, for example:

- 7.1.1.1 Labor costs paid
- 7.1.1.2 Salaries (supervisory and administrative)
- 7.1.1.4 Fringe benefits and taxes
- 7.1.2 Subcontractors' costs
- 7.1.4.2 Rental of equipment
- 7.1.4.5 Travel
- 7.1.5.2 Taxes
- 7.2.3 Losses not covered by insurance

The need for this article is to prevent unauthorized charges from being entered against the project; only those agreed to as cost of the work are applied. The Stipulated Sum contract, on the other hand, does not contain this article because the cost of the work is one number, a lump sum. That is to say that if the contractor wants to buy a new car and charge it to the project, it is no one's business except his (and that of the IRS if he is audited). He need not answer to the owner with an itemized expenditure, but only as a percentage of the work as completed.

Article 8. Cost Not To Be Reimbursed

This article has a bearing on the savings provisions and the disbursement of funds. Along with Article 7, stipulating what *is* chargeable, Article 8 stipulates what *is not* chargeable. Following are some examples of Article 8 items:

- 8.1.1 Office employees at main office
- 8.1.2 Main office expenses
- 8.1.3 Overhead
- 8.1.7 Anything not listed in Art. 7
- 8.1.8 Cost beyond the GMP amount

As with Article 7, the Stipulated Sum Contract contractor need not identify individual items expended; therefore, there is no need for Article 8 in a Stipulated Sum contract.

Article 9. Discounts

This article clarifies the matter of cash discounts. These discounts are kept by the contractor because he paid the bill in advance of payment from the client. If the client establishes a fund for this purpose, then the discounts can be accrued by the client. For the Stipulated Sum Contract, however, the cash discounts (similar to all other savings) remain with the contractor; the owner's advancement would not be relevant. If discounts were incorporated into a Stipulated Sum contract, they would certainly not be a contractual requirement, and could only be done with the consent of the contractor – a rare circumstance.

Article 10. Subcontract and Other Agreements

This article continues the shared information and approval approach by requiring the agreements to be approved by the client. This approval, with the advice of the design/build contractor, is necessary for the control of expenditures and the maintenance cost. The Lump Sum Bid contractor need not seek this approval and is solely responsible for securing the subcontractors and executing subcontract and other agreements. (This clause is usually modified to reflect the design/build company's relationship to the architect.)

Article 11. Accounting Records

This article describes the cost accounting method and provides for the owner's access to records. Stipulated Sum contract agreements do not incorporate this article because it is not the owner's concern how the funds are expended.

Article 15. Termination of Contract

The provisions for termination of the contract are needed in a design/build contract because of the formula of cost disbursement and its relationship to other articles in the agreement. The Lump Sum contract refers to the Standard General Conditions for this requirement. The Standard AIA General Conditions, AIA A201, is always included as part of the contract documents for Stipulated Sum contracts.

Standard Form Contract

A variety of organizations have issued printed, standard-form contract documents. See the list below.

AIA	American Institute of Architects
AGC	Association of General Contractors
ACEC	American Consulting Engineers Council
ASCE	American Society of Civil Engineers
CSI	Construction Specifications Institute
EJCDC	Engineers' Joint Contract Documents Committee
NSPE	National Society of Professional Engineers

The printed standard contract (AIA or other) is usually modified. Sometimes the modifications contain more language, word for word, than the original document. This is because of the unique nature of each project and the particular relationship between the design/build company and the client. Some design/build companies offer the client the standard form of printed contract along with their own "standard" printed modifications to that contract. The reasons for this approach are clear when explained in a legal context. The printed standard form contract has both pros and cons, as follows:

PROS The standard document . . .

- is carefully designed.
- is based on a collective body of legal experience.
- can be obtained cheaply.
- may carry more weight in court due to its legal foundation and extensive use.
- will be more familiar to the owner.

CONS The standard document . . .

- is slanted in favor of the organization that drafted it.
- is based on assumptions which may not be valid in all cases.
- contains clauses that reflect typical roles between parties; these roles may vary widely from contract to contract.

Because of the familiar nature of the printed contract and the prestige of the organization that originated it, many design/build companies opt for the standard form and rely on the modifications to tailor it to the particular needs of the project.

Original Drafted Contract

The modifications for some projects can become so extensive that it probably is a better idea to create an original contract. Such a document will, of course, be unfamiliar to a client, and will require a more thorough review by the client's attorney than a printed form – even one that has been heavily modified. Some of the pros and cons of an *original draft contract* are as follows:

PROS An original draft contract . . .

- will be more precisely designed to the project at hand.
- can implement specific project controls.
- may better reflect roles and responsibilities.
- will contain terms and conditions specific to the agreement.
- can be impartial to both parties in terms of the language used.
- will match the documents more closely.

CONS An original draft . . .

- will be more expensive.
- may contain untested clauses.
- may contain unenforceable clauses.
- is unfamiliar to owners.
- could be poorly written.

What Constitutes a Contract?

A contract should be fair and reasonable. It should also incorporate the following attributes: it should be well defined and priced, protective of both parties' interests, and should include enforceable statements for control and compliance. The contract must be capable of being administered properly, and should be complete in: terms and conditions, scope of work, contract documents, and form of agreement. A contract incorporates three defined categories; they are: a definition of work and terms of the Agreement, payment terms, and performance criteria.

Within the body of these categories are 25 points which can constitute a contract. Each of the points is unique to a category and some overlap into more than one category. The contract points are as follows:

1. Definition of the parties to the contract
2. By authority of (contract with a third party)
3. Type of work

4. Amount of contract
5. Scope of the work – included items
   – excluded items
6. Contract documents
7. Insurance requirements
8. Statement on progress payments
9. Statement on final payment
10. Bond statement
11. Guarantees (work and workmanship)
12. Warranty – Statement that the material is new, free from defects and faults, and in conformance with the plans and specifications.
13. Changes in the Work – procedure for processing change orders
    - Mark-up allowed for contractors, subcontractors, sub-subcontractors.
    - notification
14. Submittals
    - form
    - processing and submittal
    - material stored on-site & off-site
15. Approvals
16. Time of completion
17. Indemnification
18. Assignment
19. Backcharges
20. OSHA – Occupational Safety and Hazards Administration
21. EEO – Equal Employment Opportunity
22. Protection and clean-up
23. Arbitration statement
24. Terms of contract
25. Signature

The actual language included might be composed by the author and reviewed by his attorney, or the entire document should be drafted by the attorney. In *all* cases, legal documents and material which may have legal implications should be reviewed by an attorney. One important clause is Number 24 – Terms of Contract. This clause refers to the acceptance of other forms of agreement which may be part of the body of data included in the contract. This clause is necessary to prevent the inadvertent agreement to a statement included for purposes of scope. The Terms of Contract clause might read as follows:

> "This Agreement expressly limits acceptance to the terms of this contract; any additional or different terms proposed by any document or contractor or subcontractor, included in this document for any reason, are rejected unless expressly assented to in writing by the participants of this contract."

The degree of formality in contracts is based on the requirements of the participants. Some say that there was a time when a handshake was all that was needed, but the relationship between a client and a design/build, or any other type of construction company has almost always been through written documents. This approach is intended to protect both parties.

Providing neither is a litigant and the project does not run into a problem, simple forms are acceptable. There is, however, nothing wrong with protecting one's interests, being careful to create an equitable instrument that deals with the business relationship between parties; this is only good business.

Specific Clauses

Certain other clauses (in addition to the terms of the contract) also need to be written with particular care. Clause no. 5, Scope of the Work, involves the definition of the project. Since design/build has little "hard" documentation in the form of completed plans and specifications, the material that is used to define the project has to be clear and complete. The outline specifications define the scope of both included and excluded work. The concept plan, site plan, conceptual estimate, and any other documents are particularly defined in this section and in Clause no. 6, Contract Documents.

Clause no. 8, Progress Payment, and Clause no. 9, Final Payment, need to be specific in stipulating a schedule of payment for the design portion of the work, the fee disbursement, and the actual cost of work completed. As a rule, the design portion is paid at a percentage rate in accordance with the different levels of the design without retainage. (Retainage is that money, usually 5% to 10% of the work completed, that is withheld from payment. This is done to insure the completion of the work.) The fee is to be paid as a lump sum but is reduced to a percentage for the purpose of progress payment; this percentage then becomes a multiple of the percentage amount completed to date. The reconciliation (assuming a 5% fee) is shown in the following example based on the case study:

Construction cost	$2,800,000.00
Design cost	200,000.00
Fee	150,000.00
Total Project cost	**$3,150,000.00**

Amount of work completed to date $ 1,000,000.00

Percentage to date:

$$\frac{\text{amount completed to date}}{\text{cost of work}} \times 100 = \%\ \text{completed}$$

$$\frac{\$1{,}000{,}000.00}{2{,}800{,}000.00} \times 100 = 36\%$$

Fee: 36% x $150,000.00 =	$54,000.00
Total contract to date	54,000.00
less retention (5%)	2,700.00
Net amount	51,300.00
Less previous payment	20,500.00
Total payment this period	$30,800.00

This method allows for a fair disbursement of the fee; it must be stipulated as the chosen method in order to avoid any confusion or the introduction of other methods. Final payment also must be clearly stated and is usually considered to be paid 30 days after substantial completion.

Clause 13, Changes in the Work, only requires that the method of communication and the percentage multipliers for overhead and profit be clear.

Clause 17, Indemnification, may be unclear to contract participants. However, the vulnerability to third party suits is great in the construction industry. Architects, engineers, and owners are subject to claims for damages arising out of construction operations. For protection, most contracts include a "Hold Harmless" Clause. This clause basically states,"One party compensates a second party for a loss that the second party would otherwise bear." There are three types of Indemnity, or Hold Harmless clauses:

1. *Limited Form Indemnification*. This type holds the owner and the design/build contractor/architect harmless against claims caused by negligence of the subcontractors.
2. *Intermediate Form Indemnification*. This form includes not only claims caused by the subcontractor, but also in which the owner and/or the design/build contractor/architect may be jointly responsible.
3. *Broad Form Indemnification*. This type of clause indemnifies the owner and the design/build contractor/architect – even when they are solely responsible for the loss.

Type 3 (broad form) would take the place of the errors and omissions insurance carried by the design/build contractor/architect, reducing significantly the requirements for liability insurance. Since the terms of the owner/design/build company contract are duplicated in the subcontracts, the burden is then on the subcontractors, whose insurance costs will go up, providing they are able to obtain insurance. This Type 3 form of indemnification is illegal in most states.

Type 2 (intermediate) is the form of indemnification most used. It is the form found in the AIA Standard General Conditions A201 (1976 edition).

Clause 23, *Arbitration*, is a common consent by the disputants to have their differences settled in other than a court. Court action can impose delay, expense, and inconvenience. Arbitration offers a settlement that is prompt, private, convenient, and economical. The rules are as prescribed by the American Arbitration Society. The decision is final and only if both parties agree, can a hearing be reopened. Court decisions, by contrast, are open to lengthy appeals.

Guaranteed Maximum Price (GMP)

The price for the design/build project is guaranteed. That means that the cost is fixed and will not exceed that price as a cost to the client. If the cost does exceed the guarantee, then the design/build company guarantees to complete the building using its own financial resources, if necessary. The only condition of change to this GMP is if the owner institutes changes, or if governing bodies alter the codes and require retroactive compliance. This latter occurrence is unusual. Normally, the code compliance is, in effect, tied to the building code revision when a building permit is issued. However, catastrophic occurrences, such as disastrous fires, have prompted such official measures.

Again, one of the big differences between design/build and other types of construction is the point at which the GMP is given. If one examines the stages of the design process, it normally progresses as shown in Figure 10.1.

Design/build provides a GMP at a point between Phase 1 and Phase 2. In a construction management/contractor situation, the GMP is given at a point between Phase 3 and Phase 4. In the case of professional construction management (acting solely as a consultant), a GMP is never provided. A lump sum contract situation provides for the GMP only after the completion of Phase 4. With a potential time span separation between Phase 1 and Phase 4 of three to six months or more, the significance of the design/ build method becomes quite apparent. The problem, however, is the amount of documentation available at that (GMP) point in time. This documentation is developed as explained earlier through the outline specifications, the concept plans, and the conceptual estimate. Further, any other documentation is to be added to these three basic elements to clarify or explain the project as fully as possible. That is why the description of the work and the documentation requires careful compilation and definition.

Included in the description are allowance items. Previously it was explained that those items of work that are uncertain cannot be guaranteed, and are either carried as an allowance or listed as part of *items not included*. It is preferable to "guesstimate" an item and include it in the cost rather than to omit it entirely. It is the 80% rule, which states that it is better to be at 80% of the cost than at 0, because the differences are less traumatic. This "guesstimate" amount then becomes a value that will be adjusted once solid cost information is known.

There are two types of allowances. One is the cash allowance, which is simply a sum of money to accomplish an item of work. The amount of the cash allowance may have been developed from a computation, quantification, and unit costing. Regardless of the basis used to arrive at this amount, it is still a cash allowance. Examples of cash allowance items are hardware and landscaping.

The second type of allowance is a *scope allowance*. A scope allowance defines a particular amount of work and assigns a unit cost to it. The unit cost can be adjusted, but the quantity is not adjusted. An example of a scope allowance is carpet carried at a unit cost per square yard times a defined number of square yards. Brick material cost is commonly carried as an allowance until selected. This allowance is at the rate of cost per thousand brick delivered to the job. The quantity of the brick does not change and the adjustment is made only on the unit cost.

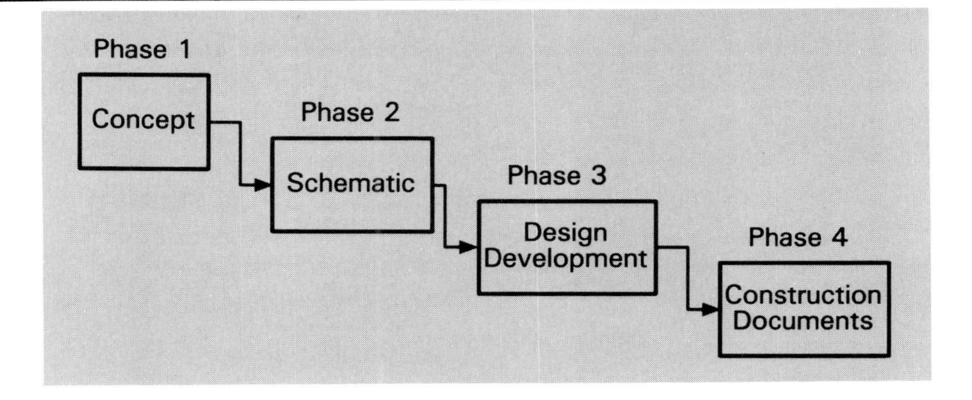

Figure 10.1

Part Three

DESIGN AND PRECONSTRUCTION ACTIVITIES

Chapter Eleven

DESIGN REVIEW

Design Levels

The design of a building construction project, with its complex and interdependent systems, is usually accomplished through various stages of development. Figure 11.1 illustrates the major stages.

During the *concept* stage, the basic parameters of the building are conceived. The basic elements of the building and their relationship evolve during the *schematic* phase. At the *design development* stage, the details are worked out and the design elements specifically defined. The *construction* stage represents completion of the design at the *working drawings*. Working drawings are used to construct the building. For the *bid* method, the design is complete by the start of the construction stage and the work can commence immediately upon award. For the

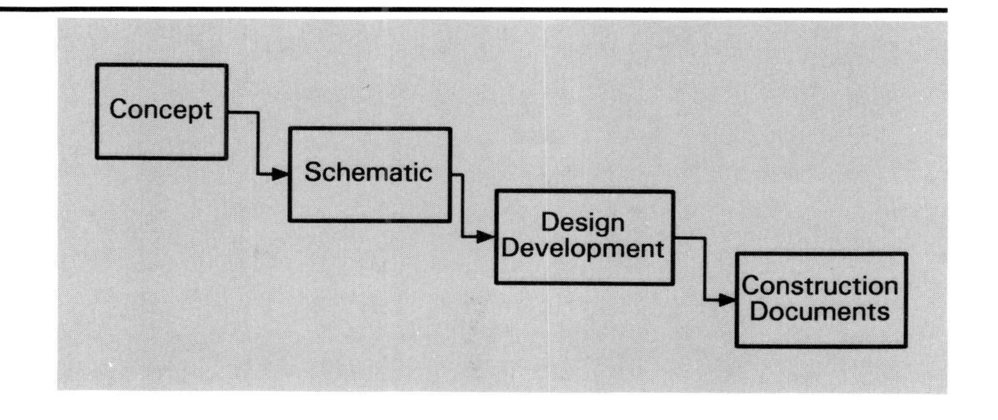

Figure 11.1

construction management method, less design work must be accomplished in order to complete the construction drawings. Construction may proceed in advance of completed design in a programmed way, called the *fast track* method, which involves bid packages.

For design/build, the design must also be completed through the construction drawings. The difference is time. In the construction management method, various alternatives are explored and examined in the course of arriving at a final design. These studies consist of sketches and revisions to large scale drawings. Changes to the design continue to and beyond the issuance of a guaranteed maximum price (GMP).

Design/build fixes the design at the concept level and then allows the process to continue with that fixed concept. Any changes tend to be in the quality/quantity areas of the design through the process of *value engineering*. Some may claim that this fixing inhibits creativity, stagnates the normal design process, and discourages design alternatives. It is true that once the design is fixed, no further changes in design concept can be tolerated. The design concept cannot change because both the GMP and the contract are predicated upon that design. The quality, quantity, and costing value engineering process continues. The involvement of all the participants (including designers, subcontractors, and the client) is encouraged.

The design/build process does not inhibit creativity or discourage design alternatives; it tests variations to the design and system concepts in the initial concept stage. This testing attempts to incorporate the client's wants into a practical, cost-effective design. The "money is no object" client does not usually opt for design/build, but proceeds through the design in preparation for a Lump Sum Bid. Design/build clients generally have a restrictive budget, require an early guaranteed price, and need to have the project completed in as short a time span as possible.

Design/build proceeds from the concept phase to design development, or directly to the construction phase, depending on the degree of difficulty of the project. Rapid design completion enhances the fast track method. The need for bid packages is the same for design/build as it is for construction management, except that it is carried out more quickly since the entire process is accelerated.

Another benefit of the design/build method is the lower cost of the design work. With the construction management method, the cost is somewhat less than for the bid method because the details need not be as extensive and every item of the building need not be shown. Both the construction management and the bid method progress through the successive design levels. The design/build design process does not necessarily progress in a regimented way. This all adds up to a reduction in design time and drafting hours, which translates into less cost and faster production working drawings. The end result is, of course, faster production of the building.

Design Progress

Once the design concept is established and frozen, it is given to the designers. The conceptual estimate, outline specifications, and the conceptual plan become the definition of the work. Clearly, the architect should be involved with the project at the inception in order to be part of the conceptualization process. This partnership and collaboration makes the transition from *concept* to *construction* (or working drawings) much easier.

The general practice in building design is for the architect to furnish the other design disciplines with the background plans onto which they can superimpose their own design efforts. In design/build, this procedure is the same. The architect converts the *concept* drawings into *design development* or *construction* level drawings and gives them to the other designers. Then the architect proceeds with the completion of the design. The other designers need not be design engineers, but could also be design/build subcontractor/designers. These subcontractors would complete the design in the same manner.

During the course of the design, all of the plans are critiqued on a periodic basis by the design/build company for review and verification of content and scope. When the design is complete and has received a final review by the design/build company, it is reviewed by the owner.

Sign-Off with Owner

As the design progresses (and before bidding of the work), it must be verified by the owner to make sure that it is in keeping with his concept of the building (in conformance with the presentation documents and the contractual agreement). When the owner has reviewed the construction-level documents, he must signify acceptance by signing each document.

The design/build company project manager reviews the plans with the client, explaining how what is shown is the same as what was proposed. When the client has been thoroughly briefed and is satisfied that the completed design and the conceptual proposal are the same, then he is asked to sign a record set of drawings to signify that fact. If the client is not convinced that what is drawn on the plans is the agreement, then the review and/or the design must be worked through again until there is unanimity. The progress of the project cannot go forward until such issues are resolved. The agreement of the client is key to the design/build construction method.

At this stage, the outline specifications and the conceptual estimate provide the qualitative and quantitative definitions of the project, and the conceptual plan provides a visual definition. These documents have translated the early, conceptual view of the project into formal plans which require some experience to read and understand. These plans do not resemble the concept plans, except for the elevations. The client, therefore, requires sufficient explanation in order to understand that he is, in fact, the recipient of the building for which he contracted. The importance of this explanation – and the owner's careful review – is evident, to head off misunderstandings. For example, the plumbing section of the outline specifications states that three urinals, ten lavatories, and eight water closets are part of the work. All that the client needs to do is to count the items on the plumbing plan to verify. He should not be obliged to determine pipe sizes for adequacy since the code stipulates good design and the plans would be reviewed by a building official. He should be cognizant of the quality of these items, and can check this information by comparing the outline specification and the construction specification.

The client should be asked to review the finish schedule and the specifications to verify the type of ceiling, flooring, and wall construction. He should be shown the types of walls and where the various finishes of those walls are to be applied. The client does not need to examine the gauge of the metal drywall stud, nor be preoccupied with structural connections. He only needs to know that the structure is of specified structural steel and not unspecified masonry bearing.

The client's review is quite broad and covers only the areas he can verify. The client can always select a consultant to review the plans and make recommendations on design and content. Most clients will do just that, covering themselves with this extra measure of insurance.

As with any other step in the process, the design/build company needs to do this job well to maintain the client's trust and confidence. A review of the construction set of documents is among the most important steps in the complete process.

Bid Package Format

The construction level design proceeds at as rapid a pace as possible, but with a predetermined direction. This predetermined direction is called the *bid package* list. The data for the schedule of bid packages comes from the management schedule. The bid packages are design packages for specific items of work. They are generally determined by three factors, as follows:

1. Long delivery time
2. Critical item of work
3. Coordination

The management schedule determines which item of work needs to be procured early as a consequence of: the date the project needs the item, the length of time it takes to receive the item, and at what stage the construction level design will progress. Figure 11.2 is a segment of a simple bar chart construction schedule.

The design period represents the total time required to complete the construction phase of the design activity. However, the master schedule shows that to have a certain item of work available on time, the design of that item of work needs to be completed in advance of the total design, at a point designated as "A". That means that in order to meet the construction schedule for structural steel, the design of that item must be segregated and completed at an earlier time. This is called *Bid Package No. 1*. At point "B" in the design schedule, the design for another item of work must be completed in order to be released for bid. That item is to receive the same design treatment as "A", and is designated as *Bid Package No. 2*.

The list of bid packages continues through "C" & "D". At point "E", the list stops. The reason is that the design is complete at this point. It is no longer necessary to design out of sequence in order to satisfy job needs. All of the remaining items of work could then be bid together whether or not they are long lead items. Long lead items require a long period of time, in relation to other items, to proceed from the bid to the actual delivery of the item to the job. Examples of long lead items of work are structural steel (A) and elevators (B). Bid packages are not necessarily required because of a long lead item. Some bid packages are required because the work is critical to the success of the project.

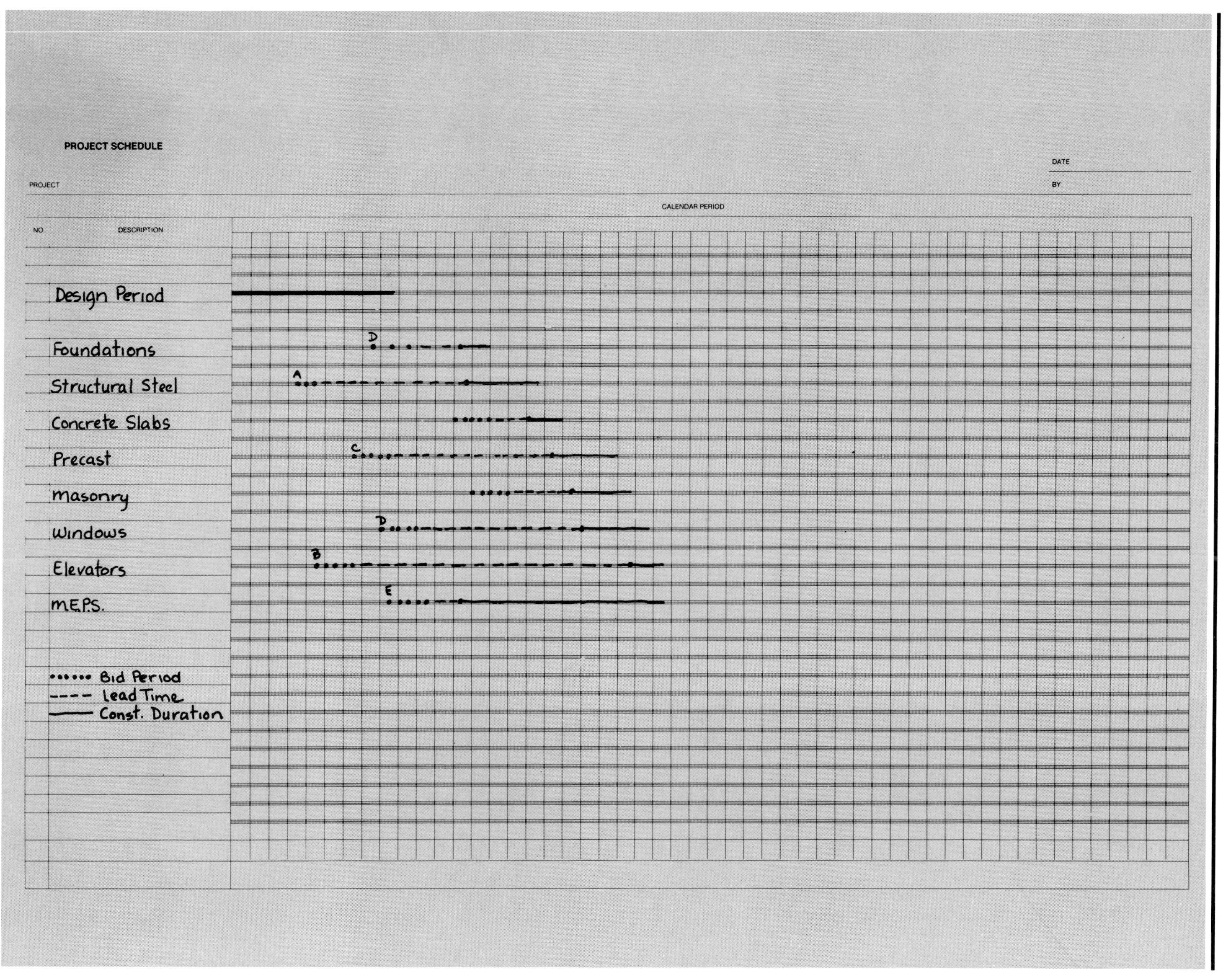
PROJECT SCHEDULE
PROJECT
DATE
BY
CALENDAR PERIOD
NO
DESCRIPTION
Design Period
Foundations
Structural Steel
Concrete Slabs
Precast
masonry
Windows
Elevators
M.E.P.S.
A
B
C
D
D
E
•••••• Bid Period
---- Lead Time
—— Const. Duration

Figure 11.2

Critical items of work in CPM (Critical Path Method) scheduling are defined as having zero "float". Float is the time period between items that allows them to be accomplished at a time other than their designated time period. If an item has zero float, it can only be done at that specific point in time determined for the success of the project. If the schedule's early start indicates that the design for a particular item is needed in advance of the completed design, then that item needs to be designated for a bid package.

The third factor is *coordination*. Coordination of certain design elements or items of work would also require portions of the construction design to be completed in advance of the whole. An example is field-installed brick masonry and in-fill precast panels. The lead time associated with precast requires that it be on the bid package list (C). The masonry bid package would not normally be required early, but the coordination of these two contiguous work items makes it important that the brick masonry subcontractor be secured as early as the precast subcontractor, so that they can coordinate their work in the shop drawing stage. The window subcontractor in this example would also be required early to participate in the same coordination process. Sometimes, the coordination is with field activity such as high tension wire relocation, boring sub-surface investigation, parking lot design, and flood plain calculations for conservation approval.

Trying to start construction at a particular time of the year is also a factor in the acceleration of the design schedule. Attempting to start a project in the spring or trying to complete a project to take advantage of a sales market are two reasons for speeding construction and therefore, the design schedule.

The bid package list can include any number of items that require out-of-sequence design. The construction schedule requirements are the deciding factor. Design/build can proceed with construction before the design is completed (fast track), better than any other type of construction because of the *total control* of the design.

Design Schedule

A construction schedule of activities is based on productivity in man-hours per unit quantity. Design scheduling is also based on productivity, in design hours. Design work differs in that creativity is difficult to quantify and only a finite number of people can work on the drawings at one time. Construction work can utilize large numbers of workers doing the same kind of work in different locations in the project. Only a few people can draw a series of sheets of drawings since continuity and coordination are so important.

Design can be manipulated and prioritized to some extent in order to produce certain items that may be required first. Nevertheless, short of working night and day, seven days a week, only so much work can be accomplished in a given period.

Knowledge of design productivity is important to the design/build company because of the early commitments that must be made for this mode of construction. These commitments must reflect realistic time spans with all factors considered – not the least of which is the design process.

As stated earlier, the design of a project proceeds through various stages. By the use of a bar chart, the relationship and typical durations can be shown as in Figure 11.3.

This schedule illustrates a project for which the normal duration for design – from concept to completion of the working, or construction, documents – is about 23 weeks, or almost six months.

In a bid package for a fast track format, the design schedule for this same project would appear as shown in Figure 11.4.

This schedule shows the out-of-sequence activity necessitated by bid packages. The design period remains the same, but the specific items on the list must be worked independently. The overlap in the design development and construction documents makes the architect's work more complete.

The design/build method involves design stages in a schedule similar to that shown in Figure 11.5.

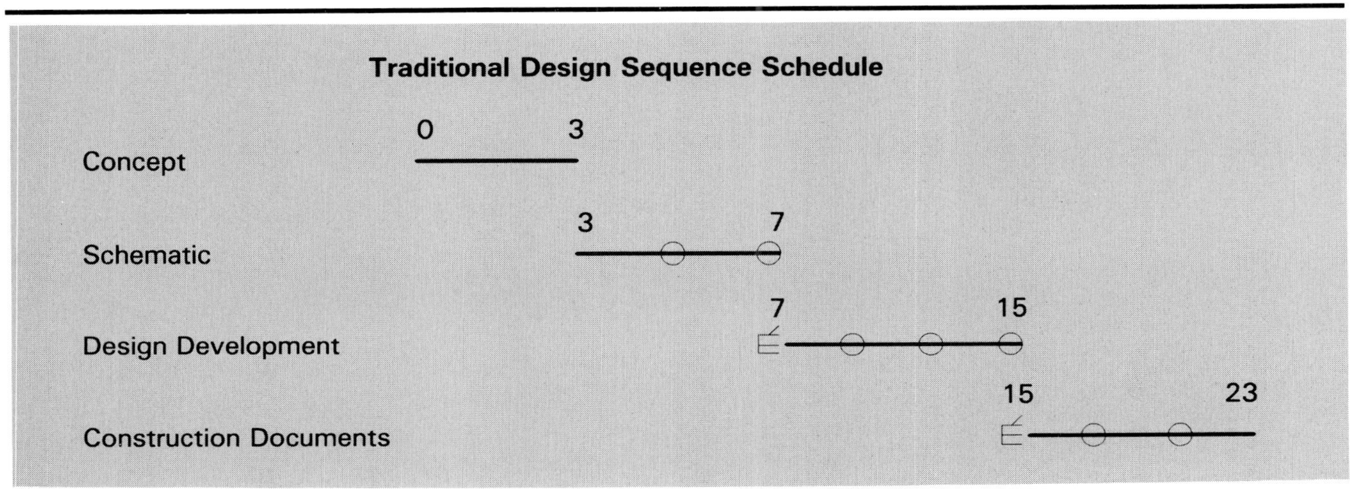

Figure 11.3

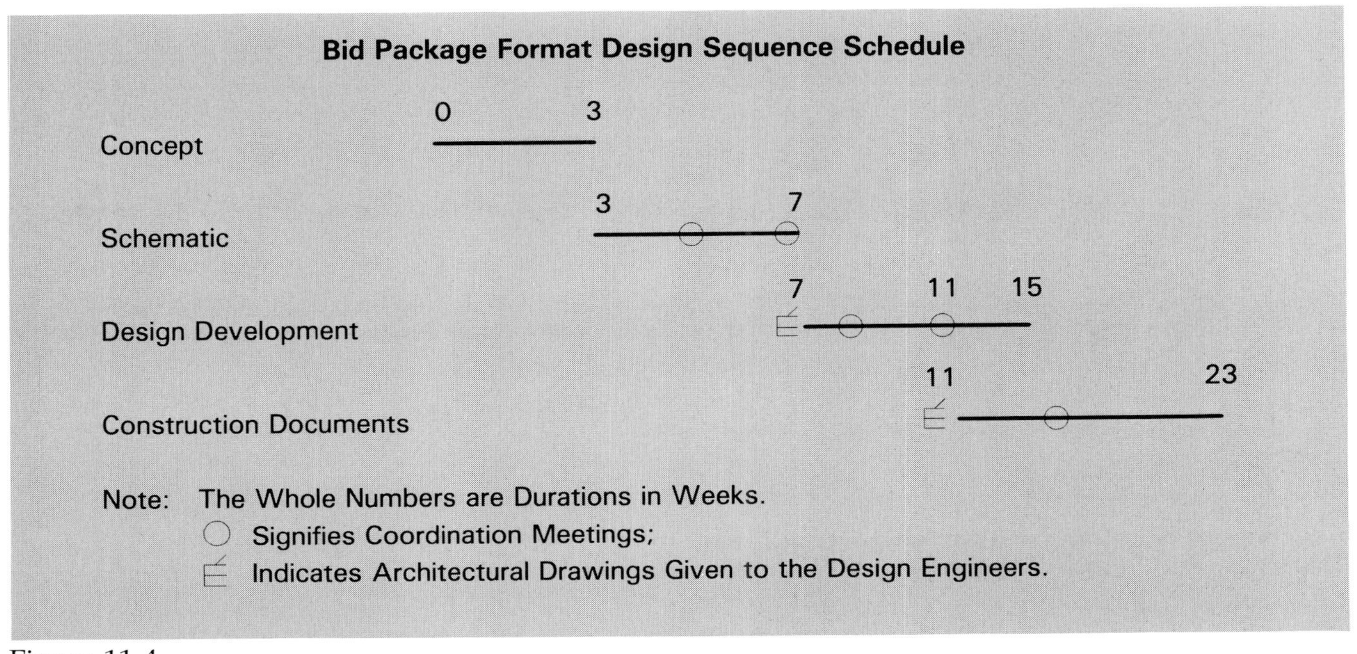

Figure 11.4

The design/build design time is shorter because the schematic stage is omitted. Further reductions of the construction stage are possible since there are fewer details on the drawings. Many of the details are known or are worked out on shop drawings. The standard design time can be reduced from 23 weeks to 19 weeks, or even as low as 15 weeks. This design time reduction, coupled with the early start of construction, goes a long way toward producing the project quickly and reducing the cost.

The traditional design sequence activities are as follows:

Concept Stage

- Site layout analysis
- Design feasibility studies
- Zoning review
- Initial material selection
- Initial M.E.P.S. designation
- Initial structure selection
- Elevation views
- Initial scale drawings

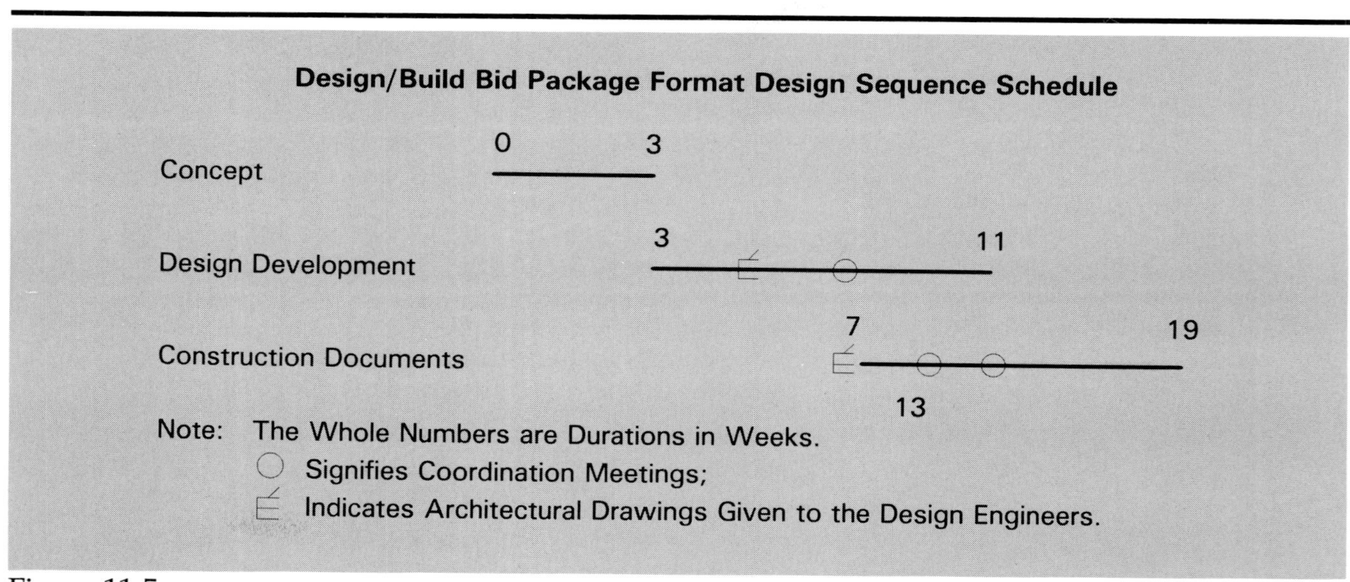

Figure 11.5

Schematic Stage

- Confirmation of program requirements
- Final site location of building
- Beginning of floor layouts
- Test pits/borings
- Preliminary site plan
- Preliminary foundation plan
- Confirmation of live loads
- Selection of framing system
- Initiation of building code review
- Egress study
- Fire/life safety study
- Utility study
- Flow tests
- M.E.P.S. selection
- Determination of building materials
- Establishing quality levels
- Product distribution study
- Revised scale drawings

Design Development

- Establish floor heights
- M.E.P.S. design
- Framing system design
- Foundation design
- Coordination of drawings between designers
- Establishing egress and fire zones
- Fixing building materials
- Fixing floor layouts
- Developing wall sections
- Developing window sections
- Developing details
- Preliminary door and finish schedules
- Specification drafts
- Scale drawings as required

Construction Documents Stage

- Complete floor layouts
- Complete elevations
- Complete M.E.P.S.
- Complete structural
- Complete specifications
- Complete coordination of design disciplines

Most of the activities normally listed under the schematic stage have to be accomplished as part of the design/build concept stage. This is important to allow the guaranteed price to be established early in the process. Once these concept/schematic items are accomplished, the design work can proceed directly to the next level. Even some of the design development stage items are accomplished in the concept stage for a design/build project. Establishing the floor heights and selecting the M.E.P.S. are two examples. In analyzing the design process, the importance of data gathering becomes quite evident – as preparation for the conceptual design and estimate.

Design Conformance to GMP

In the design/build construction method, the design is frozen at the GMP (guaranteed maximum price). No further design input can be tolerated since it is not included as part of the conceptual estimate. Review of the completed design also means the verification of design conformance. As the design progresses, and at progress points prior to the presentation to the client, it is reviewed by the design/build company.

The design/build company review includes an examination of the architectural plans to compare and verify the quality and quantity shown against the budget. If the budget required metal siding, it cannot now be designed in brick. If the building windows were of a certain height and type, they cannot now be different. Usually this kind of flagrant departure from the scope of work does not occur. More common is substitution of finish work, such as using drywall ceilings in place of acoustic lay-in, or the introduction of vinyl wall covering where only painted surfaces were specified. The option to include these features is the prerogative of the client and not the designer. Such changes can be suggested by the designer, but must not be included until accepted by the owner. Among items to be reviewed are the following:

- The structural design should be in accordance with the agreed upon concept. A "takeoff" of the pieces might be necessary to verify that concept.
- Mechanical systems need to be verified, unit heaters counted, and sizes noted, tons of air conditioning noted and checked against the conceptual estimate.
- Plumbing fixtures must be counted and the manufacturers checked against the quality level priced.
- Electrical design is also to be checked by quantity and quality.

This review procedure does not reflect an adversarial relationship between the designers and other members of the design/build team. The outline specification, conceptual estimate, and all other pertinent documents are given to the designers at the start for their use in developing the design. Much of the information generated as part of the conceptual estimate and stipulated at the presentation (and thus included in the contract and price), is furnished by the designers, and in many cases reviewed prior to presentation. What sometimes occurs is a rethinking of the concept by another party not involved in its original development. Changes for the better are not discouraged, but their economic impact must be ascertained and they must be approved by the owner. This is the ongoing value engineering process. Some upgrade changes can be accommodated at no change in cost, but they cannot be unilaterally introduced and must receive the approval of the client.

Only when the design/build company is satisfied that the design, the plans, and construction specifications reflect the conceptual estimate and data contained in the outline specifications, will this package be accepted, shown to the client and released for bidding and/or construction.

Chapter Twelve

COORDINATION

What is Coordination?

There are two forms of coordination associated with a building project. One is *design coordination*, and the other is *job coordination*. Design coordination involves the relationship of the building's systems within the complete structure. Job coordination deals with the individual items (within each system) and their physical locations in the building. Design coordination is done by the designers. Job coordination involves the installer.

Design Coordination

The design stages of *concept, schematic, design development*, and *construction documentation* are determined primarily by the architectural design. The remaining designers – whether consulting engineers or design/build subcontractors – superimpose their designs on the drawings they receive from the architect. The number of designers involved with a project may vary depending on the type of project. A representative list could include the following:

- Mechanical (A)
- Electrical (A)
- Plumbing (A)
- Sprinkler (M)
- Structural (A)
- Civil Site (M)
- Landscape Architect (M)
- Interior designer
- Consultants
    - Elevator
    - Specification (M)
    - Hardware
    - Fire Safety
    - Hotel Management
    - Food Service
    - Scheduling
    - Costing

- Construction
- Construction Management
- Hospital Services
- Historical

The particular designer's need will depend on the project use. Not all of the designers listed will be required for a given project. Those identified with (M) are required for most projects; those with (A) are *always* required – regardless of the type of project. The designs need not be developed by independent consulting engineers; they could be furnished in conjunction with design/build subcontractors. Either way, the design should be done by a qualified engineer. Having so many different disciplines requires that a single source coordinate all of their efforts.

This coordination by one individual involves examining the plans, checking to make sure that the input is correct, and relating each system to the building as a whole. The location of shafts and pipe chases, for example, must be the same on all of the drawings. The routing of piping vertically and horizontally through certain spaces should be checked against available space and for their position relative to contiguous work. The positioning of mechanical and electrical machinery and equipment in their assigned spaces needs to be checked for size; all dimensions should be verified as correct.

In the bid and construction management methods of construction, the coordination is done by the architect who has a contractual relationship with the designers and consultants. In those instances where the architect is not contractually related to the other designers, the coordination work is generally included as part of his contract. The architect is a logical choice for coordinator because it is a task that can be done in the course of his own design work. The architect is the person most familiar with the nuances of the building.

The Design/Build Approach

Design/build companies customarily enter into separate contracts with the different design disciplines, or design/build subcontractors. There is, however, no contractual relationship between the owner and these designers.

The design/build company can accomplish design coordination in several ways. One choice is to assign this task to the architect as part of his work. Alternatively, the design/build company might hire an independent consultant to do the design coordination, or they can do it themselves. Those companies that do their own coordination must attend the coordination meetings and secure a progress print from all of the designers. Each print is then critiqued based on a comparison with the architect's plans. The prints are then returned – with comments – to their originators. This critique is not limited to the designers of building systems; comments may also be made to the architect if there are deficiencies in his plans. The major concern at this point is the availability and use of space.

Initially, the architect is given the required dimensions for the mechanical-electrical-telephone rooms and space needs for major equipment and devices. The architect also must consider the vertical and horizontal disbursement of ductwork, piping, and conduit, and provide shaftways and ceiling space for those purposes. This initial data is then manipulated for the maximum efficiency of the design, to accommodate the needs, but also maximize the useable space. Such manipulation requires compromise and relocation. This is where the design coordination is important. During this stage, the routing of work items and the useable spaces and rooms for this purpose are in a constant state of change. In addition to the M/E work, the structure itself must also be considered. The structural design may require space constriction due to a brace or a connection. This revision may in turn oblige changes in areas previously assigned or dimensioned. With this state of change, common to all jobs, the necessity for the coordination is as important for design/build as it is for any other construction method.

If the designers are associated with the design/build subcontractors, the process is no different. Their efforts must be cooperative and they must proceed in the same manner as if they were independent consultants. One of the benefits of design/build is that job coordination may be a smoother process since the designers are usually also the installers.

Job Coordination

The job coordination takes place after the design is complete. It is based on the construction, or working, drawings. When these drawings have been received by all trades, meetings are set up to carry out *job coordination*.

The building design has, at this point, been reviewed to make sure that all of the proposed elements can be accommodated within the building. Much of the information and systems are defined schematically. The job coordination now involves the to-scale drafting of certain items to ensure if they can indeed fit within the building. As a rule, the space available is ample to accommodate the items and the job coordination runs smoothly. In extreme cases, building design modifications may be necessary.

Job coordination is concerned mainly with the horizontal space between the ceiling and the underside of the floor above. It is in this area, depending on the elements of the project, that most of the project's functional items are installed. Within this space are the sprinkler system, mechanical system, electrical work, plumbing, and the structural skeleton of the building.

The sprinkler system consists of horizontal piping and vertical extension of the piping to the ceiling line, as well as valves. Sprinkler components must traverse to every part of the floor.

The mechanical system includes piping and ductwork. The ductwork, which takes up the most room in this space, must be positioned carefully so that other elements may go around it, along side of it, or, if totally unavoidable, through it. Fans, heaters, and heating/cooling units also need to be accommodated in the space.

The electrical work takes much the same path as the sprinkler piping, but involves electrical conduit. This conduit will be located in most of the parts of the floor. The space must also allow room for the light fixtures, usually from three to six inches in depth.

The plumbing work is of less concern because it is usually in specific locations and not in every segment of the floor. When it is routed horizontally, it does not require much space. Its vertical position is "stacked", like the roof drain leaders, so it requires little space.

Also sharing this space is the structural skeleton for the building. If the system is shallow, as in the case of reinforced concrete, then the space is reasonably free of obstructions. If the system is structural steel, then the members of that system need to be particularly considered in the job coordination process. The coordinated items will be located mostly below the steel structure. If necessary, some of the elements will go through the structure. In the case of bar joists, there is often adequate space available for small duct, piping, and conduit. (See Figure 12.1.) If section beams are used, then piping and conduit and, in some cases, the ductwork will be installed by cutting or burning holes in the structural members.

Job Coordination Meetings

The first meeting is to resolve basic layout questions and space requirements. All of the other trades must defer somewhat to the mechanical contractor due to the nature of his work. The lead work in job coordination is with the mechanical contractor. If the system is an air heating/cooling design, then there is a lot of work in layout of the major duct runs and lateral distribution. Once this is accomplished, then the sprinkler contractor usually superimposes his layout on the duct layout drawing and adjusts the piping runs (where possible) to avoid the duct.

The sprinkler subcontractor actually prepares two drawings. The first is a complete layout drawing showing all of the sprinkler heads in their exact location and also showing all of the piping, completely dimensioned and drawn to scale inclusive of all the fittings. This drawing is the one that is sent to the insurance underwriters and to the fire marshal for concurrence. The sprinkler subcontractor also locates his

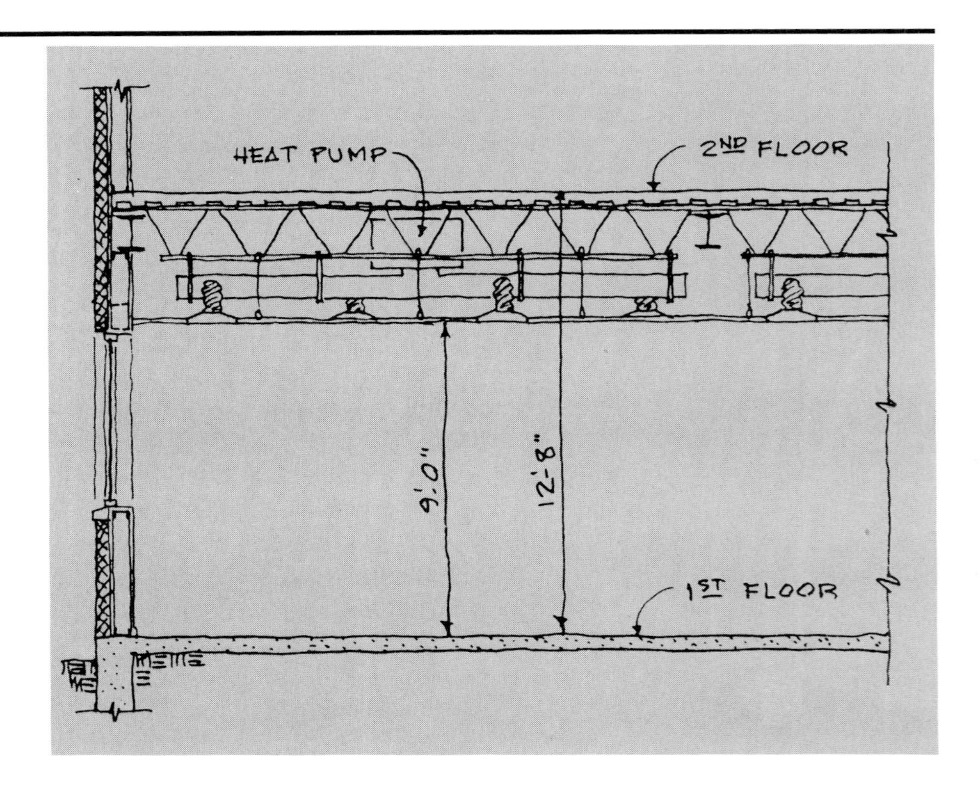

Figure 12.1

piping on the coordination drawing and, once approved, will include any changes on his original drawings. It is from that original drawing that the piping will be cut and installed on the job.

Next is the electrical subcontractor who also lays out his lines on the duct drawings in the designated spaces. The electrician must also be aware of the location of equipment, but usually installs his conduit right to the structure and can "sweep" past most obstacles.

Finally, the plumbing subcontractor adds piping and other plumbing items to the plan. His task may be considered the least complicated in the process, except for drain piping that requires sloping or pitch to allow for gravity flow.

The process of superimposing the layouts of other systems on the mechanical layout drawings is illustrated in Figure 12.2.

After all of the trades have introduced their work onto the drawing, it is reviewed by all interested parties. The architect judges the drawing based on conformance to the design. The design/build company makes sure that no added work has been introduced and confirms conformance to their agreement with the client. The structural engineer checks any changes affecting the structural beams and loads. Anyone else who might have an interest also has an opportunity to review the drawing at this point.

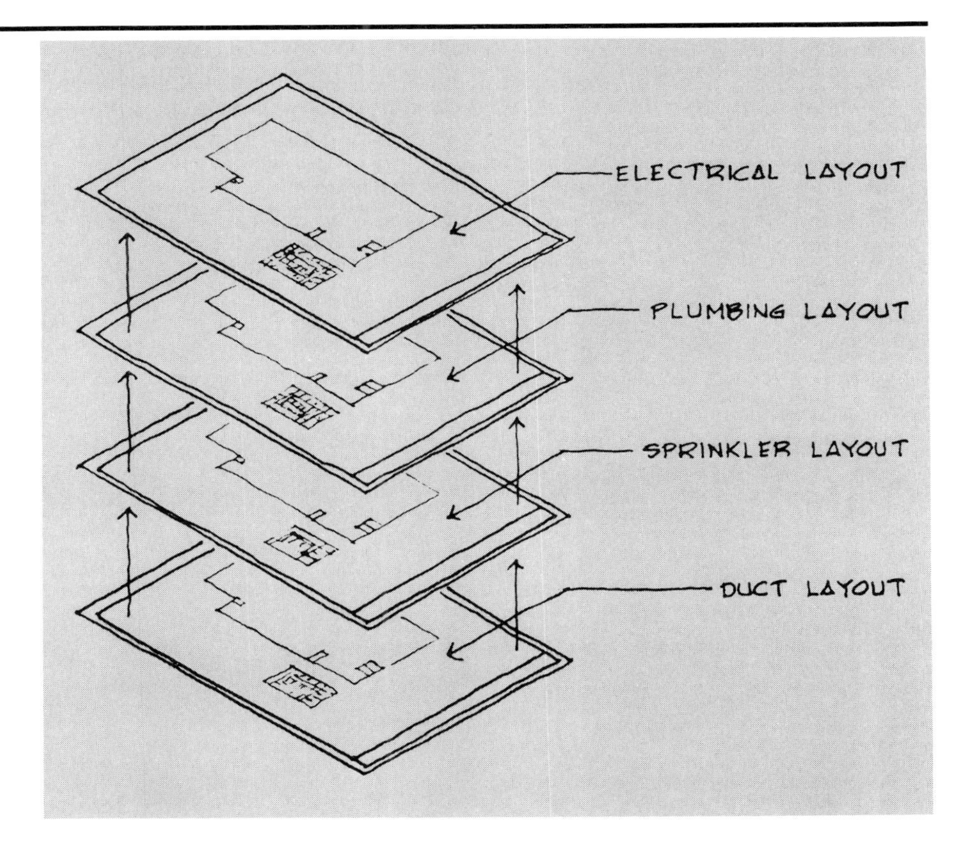

Figure 12.2

The drawing is then reproduced for all of the trades so that they can begin installing their work in the designated areas. Actually, the construction work has already begun by the time the job coordination drawing is approved. Any variation is field adjusted.

Job coordination does not always have to be on such a large scale as in the case described above. Some jobs are not so complex or the available space may be such that there is plenty of room to accommodate all systems. Nevertheless, control of the trades and their methods of installation is always a good idea.

Design/Build Services

It is important that the practitioner of design/build services be aware that coordination is part of the service and is to be considered accordingly. The cost imposed for the design has to take into consideration all of the requirements. These include services for:

- Architectural
- Civil
- Structural
- Electrical
- Mechanical
- Plumbing
- Landscaping
- Construction
- Management

The construction services include the job coordination and construction drawing review. The management services include design coordination and the design drawing review.

In addition to design, there are testing and inspection services. The cost of these services is usually carried as part of the General Conditions. The testing contract is drawn up with an independent consultant to perform the testing of all materials required, such as concrete, asphalt, and soils. The inspection work is also performed by an independent consultant and can include concrete, asphalt, soils, structural steel, reinforcing steel, roofing and masonry work.

All of the activities normally included as part of the architect's work (in other methods of construction) become part of the design/build company's requirements in design/build construction. All of the responsibility is now with one entity, the design/build company, and is not fragmented as it is in other methods.

Chapter Thirteen

Cost Accounting

What is Cost Accounting?

Cost accounting is the method used to gather job costs. Costs can be accounted for in the following ways: in aggregate under broad categories for the company, in aggregate under broad categories by project, or as finite numbers for the company or for each project. The application of these costs will vary depending on the requirements of the business and the accounting system being used.

Cost accounting is necessary for many reasons, including:

- for tax purposes
- for insurance purposes
- information for the Bonding company
- for historical information
- for testing and upgrading unit prices
- to monitor the progress of the job as it relates to the estimate
- for contractual obligations

Taxes

The maintenance of cost records for tax purposes is a clear necessity; it is the law. Companies doing construction work may be on an accrual tax basis. The accrual system takes the billing and subtracts the expenses to determine net income. Taxes are then figured against this resultant income regardless of the receipt of payment. In construction in general or design/build in particular, taxes are paid on the basis of totals; it is not necessary to keep separate project records to demonstrate the amount to the IRS. Many businesses that sell one or more products are on an accrual tax basis and keep total records. Construction companies could also do this if their only concern was the IRS.

Insurance

Worker's Compensation insurance depends upon a cost accounting system. Worker's Compensation contributions are based on the straight time labor expended. Different rates are applied to different types of work categories. Insurance costs for laborers, for example, would be at a different rate than those for ironworkers, and different still from masons' insurance costs. The reason for the different rates is the differing risks associated with the individual trades. The accounting system need only gather payroll costs into the appropriate codes to coincide with the Worker's Compensation risk category list. The list would only address the type of labor directly employed by the design/build company. Subcontractor labor and related expenses (such as Worker's Compensation) would be carried as part of the subcontractor's payroll.

Unemployment insurance payments to the federal and state governments must also be included in project costs. Like Worker's Compensation, unemployment insurance is also considered a labor cost. It is assessed on the basis of layoff experience. Companies with a high rate of personnel turnover will have a higher cost for unemployment insurance. The assigned rate is multiplied by the gross earnings – usually at about the first $7,000.00.

Bonding

Bonding is a necessity of the design/build method of construction. Most clients would insist that bonding be available and then exercise the option to require it. Bonding companies require three years of certified audits as part of the approval process. Most audits are not certified, but only state that the financial information has been reviewed. A certified audit requires a more intense accounting procedure. The records for this type of audit are cumulative and must reflect the performance of the company as a whole. Once accepted by a bonding company, the actual financial performance of a company is monitored, and compared to its projected performance. This performance involves three areas: the management of the company, the actual amount of the awarded contracts vs. the projected sales, and the amount of the actual profits vs. the projected profits. If all of the jobs were bonded, the bonding company might be interested in a total statement. However, since all of the jobs are usually *not* bonded and the bond rate varies for those that are bonded, then it is worthwhile keeping separate cost records by job. The cost of the bond is on a sliding scale based on the value to be bonded. The final cost of the bond will be based on the total amount of payment received for each job. This amount is verified by the bonding company audit at the conclusion of the project.

Historical Records

Job performance records are historically useful in the review of conceptual estimates for future jobs. If six prior awarded jobs cost a certain amount and the conceptual estimate of the current project is outside of that range, then the reason for this deviation should be understood. A comparison of the historical building system to the currently estimated building system focuses on that difference. Such an analysis is important so that the price for the project is not developed beyond the reasonable market cost. Also, the system costs that go beyond the historical parameters can be used for redesign or can be highlighted to show the client that they represent an exceptional cost or design requirement. This attention to detail makes the presentation more professional, builds confidence in the design/build company, and helps to increase the likelihood of the sale. These advantages would not be possible if good accounting records had not been maintained.

Monitoring Job Progress

In reviewing the different reasons for a cost accounting system, most do not require separate records for different jobs. The IRS is interested only in the totals. Worker's Compensation insurance is based only on the labor data gathered as part of the company's own cost code system (in most instances, two cost code systems are required for cost accounting in order to satisfy this separate need). The bonding company is interested in the company's overall performance and in the individual job for billing. If problems arise, the bonding company becomes interested in the individual project accounting. Historically, the information by job could be useful, but not important enough to be the sole purpose for a separate job cost accounting system. While the testing and upgrading of composite unit costs is a good reason to introduce that level of reporting, this kind of updated information (which concerns the changes in productivity) can also be monitored in other, less involved ways; the unit cost update can be accomplished through new data on labor wages, benefits, and material costs. To monitor job progress as it relates to the estimate is a better reason for a cost accounting system.

Good job or project management tries to forecast a problem in advance of its occurrence. An accurate forecast allows time to make corrections and develop alternatives. The information base necessary to do this forecasting is a cost accounting system by job. Of course, the system alone will not be a problem solver, but monitoring and sampling the data presented, and reporting and projecting from it, can be very constructive.

Maintaining good job accounting records is, above all, sound business practice. Such a system, once implemented, may address many other needs, as it enhances the company's data base. Not withstanding these many benefits, one other reason for maintaining a job cost accounting system is that it may be required by some contracts.

Contractual Obligation

The contractual obligation regarding cost accounting and cost records is stipulated within the body of the agreement. The standard form contract agreement AIA Document A111 – *Standard Form of Agreement Between Owner and Contractor, Cost of the Work Plus a Fee*, Article 11 states, "The Contractor shall keep such full and detailed accounts......as may be necessary for proper financial management......". It further states, "The Owner......shall be afforded access to all of the contractor's records, books,..........and other data relating to this contract........" Similar terms are usually found in other forms of contracts with design/build companies. Furthermore, items to be included in and excluded from the work are also stipulated in Articles 7 and 8 of AIA Document A111 and in other forms of design/build company contracts.

One reason for this condition of agreement is the savings provision that is usually included in the contract. In order that an owner may be assured that the cost of the work is as reported, audit privileges are granted, though this privilege is not always exercised. Nevertheless, costs must be separated by job and not absorbed into a company general fund, as is the case with some other systems.

Another reason for a cost accounting condition in the contract relates back to Articles 7 and 8 and the Guaranteed Maximum Price (GMP). The GMP states that the cost of the work will not exceed the total cost stated. Individual items of work are not guaranteed. In the construction process, changes occur in personnel, equipment, materials, and even tradesmen. All of these changes are made to enhance the project's success, but they may not be specifically identified in the GMP.

The changes are an accepted part of the procedure, but they must be controlled. Indiscriminant charges to the job might compromise the project and the cost. For example, the need for the president of the design/build company to attend a meeting in order to achieve a certain objective (per Articles 7 and 8) is not an acceptable cost charged to the work. This expense is instead part of the fee. Sometimes this executive effort may be allowed as a consulting activity. Other time contributed by executives would also be normal in the course of a project and therefore not allowed as an added cost. Company car expenses (except those provided for on-the-job use) would not be accepted. None of these qualifications would be particularly stated in the GMP, but are a consequence of the contract. The audit privilege, therefore, is a way of verifying the accepted charge costs and rooting out those that are not acceptable.

Accounting Methods

Cost Coding

Many of the computer package cost accounting systems have cost codes which are based on the CSI (Construction Specifications Institute) MASTERFORMAT. This format is a standard used for specification section designations and lends itself to the job costs as well. The division headings are as follows:

CSI MASTERFORMAT DIVISIONS

1. General Requirements
2. Site Work and Improvements
3. Concrete
4. Masonry
5. Metals
6. Wood & Plastics
7. Thermal & Moisture protection
8. Doors & windows
9. Finishes
10. Specialties
11. Equipment
12. Furnishings
13. Special Construction
14. Conveying Systems
15. Mechanical
16. Electrical

In addition, two other divisions may be added for control. They are:

Division 0	Design and Development
Division 17	Change Orders

A cost coding system to further define these broad categories would have to be established. This system could be based on the MASTERFORMAT designations, with additional cost categories as needed. Using alpha/numerics may be one way to create these categories. In the MASTERFORMAT, the first two digits define the division section. The next three digits define the category within the division. If more digits are required by the company's cost coding system, the MASTERFORMAT five-digit scope format could be modified to a six-digit format as follows.

For example, 010240 might signify the Superintendent category. The 01 stands for Division 1 and 0240 is the code for a superintendent. 090900 could be Painting, in which case the 09 stands for Division 9, and the 0900 is the code for Painting.

The precise numerical designation for an item within a division can vary greatly. Some codes strive for consistency as an aid to memory. Temporary heat carried as part of different work categories may be coded as follows:

Concrete	030500
Masonry	040500
Painting	099500
Drywall	092500
Plastering	092500

Noting the same last three digits as a constant and varying the division for the particular work makes this redundant code easy to remember and to double check against miscoding. For similar work in the same division, the unused digit can provide a more precise code. For all of the finish work, temporary heat is in Division 09; they are all coded as 500, but:

Plastering	is 09**1**500
Drywall	is 09**2**500
Painting	is 09**9**500

The 1, 2, and 9 further define the information. With fewer digits, the description must be broader; more digits allow a more precise definition.

Codes then need to distinguish between Labor, Material, Equipment, and Subcontractor. This can be accomplished by inputting to specified columns or the identification can be as part of the cost code, for example:

Survey Crew	01050L
Survey Material	01050M
Survey Equipment	01050E

The computer, which has been programmed to accept this coding, would then assign the charged cost to the proper item (and in the proper column if in columnar form).

There are two types of cost codes: *Project* cost codes and *Standard* cost codes. The standard cost codes pertain to the classification of types of projects. Project cost codes, illustrated previously, pertain to the classification of all work items within a particular project. The standard cost code can be developed to generate information other than a numerical tag to a project. Standard cost codes may contain six digit numbers, such as:

061787

The digits in this code represent the following:

06 suburban office building
17 job name
87 contracting or sales year

092387

This number would indicate:

09 industrial building
23 job
87 contracting or sales year

The variations of building type could be limited to a maximum of nine. The remaining five numbers might then be used to provide information. The benefit of multi-information codes is that the data can be gathered in different forms.

Conceptual Estimate Conversion

The conceptual estimate contains quantities and unit prices that can be directly cost coded and input into a cost account. Some quantities and costs cannot be directly input, and must be converted. The quantity is a composite of individual costs. The unit price is in the same format. This format is useful in producing a conceptual estimate, but it cannot be used in the construction of the project. The individual items of work must be collected and represented together in order to establish the value of each work item. Before the universal use of computers, the gathering of similar items after extending the quantities was accomplished by hand using an accounts spreadsheet. With the use of computers, this laborious task is no longer a project manager's duty, and can instead be accomplished by a data processing person.

The composite price for the example in Figure 13.1, cast-in-place concrete floor slab, is made up of unit prices, all from Division 3. Each item (unit price) also has a unique cost code (see Figure 13.2). The first two digits represent the unit price division. The following three digits represent the task. The "L" or "M" specify the task as *labor only* or *material only*.

If the C.I.P. flat plate of a project was 75,500 square feet, then that number (for concrete) is multiplied by the quantity for all the items. The conceptual estimate would look like this:

75,500 S.F. x $5.88/S.F. = $443,940.00

In order to construct a project, materials must, at some point, be ordered and purchased. This process is commonly referred to as the project "buy-out". Using the composite cost approach, the individual unit costs have associated *required quantities* per unit of measure of the composite cost. (See Figure 13.3.) To find out the total required quantities of the individual units, it is necessary to multiply the total number of composite units by the associated individual required quantities per composite unit. If a cost code is assigned to each of the individual unit costs, the computer can identify and assemble like cost codes.

Many times the composite unit cost involves more than one division, such as furnishing and installing a metal door and frame as shown in Figure 13.4. (See Figure 13.5 for an illustration.)

The end result is that the composite estimate, in a form handy for developing the estimate, has now been converted to a form that is handy for building the project.

FLOORS	B3.5-150	C.I.P. Flat Plate

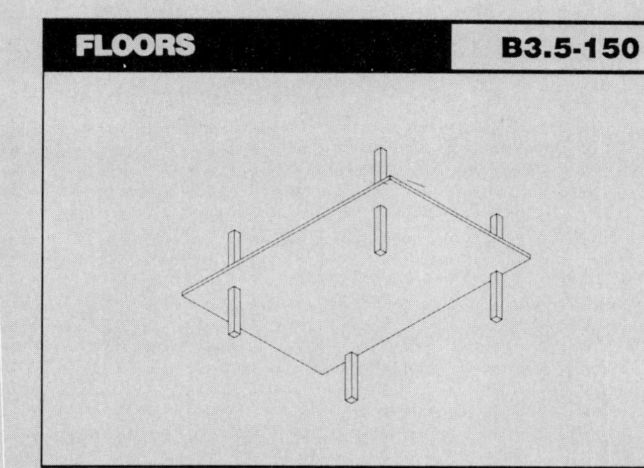

General: Flat Plates: Solid uniform depth concrete two way slab without drops or interior beams. Primary design limit is shear at columns.

Design and Pricing Assumptions:
Concrete f'c to 4 KSI, placed by concrete pump.
Reinforcement, fy = 60 KSI.
Forms, four use.
Finish, steel trowel.
Curing, spray on membrane.
Based on 4 bay x 4 bay structure.

3

System Components	QUANTITY	UNIT	COST PER S.F. MAT.	COST PER S.F. INST.	COST PER S.F. TOTAL
SYSTEM 03.5-150-2000					
15'X15' BAY 40 PSF S. LOAD, 12" MIN. COL.					
Forms in place, flat plate to 15' high, 4 uses	.992	S.F.	.53	2.60	3.13
Edge forms to 6" high on elevated slab, 4 uses	.065	L.F.	.02	.12	.14
Reinforcing in place, elevated slabs, #4 to #7	1.706	Lb.	.49	.31	.80
Concrete ready mix, regular weight, 3000 psi	.459	C.F.	.95		.95
Place and vibrate concrete, elevated slab less than 6", pump	.459	C.F.		.34	.34
Finish floor, monolithic steel trowel finish for finish floor	1.000	S.F.		.46	.46
Cure with sprayed membrane curing compound	.010	C.S.F.	.02	.04	.06
TOTAL			2.01	3.87	5.88

3.5-150 Cast In Place Flat Plate

	BAY SIZE (FT.)	SUPERIMPOSED LOAD (P.S.F.)	MINIMUM COLUMN SIZE IN.	SLAB THICKNESS (IN.)	TOTAL LOAD (P.S.F.)	COST PER S.F. MAT.	COST PER S.F. INST.	COST PER S.F. TOTAL
2000	15 x 15	40	12	5-½	109	2.01	3.87	5.88
2200	(17) (18)	75	14	5-½	144	2.02	3.86	5.88
2400		125	20	5-½	194	2.13	3.90	6.03
2600		175	22	5-½	244	2.18	3.91	6.09
3000	15 x 20	40	14	7	127	2.34	3.91	6.25
3400		75	16	7-½	169	2.52	3.98	6.50
3600		125	22	8-½	231	2.79	4.09	6.88
3800		175	24	8-½	281	2.82	4.09	6.91
4200	20 x 20	40	16	7	127	2.33	3.90	6.23
4400		75	20	7-½	175	2.54	3.99	6.53
4600		125	24	8-½	231	2.80	4.07	6.87
5000		175	24	8-½	281	2.83	4.09	6.92
5600	20 x 25	40	18	8-½	146	2.76	4.09	6.85
6000		75	20	9	188	2.88	4.13	7.01
6400		125	26	9-½	244	3.14	4.24	7.38
6600		175	30	10	300	3.27	4.29	7.56
7000	25 x 25	40	20	9	152	2.87	4.13	7
7400		75	24	9-½	194	3.06	4.22	7.28
7600		125	30	10	250	3.28	4.30	7.58
8000								

For expanded coverage of these items see *Means Concrete Cost Data 1987*

Figure 13.1

Code	Description	Quantity	Unit	M	L	T
03040M	Forms in Place	0.992	SFCA	0.53		0.53
03045L	Forms in Place	0.992	SFCA		2.60	2.60
03060M	Edge Forms	0.065	LF	0.02		0.02
03060L	Edge Forms	0.065	LF		0.12	0.12
03050M	Reinforcing in Place	1.706	LBS	0.49		0.49
03050L	Reinforcing in Place	1.706	LBS		0.31	0.31
03010M	Concrete Ready Mix	0.459	CF	0.95		0.95
03018L	Place and Vibrate	0.459	CF		0.34	0.34
03040L	Finish Floor	1.000	SF		0.46	0.46
03090M	Cure Floor	0.010	CSF	0.02		0.02
03090L	Cure Floor	0.010	CSF		0.04	0.04
	Total	**1.000**	**SF**			**5.88**

Note: S.F.C.A.: square foot contact area
L.F.: linear foot
C.F.: cubic feet
C.S.F.: hundreds of square feet
LBS.: pounds

Figure 13.2

Quantities Required					Dollars Required		
Cost Code	Description	Quantities Per Composite Unit	Total Composite Units	Required Individual Quantities	Total Composite Units	Individual Unit Cost Per Composite	Total Dollars
03045M	Forms in Place	0.992	75,500	74,896 S.F.	75,500	$ 0.53/S.F.C.A.	$ 40,015
03045L	Forms in Place	0.992	75,500	—	75,500	2.60/S.F.C.A.	196,300
03060M	Edge Forms	0.065	75,500	4,908 S.F.	75,500	0.02/L.F.	1,510
03060L	Edge Forms	0.065	75,500	—	75,500	0.12/L.F.	9,060
03050M	Reinf. in Place	1.706	75,500	128,803 lbs.	75,500	0.49/lb.	36,995
03050L	Reinf. in Place	1.706	75,500	—	75,500	0.31/lb.	23,405
03010M	Concrete Ready-Mix	0.459	75,500	34,655 C.F.	75,500	0.95/C.F.	71,725
03018L	Place & Vibrate	0.459	75,500	—	75,500	0.34/C.F.	25,670
03040L	Finish Floor	1.000	75,500	—	75,500	0.46/S.F.	34,730
03090M	Cure Floor	0.010	75,500	755 C.S.F.	75,500	0.02/C.S.F.	1,510
03090L	Cure Floor	0.010	75,500	—	75,500	0.04/C.S.F.	3,020
							$443,940

Figure 13.3

System Components	QUANTITY	UNIT	COST EACH MAT.	INST.	TOTAL
SYSTEM 06.4-220-1200					
STEEL DOOR, HOLLOW, 20 GA., HALF GLASS, 2'-8"X6'-8", D.W.FRAME, 4-⅞" TH					
Steel door, flush, hollow core, 1-¾" thick,half glass, 20 ga.,2'-8"x6'-8"	1.000	Ea.	170.50	24.50	195
Steel frame, KD, 16 ga., drywall, 3-⅞" deep, 2' x 8" x 6' x 8", single	1.000	Ea.	77	33	110
Float glass, 3/16" thick, clear, tempered	5.000	S.F.	20.65	18.35	39
Paint exterior door & frame one side 2'-8" x 6'-8", primer & 2 coats	1.000	Ea.	2.59	25.41	28
TOTAL			270.74	101.26	372

Figure 13.4

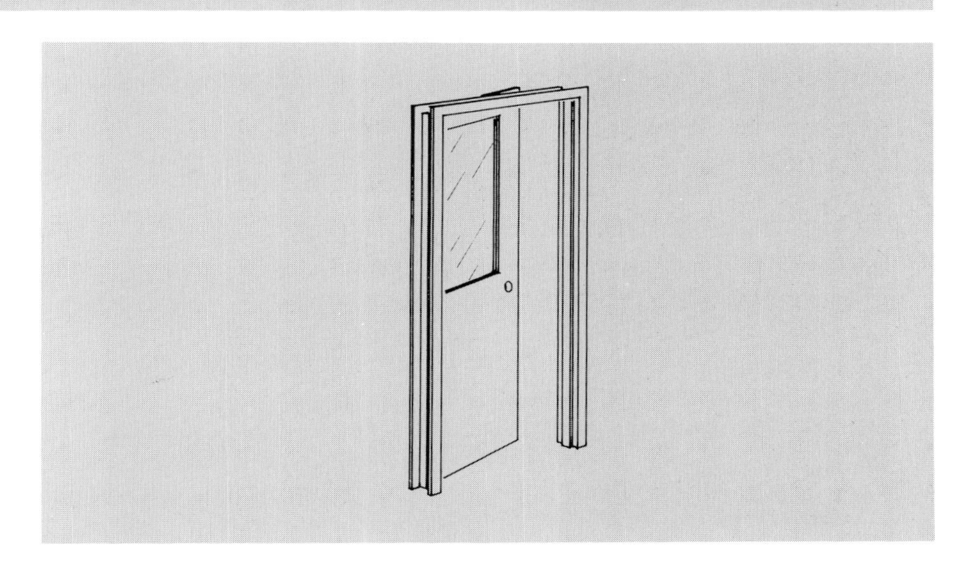

Figure 13.5

Reporting Forms and Formats

The methods of reporting costs vary from company to company. A good basic reporting form is one that will allow for adjustment and facilitate user input. Such a form might include headings as shown in the example form in Figure 13.6.

The variance is derived by the following formula:

COL. 9 = COL. 5 – COLS. (7 + 8)

where 8 is manually input by the project manager based on his knowledge of the project.

Such a form should provide an overview of the project. The information for this form is developed from a more detailed spreadsheet which identifies each estimated quantity and shows the progress by insertion of job performed quantities. An example of such a form with suggested headings is shown in Figure 13.7.

In this format, one or more horizontal lines are used for both the original and the updated estimate. Any number of other horizontal lines (as needed) in the same cost code section are used for field information and "drawn down", or subtracted from the total estimated amount for each item of work. This form summarizes by cost code, by division and by division summary.

Other reporting forms might require information on disbursement of funds through a vendor disposition spreadsheet. This sheet could list all of the material suppliers and subcontractors by their own vendor or pay number. The listing would track by job all of the payments made to that vendor. This information contributes to the *cost to date* column on the main or management reporting sheet. See Figure 13.8 for suggested headings for this type of form, which could be summarized by job or by vendor.

Purchasing

Another reporting form could be created for recording purchasing activities. The subcontract and material items of work to be purchased could be extracted from the estimate and superimposed upon this form. The headings for this form could be as shown in Figure 13.9.

The precise type of format a design/build company will use for cost accounting is a matter of preference and need. The only requirement is that the system provides for all the business, estimating, and contractual needs dictated by this method of construction.

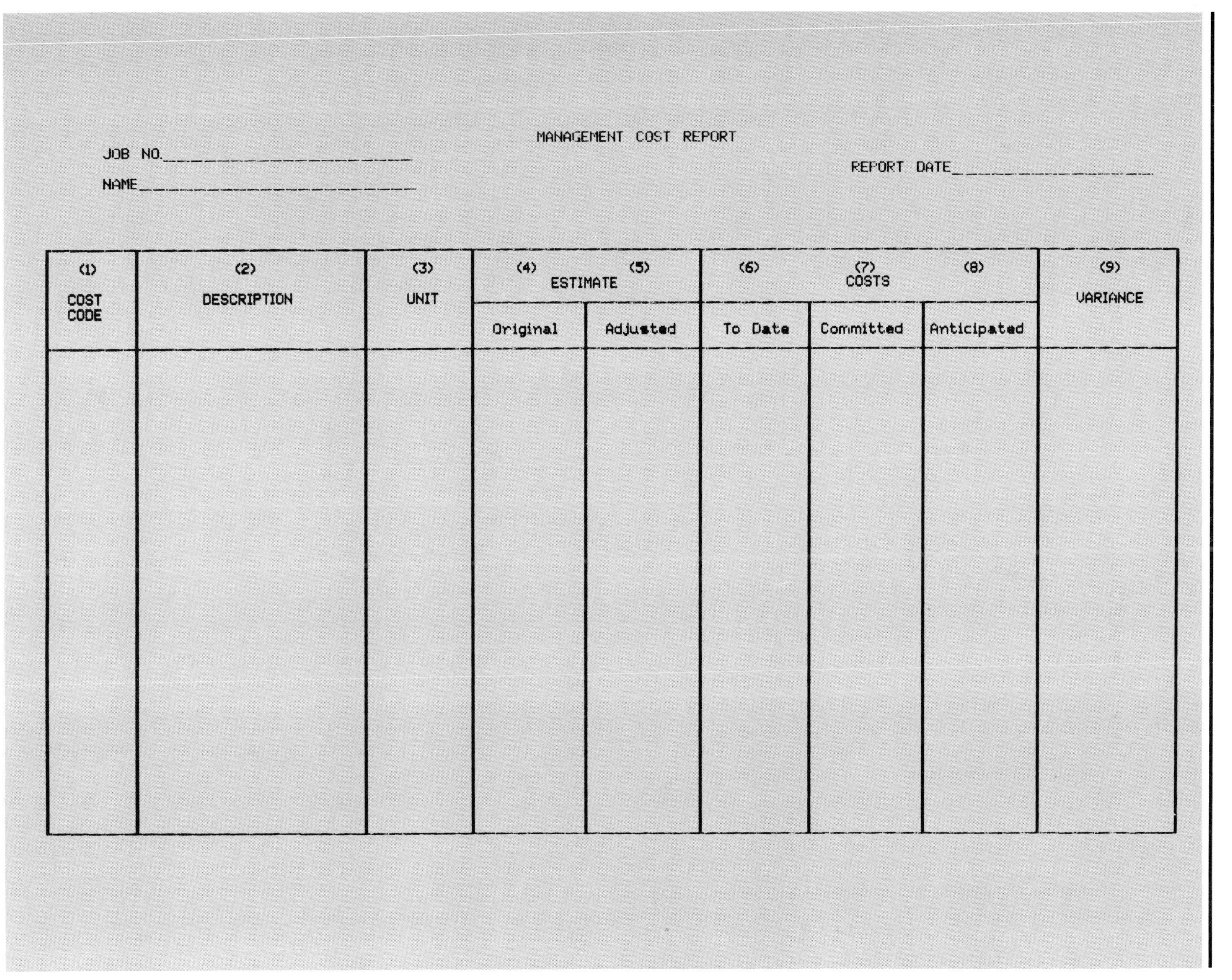

MANAGEMENT COST REPORT

JOB NO.______________

NAME______________

REPORT DATE______________

(1) COST CODE	(2) DESCRIPTION	(3) UNIT	(4) ESTIMATE	(5)	(6) COSTS	(7)	(8)	(9) VARIANCE
			Original	Adjusted	To Date	Committed	Anticipated	

Figure 13.6

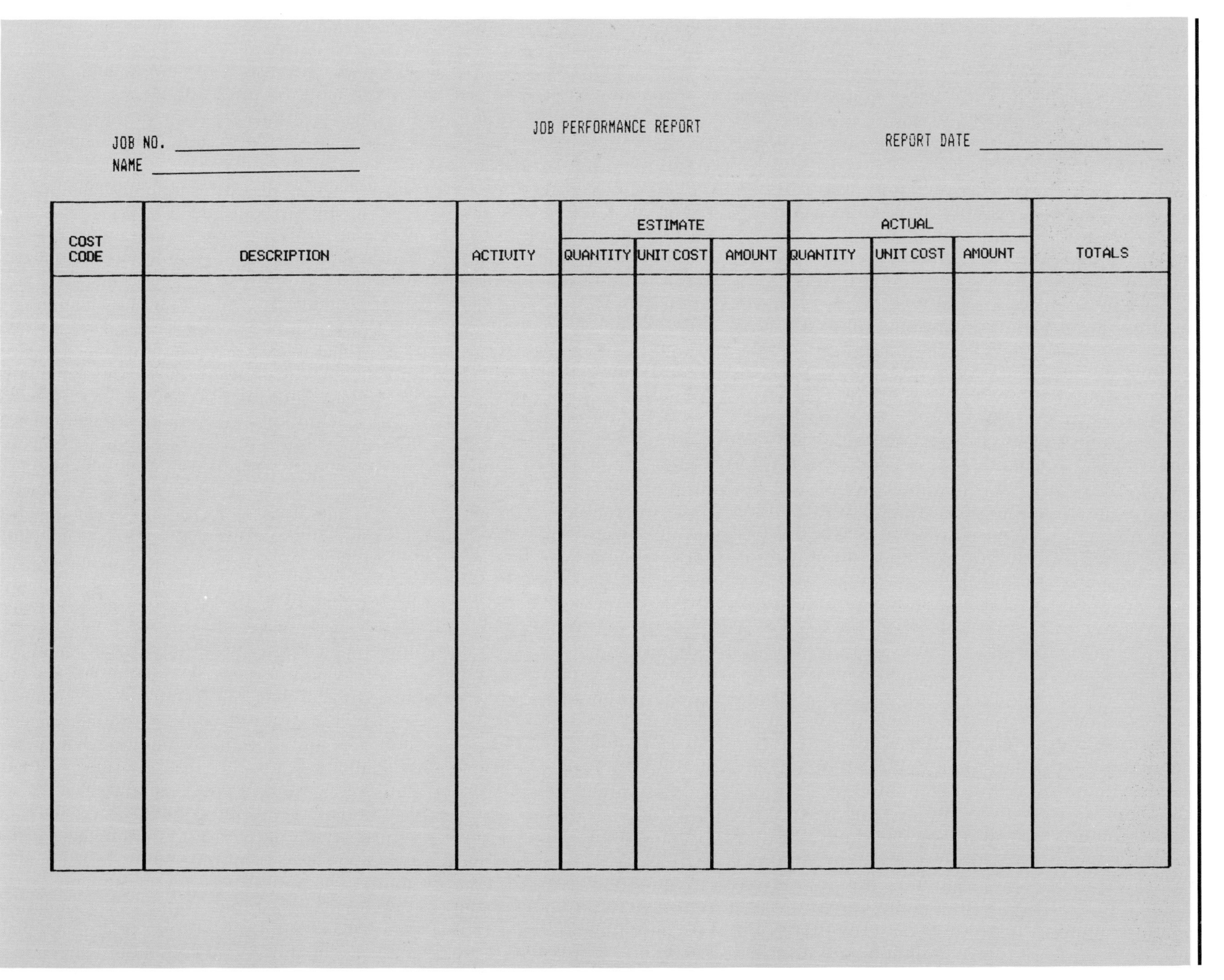

JOB PERFORMANCE REPORT

JOB NO. ________
NAME ________

REPORT DATE ________

COST CODE	DESCRIPTION	ACTIVITY	ESTIMATE			ACTUAL			TOTALS
			QUANTITY	UNIT COST	AMOUNT	QUANTITY	UNIT COST	AMOUNT	

Figure 13.7

VENDOR STATUS REPORT

VENDOR NUMBER ________________ VENDOR NAME ____________________

WORK ITEM ____________________

JOB NUMBER	JOB NAME	AMOUNT CONTRACT	AMOUNT PAID	RETAINED

Figure 13.8

PURCHASING REPORT

REPORT DATE ____________

COST CODE	DESCRIPTION	SUBCONTRACTOR	P.A.	JOB NO.	DATES			AMOUNT		DIFF.
					DESIGN	BUY	ON JOB	ESTIMATED	BUY	

Figure 13.9

Chapter Fourteen

SCHEDULING

Scheduling as a Management Tool

Scheduling of activities is a method used in construction to plan and control a project. In design/build construction, it is not enough to control the construction activities; it is also necessary to control all of the activities associated with a project to completion. The reasons behind this need for control are the fast track construction of the project, and the design/build company's responsibility for all of the elements of the design. A master control schedule is the means to achieve this goal. The master control schedule, also called a management schedule, incorporates all of the activities that need to occur for the successful accomplishment of the project. The management schedule includes and isolates the activities of the design/build contractor in design, procurement, and construction. This schedule would also include the designer's role in the overall design process – both by individual designers and by the collective group. The management schedule would also show the owner's involvement in the project; dates would be indicated for the exchange of information, decision points, securing of the land, securing of the financing, and sign-off of the design. The management schedule for the case study is shown in Figure 14.1.

A management schedule commonly starts with the construction activity. The lead time duration and the bid time duration for each of the items is introduced. The design schedule is then overlaid and design bid packages indicated. Since the goal of any construction project is the timely production of the building, everything must be directed to that end. As the other activities are input to the left against the construction schedule to the right, it becomes apparent which items are critical (those most to the left). The illustrated management schedule bar chart (Figure 14.1) shows construction activities to the far right, lead time next, with negotiation, bid assembling, scoping and design (in that order) from right to left. This procedure does not preclude the use of CPM (Critical Path Method) scheduling techniques which will be discussed later.

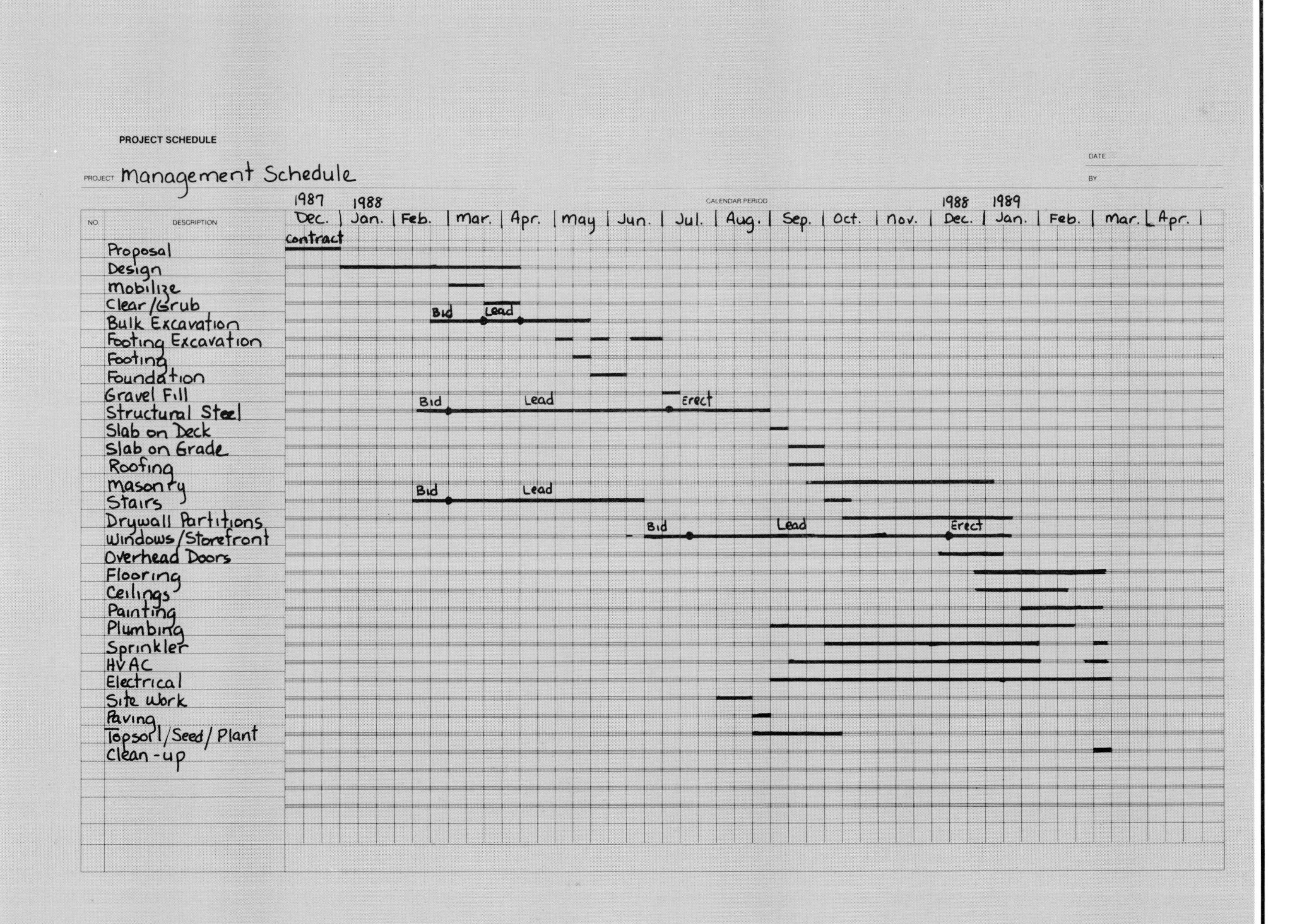

PROJECT SCHEDULE
PROJECT Management Schedule
DATE
BY
CALENDAR PERIOD
NO.
DESCRIPTION
1987
Dec.
1988
Jan.
Feb.
Mar.
Apr.
May
Jun.
Jul.
Aug.
Sep.
Oct.
Nov.
1988
Dec.
1989
Jan.
Feb.
Mar.
Apr.
Contract
Bid
Lead
Erect
Proposal
Design
Mobilize
Clear/Grub
Bulk Excavation
Footing Excavation
Footing
Foundation
Gravel Fill
Structural Steel
Slab on Deck
Slab on Grade
Roofing
Masonry
Stairs
Drywall Partitions
Windows/Storefront
Overhead Doors
Flooring
Ceilings
Painting
Plumbing
Sprinkler
HVAC
Electrical
Site Work
Paving
Topsoil/Seed/Plant
Clean-up

Figure 14.1

Design Schedule

In a previous chapter, it was explained that the design stages involved concept; schematic; design development; and construction documents. The relationship of the design to the management schedule involves bid packages. The bid packages represent the out-of-sequence design of certain trade work or material items that are needed in advance of the total design. This advance need is what is forecast and shown in the management schedule. The design of an item of work requires many activities. The concept needs to be stated, then the space blocked out on sheets of drawings, the details developed, the dimensions verified, and then the support elements (such as structural, electrical, mechanical, plumbing and sprinkler designs) must be superimposed. This procedure repeats for individual elements to arrive at the total design. Durations for each element of the design sequence vary greatly.

Doing just design package elements and not completing the design will take longer. In order for the design to move along as rapidly as possible, the bid packages must be kept to a minimum. This approach is easier with design/build than with other methods because the design concept is fixed early and the remaining tasks can be given the attention they deserve. Design bid packages are then used to allow the construction activities to start early. The question to ask is, "When will I receive the total design drawings, or the working drawings?". If the response is a design completion date that is later than anticipated or needed, then design bid packages are required and need to be developed.

Procurement

Procurement is the gathering of men and materials at the job site. This procedure is called *purchasing*. Purchasing requires a great attention to detail and to the elements that make up each material or trade item. To enable the purchasing to coincide with the needs of the construction, time durations must be established for scoping, bidding, and lead time for each item. See Figure 14.2 for typical purchasing activity durations.

Scoping occurs when documentation is received for the item of work to be procured. The person who is to purchase the item examines the plans carefully to make sure that all of the work required is shown and complete. Any segment of this work that is not shown should be listed by the purchasing agent to insure that the information is available at the time of bidding. The time period for scoping of the work ranges from one day to as much as a week, depending on the item of work and the degree of difficulty of the project.

All of the bid documents are then assembled into a "bid package", copies of which are distributed to bidders. The amount of time necessary to assemble documents depends on the printing facilities, number of documents, and the number of bidders. Usually a week is plenty of time to allow for this activity.

Bidding occurs next and involves distribution of the bid documents, conducting interviews and information meetings, and receiving the bids. The bids are then evaluated, negotiations are held with the low bidder or bidders, and final contract terms are decided. The time allotted for bidding is seldom less than two weeks and may require as much as six weeks depending on the scope of work and the type of project.

The lead time period is that time necessary to draw shop drawings; submit them for approval; receive approval or resubmit revised shop drawings, if necessary; gather raw material; fabricate the product, and ship it, ready for installation, to the job site. The time period allowed for lead time is based on the nature of the product, the degree of difficulty in detailing and fabricating that product, and the demand for the product. A product with a normal lead time of four weeks could stretch to twelve weeks if there is a great demand. Some products just cannot be secured in less than ten weeks and may even require two to three times that duration if the project is difficult, the product used is not modular, and it is in great demand. See Figure 14.3, a representative list of structural steel lead time activities.

The cumulative time needed for the activities of scoping the work, assembling documents, bidding the work, negotiating the bid, providing lead time in the production of the item, and shipping it to the site varies from item to item. Each element is included as part of the management schedule. Structural steel, for example, may involve the duration shown in Figure 14.4, whereas painting may have a duration as shown in Figure 14.5.

Scope Time, Bid Times and Lead Times
(All Durations Are in Weeks)

Item	Times		
	Scope	Bid	Lead
Excavation/Backfill	–	3	2
Piles	–	3	8
Foundations	–	2	5
Masonry	1	3	9
Structural Steel	2	5	23
Steel Stairs	1	4	15
Roofing	1	3	9
Windows	2	4	21
Drywall	1	3	8
Ceiling	1	3	8
Flooring	1	3	13
Painting	–	2	3
Mechanical	2	4	4
Electrical	2	4	4
Sprinkler	2	4	18
Plumbing	1	3	4

Figure 14.2

Lead Time Activities

1. Receive Contract
2. Prepare Initial Plant Cutting List
3. Prepare Erection Drawings
4. Submit Erection Drawings
5. Start Piece Drawings
6. Approval of Erection Drawings
7. Submit Final Plant Rolling List
8. Submittal of Phased Piece Drawings
9. Phased Approval of Piece Drawings
10. Receipt of Raw Material
11. Start Fabrication of Members
12. Store Members
13. Start Shipment to the Job
14. Unload and Sort Members
15. Start Erection

Figure 14.3

1. Scoping	1/2 Week
2. Assembling Documents	1/2 Week
3. Bidding	5 Weeks
4. Lead Time	23 Weeks
5. Ship to Job	1/2 Week
Total	29½ Weeks

Time Span: 29½ Weeks

Figure 14.4

1. Scoping	0 Weeks
2. Assembling Documents	1/2 Week
3. Bidding	2 Weeks
4. Lead Time	3 Weeks
5. Ship to Job	1/2 Week
Total	6 Weeks

Time Span: 6 Weeks

Figure 14.5

The Construction Schedule

The construction portion of the master schedule describes the order in which the project will be constructed. It establishes the sequence of individual activities and applies durations to each item of work. Durations are based on the amount of work required and the productivity of the workers.

The method used to prepare a construction schedule may depend on the size and complexity of the project, as well as the level of detail required in planning. A bar chart may be used for small projects, whereas a network schedule may be more appropriate for larger, more complex projects, where more detail and control are required. See Figures 14.6 and 14.7 for examples of a bar chart and a network (CPM) schedule, both based on the case study.

The activity durations used in the construction schedule may be developed from historical data or from available man-hour or daily output publications. Quantities may be segmented or broken down into separate activities such as form footings, reinforce footings, and place concrete. Or, the activity may be shown simply as *place continuous footings*. Quantities of the activities should be available from the bid recap.

The following are examples of duration time development using the R.S. Means annual cost data publication, *Building Construction Cost Data*, 1987, and quantities from the example project's plans. Figure 14.8 shows a detail of a footing and foundation wall. Appropriate quantities are listed below:

Quantities	
Foundation Wall	860 L.F.
Continuous Footing Forms	1720 S.F.
Wall Form	6880 S.F.
Footing Reinforcing	5 Tons
Wall Reinforcing (Horizontal Only)	2.4 Tons
Concrete Continuous Footing	64 C.Y.
Concrete Wall	127 C.Y.

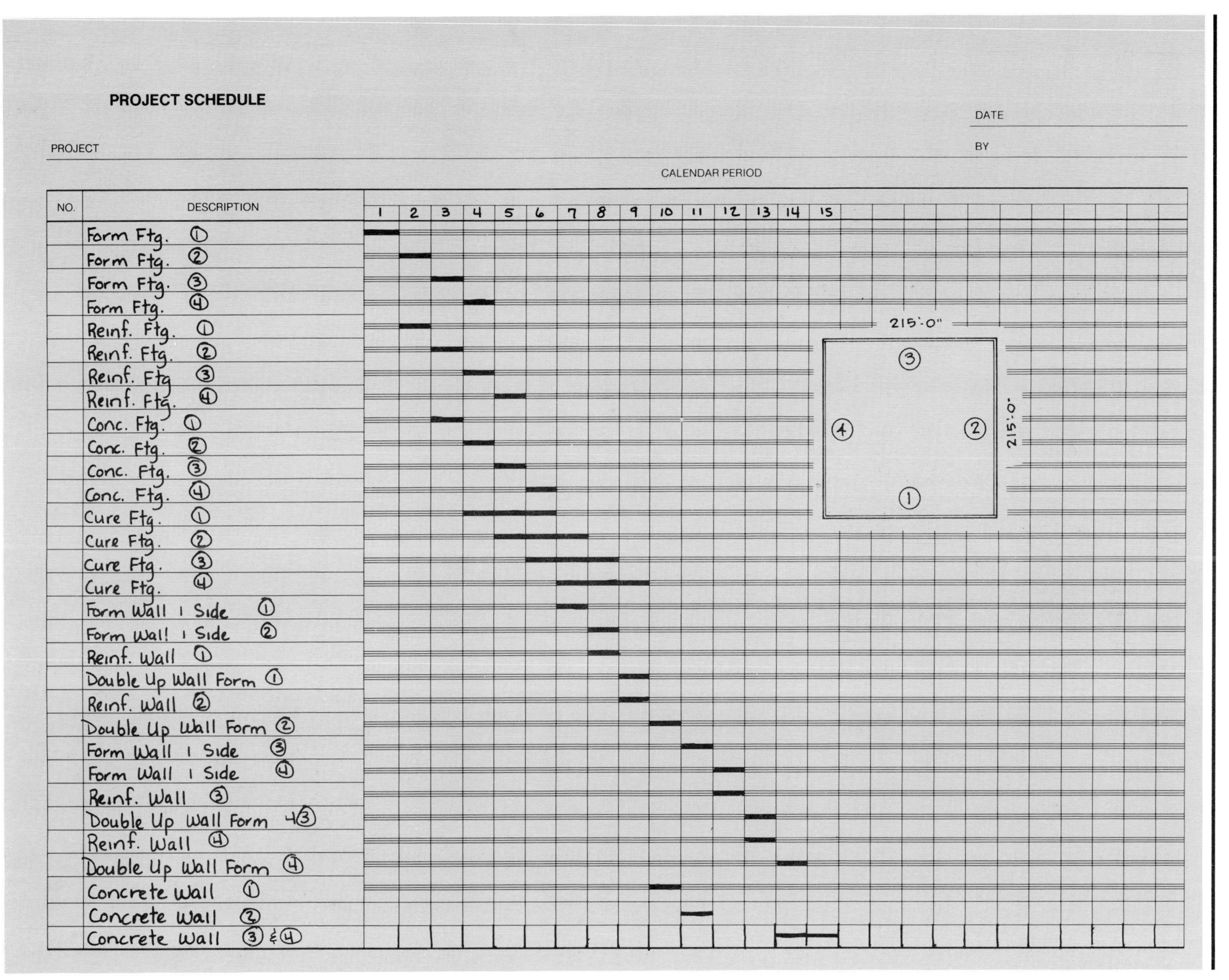
PROJECT SCHEDULE
PROJECT
DATE
BY
CALENDAR PERIOD
NO.
DESCRIPTION
1 2 3 4 5 6 7 8 9 10 11 12 13 14 15
Form Ftg. ①
Form Ftg. ②
Form Ftg. ③
Form Ftg. ④
Reinf. Ftg. ①
Reinf. Ftg. ②
Reinf. Ftg ③
Reinf. Ftg. ④
Conc. Ftg. ①
Conc. Ftg. ②
Conc. Ftg. ③
Conc. Ftg. ④
Cure Ftg. ①
Cure Ftg. ②
Cure Ftg. ③
Cure Ftg. ④
Form Wall 1 Side ①
Form Wall 1 Side ②
Reinf. Wall ①
Double Up Wall Form ①
Reinf. Wall ②
Double Up Wall Form ②
Form Wall 1 Side ③
Form Wall 1 Side ④
Reinf. Wall ③
Double Up Wall Form 4③
Reinf. Wall ④
Double Up Wall Form ④
Concrete Wall ①
Concrete Wall ②
Concrete Wall ③ & ④
215'-0"
215'-0"
①
②
③
④

Figure 14.6

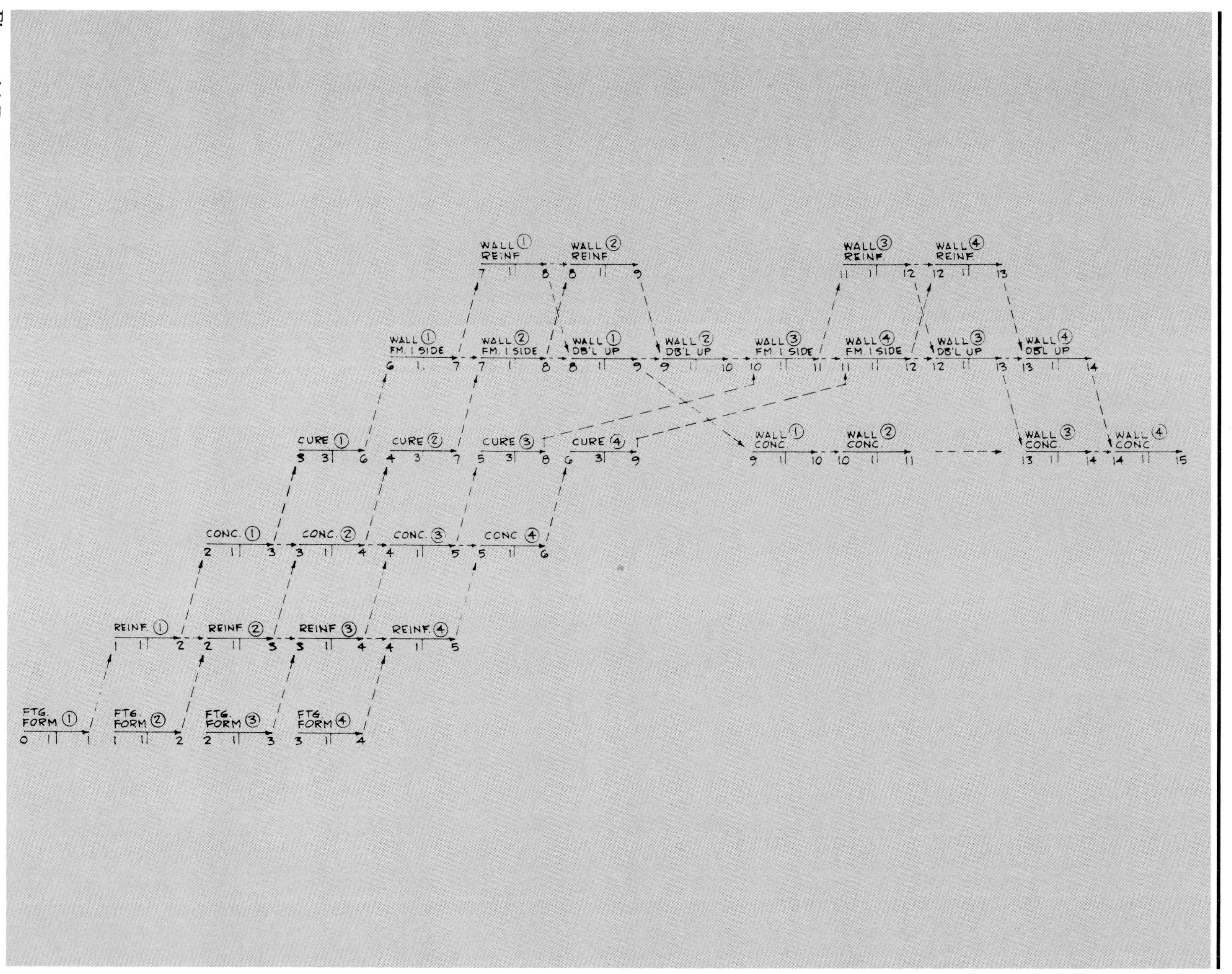
FTG. FORM ①
FTG. FORM ②
FTG. FORM ③
FTG. FORM ④
REINF. ①
REINF ②
REINF ③
REINF. ④
CONC. ①
CONC. ②
CONC. ③
CONC. ④
CURE ①
CURE ②
CURE ③
CURE ④
WALL ① FM. 1 SIDE
WALL ② FM. 1 SIDE
WALL ① DB'L UP
WALL ② DB'L UP
WALL ③ FM. 1 SIDE
WALL ④ FM. 1 SIDE
WALL ③ DB'L UP
WALL ④ DBL UP
WALL ① REINF
WALL ② REINF.
WALL ③ REINF.
WALL ④ REINF.
WALL ① CONC.
WALL ② CONC.
WALL ③ CONC.
WALL ④ CONC.

Figure 14.7

Footings

Formwork: Line No. 3.1 450 0150 (see Figure 14.9): Forms in Place, Footings Continuous Wall (4 uses)

Daily Output: 485 S.F. – C-1 Crew (see Figure 14.10)

$$\frac{1720 \text{ S.F.}}{485 \text{ S.F./day}} = 3.55 \text{ days}$$

Reinforcing: Line No. 3.2 300 0500 (see Figure 14.11): Reinforcing in-place Footings, #4 to #7.

Daily Output: 2.10 Tons Using 4 Rodmen, or .525 Tons/Rodman

$$\frac{5 \text{ Tons}}{.525 \text{ Tons/day} \times 4 \text{ Rodmen}} = 2.38 \text{ days}$$

Placing Concrete: Line No. 3.3 380 1900 (see Figure 14.12): Placing concrete and vibrating, including labor and equipment footings continuous shallow direct chute.

Daily Output: 120 C.Y. C-6 Crew (see Figure 14.10)

$$\frac{64 \text{ C.Y.}}{120 \text{ C.Y./day}} = .53 \text{ days}$$

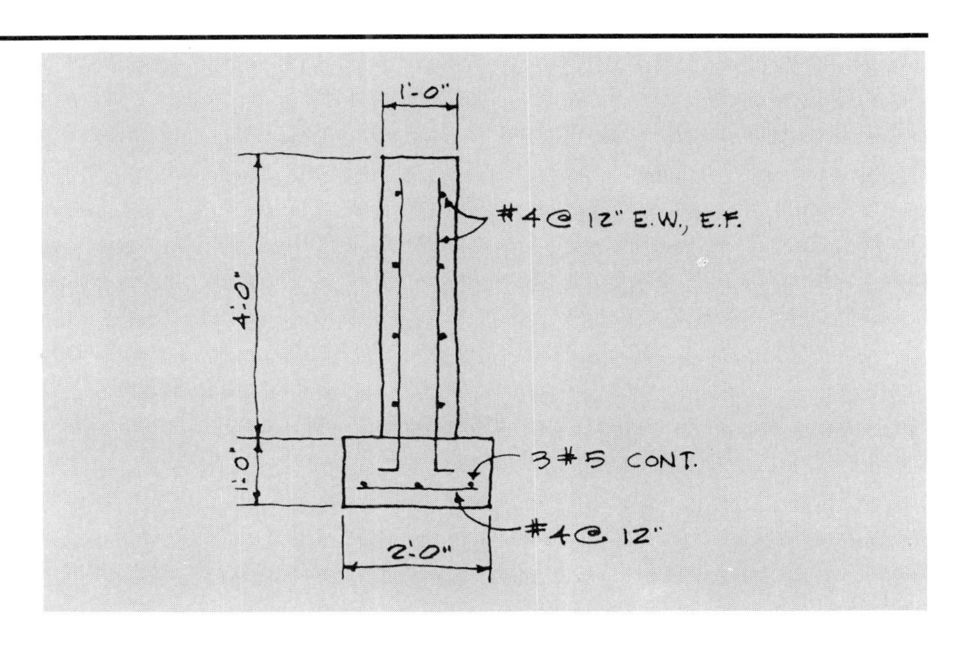

Figure 14.8

		3.1 Formwork	CREW	DAILY OUTPUT	MAN-HOURS	UNIT	BARE COSTS MAT.	BARE COSTS LABOR	BARE COSTS EQUIP.	BARE COSTS TOTAL	TOTAL INCL O&P	
350	4100	3 use	C-2	465	.103	S.F.	1.05	2.08	.09	3.22	4.31	350
	4150	4 use		495	.097		.91	1.95	.09	2.95	3.96	
	4500	With 30" fiberglass domes, 1 use		405	.119		2.32	2.39	.10	4.81	6.15	
	4520	2 use		450	.107		1.44	2.15	.09	3.68	4.84	
	4530	3 use		460	.104		1.13	2.10	.09	3.32	4.43	
	4550	4 use		470	.102		.97	2.06	.09	3.12	4.19	
	5000	Box out for slab openings, over 16" deep, 1 use		190	.253	SFCA	1.91	5.10	.22	7.23	9.80	
	5050	2 use		240	.200	"	1.13	4.03	.18	5.34	7.35	
	5500	Shallow slab box outs, to 10 S.F.		42	1.140	Ea.	6.90	23	1.01	30.91	42	
3	5550	Over 10 S.F. (use perimeter)		400	.120	L.F.	.77	2.42	.11	3.30	4.51	
	6000	Bulkhead forms for slab, with keyway, 1 use, 2 piece		500	.096		.96	1.93	.08	2.97	3.99	
	6100	3 piece (see also edge forms)		460	.104		1.30	2.10	.09	3.49	4.62	
	6500	Curb forms, wood, 6" to 12" high, on elevated slabs, 1 use	C-1	180	.178	SFCA	1.40	3.46	.18	5.04	6.80	
	6550	2 use		205	.156		.74	3.03	.16	3.93	5.45	
	6600	3 use		220	.145		.54	2.83	.14	3.51	4.90	
	6650	4 use		225	.142		.45	2.76	.14	3.35	4.71	
	7000	Edge forms to 6" high, on elevated slab, 4 use		500	.064	L.F.	.22	1.24	.06	1.52	2.14	
	7100	7" to 12" high, 4 use		350	.091	SFCA	.42	1.78	.09	2.29	3.17	
	7500	Depressed area forms to 12" high, 4 use		300	.107	L.F.	.42	2.07	.11	2.60	3.62	
	7550	12" to 24" high, 4 use		175	.183		.54	3.55	.18	4.27	6	
	8000	Perimeter deck and rail for elevated slabs, straight		90	.356		5.50	6.90	.35	12.75	16.60	
	8050	Curved		65	.492		7.55	9.55	.49	17.59	23	
	8500	Void forms, round fiber, 3" diameter		450	.071		.34	1.38	.07	1.79	2.48	
	8550	4" diameter, void		425	.075		.36	1.46	.07	1.89	2.63	
	8600	6" diameter, void		400	.080		.64	1.56	.08	2.28	3.07	
	8650	8" diameter, void		375	.085		1.05	1.66	.08	2.79	3.69	
	8700	10" diameter, void		350	.091		1.80	1.78	.09	3.67	4.69	
	8750	12" diameter, void		300	.107		2.25	2.07	.11	4.43	5.65	
	8800	Metal end closures, loose, minimum				C	25			25	28	
	8850	Maximum				"	120			120	130	
400	0010	**FORMS IN PLACE, EQUIPMENT FOUNDATIONS** 1 use	C-2	160	.300	SFCA	1.55	6.05	.27	7.87	10.85	400
	0050	2 use		190	.253		.85	5.10	.22	6.17	8.65	
	0100	3 use		200	.240		.65	4.83	.21	5.69	8.05	
	0150	4 use		205	.234		.55	4.72	.21	5.48	7.75	
450	0010	**FORMS IN PLACE, FOOTINGS** Continuous wall, 1 use	C-1	375	.085		.79	1.66	.08	2.53	3.40	450
	0050	2 use		440	.073		.43	1.41	.07	1.91	2.62	
	0100	3 use		470	.068		.32	1.32	.07	1.71	2.37	
	0150	4 use		485	.066		.27	1.28	.07	1.62	2.25	
	0500	Dowel supports for footings or beams, 1 use		500	.064	L.F.	.52	1.24	.06	1.82	2.47	
	1000	Integral starter wall, to 4" high, 1 use		400	.080		.80	1.56	.08	2.44	3.25	
	1500	Keyway, 4 uses, tapered wood, 2" x 4"	1 Carp	530	.015		.07	.31		.38	.53	
	1550	2" x 6"		500	.016		.09	.33		.42	.58	
	2000	Tapered plastic, 2" x 3"		530	.015		.40	.31		.71	.90	
	2050	2" x 4"		500	.016		.50	.33		.83	1.03	
	2250	For keyway hung from supports, add		150	.053		.50	1.10		1.60	2.16	
	2260											
	3000	Pile cap, square or rectangular, 1 use (33)	C-1	290	.110	SFCA	1.30	2.14	.11	3.55	4.70	
	3050	2 use		346	.092		.73	1.80	.09	2.62	3.54	
	3100	3 use		371	.086		.53	1.68	.09	2.30	3.13	
	3150	4 use		383	.084		.47	1.62	.08	2.17	2.99	
	4000	Triangular or hexagonal caps, 1 use		225	.142		1.48	2.76	.14	4.38	5.85	
	4050	2 use		280	.114		.82	2.22	.11	3.15	4.28	
	4100	3 use		305	.105		.62	2.04	.10	2.76	3.79	
	4150	4 use		315	.102		.52	1.97	.10	2.59	3.58	
	5000	Spread footings, 1 use (33)		305	.105		.92	2.04	.10	3.06	4.12	
	5050	2 use		371	.086		.50	1.68	.09	2.27	3.10	
	5100	3 use		401	.080		.39	1.55	.08	2.02	2.79	
	5150	4 use		414	.077		.34	1.50	.08	1.92	2.66	
	6000	Supports for dowels, plinths or templates, 2' x 2'		25	1.280	Ea.	2.70	25	1.27	28.97	41	
	6050	4' x 4' footing		22	1.450	"	5.80	28	1.45	35.25	49	

64

For expanded coverage of these items see *Means Concrete Cost Data 1987*

Figure 14.9

CREWS

Crew No.	Bare Costs Hr.	Bare Costs Daily	Incl. Subs O & P Hr.	Incl. Subs O & P Daily	Cost Per Man-hour Bare Costs	Cost Per Man-hour Incl. O&P
Crew B-89	**Hr.**	**Daily**	**Hr.**	**Daily**	**Bare Costs**	**Incl. O&P**
1 Skilled Worker	$20.80	$166.40	$30.70	$245.60	$18.45	$27.17
1 Common Laborer	16.10	128.80	23.65	189.20		
1 Cutting Mach. & Access.		27.70		30.45	1.73	1.90
16 M.H., Daily Totals		$322.90		$465.25	$20.18	$29.07
Crew B-90	**Hr.**	**Daily**	**Hr.**	**Daily**	**Bare Costs**	**Incl. O&P**
1 Labor Foreman (outside)	$18.10	$144.80	$26.55	$212.40	$17.43	$25.50
3 Highway Laborers	16.10	386.40	23.65	567.60		
2 Equip. Oper. (light)	19.60	313.60	28.65	458.40		
2 Truck Drivers (heavy)	16.95	271.20	24.60	393.60		
1 Road Mixer, 310 H.P.		1364.00		1500.40		
1 Dist. Truck, 2000 Gal.		212.80		234.10	24.63	27.10
64 M.H., Daily Totals		$2692.80		$3366.50	$42.06	$52.60
Crew B-91	**Hr.**	**Daily**	**Hr.**	**Daily**	**Bare Costs**	**Incl. O&P**
1 Labor Foreman (outside)	$18.10	$144.80	$26.55	$212.40	$18.78	$27.48
2 Highway Laborers	16.10	257.60	23.65	378.40		
4 Equip. Oper. (med.)	20.75	664.00	30.35	971.20		
1 Truck Driver (heavy)	16.95	135.60	24.60	196.80		
1 Dist. Truck, 3000 Gal.		221.20		243.30		
1 Aggreg. Spreader, S.P.		367.60		404.35		
1 Roller, Pneu. Tire, 12 Ton		157.60		173.35		
1 Roller, Steel, 10 Ton		173.60		190.95	14.37	15.81
64 M.H., Daily Totals		$2122.00		$2770.75	$33.15	$43.29
Crew B-92	**Hr.**	**Daily**	**Hr.**	**Daily**	**Bare Costs**	**Incl. O&P**
1 Labor Foreman (outside)	$18.10	$144.80	$26.55	$212.40	$16.60	$24.37
3 Highway Laborers	16.10	386.40	23.65	567.60		
1 Crack Cleaner, 25 H.P.		63.40		69.75		
1 Air Compressor		52.80		58.10		
1 Tar Kettle, T.M.		33.40		36.75		
1 Flatbed Truck, 3 Ton		68.00		74.80	6.80	7.48
32 M.H., Daily Totals		$748.80		$1019.40	$23.40	$31.85
Crew B-93	**Hr.**	**Daily**	**Hr.**	**Daily**	**Bare Costs**	**Incl. O&P**
1 Equip. Oper. (med.)	$20.75	$166.00	$30.35	$242.80	$20.75	$30.35
1 Feller Buncher, 50 H.P.		263.80		290.20	32.97	36.27
8 M.H., Daily Totals		$429.80		$533.00	$53.72	$66.62
Crew C-1	**Hr.**	**Daily**	**Hr.**	**Daily**	**Bare Costs**	**Incl. O&P**
3 Carpenters	$20.55	$493.20	$30.15	$723.60	$19.43	$28.52
1 Building Laborer	16.10	128.80	23.65	189.20		
Power Tools		31.80		35.00	.99	1.09
32 M.H., Daily Totals		$653.80		$947.80	$20.42	$29.61
Crew C-2	**Hr.**	**Daily**	**Hr.**	**Daily**	**Bare Costs**	**Incl. O&P**
1 Carpenter Foreman (out)	$22.55	$180.40	$33.10	$264.80	$20.14	$29.55
4 Carpenters	20.55	657.60	30.15	964.80		
1 Building Laborer	16.10	128.80	23.65	189.20		
Power Tools		42.40		46.65	.88	.97
48 M.H., Daily Totals		$1009.20		$1465.45	$21.02	$30.52
Crew C-3	**Hr.**	**Daily**	**Hr.**	**Daily**	**Bare Costs**	**Incl. O&P**
1 Rodman Foreman	$24.10	$192.80	$37.70	$301.60	$20.53	$31.50
4 Rodmen (reinf.)	22.10	707.20	34.60	1107.20		
1 Equip. Oper. (light)	19.60	156.80	28.65	229.20		
2 Building Laborers	16.10	257.60	23.65	378.40		
Stressing Equipment		31.80		35.00		
Grouting Equipment		52.75		58.05	1.32	1.45
64 M.H., Daily Totals		$1398.95		$2109.45	$21.85	$32.95

Crew No.	Bare Costs Hr.	Bare Costs Daily	Incl. Subs O & P Hr.	Incl. Subs O & P Daily	Cost Per Man-hour Bare Costs	Cost Per Man-hour Incl. O&P
Crew C-4	**Hr.**	**Daily**	**Hr.**	**Daily**	**Bare Costs**	**Incl. O&P**
1 Rodman Foreman	$24.10	$192.80	$37.70	$301.60	$22.60	$35.37
3 Rodmen (reinf.)	22.10	530.40	34.60	830.40		
Stressing Equipment		31.80		35.00	.99	1.09
32 M.H., Daily Totals		$755.00		$1167.00	$23.59	$36.46
Crew C-5	**Hr.**	**Daily**	**Hr.**	**Daily**	**Bare Costs**	**Incl. O&P**
1 Rodman Foreman	$24.10	$192.80	$37.70	$301.60	$21.60	$33.25
4 Rodmen (reinf.)	22.10	707.20	34.60	1107.20		
1 Equip. Oper. (crane)	21.20	169.60	31.00	248.00		
1 Equip. Oper. Oiler	17.55	140.40	25.65	205.20		
1 Hyd. Crane, 25 Ton		407.60		448.35	7.27	8.00
56 M.H., Daily Totals		$1617.60		$2310.35	$28.87	$41.25
Crew C-6	**Hr.**	**Daily**	**Hr.**	**Daily**	**Bare Costs**	**Incl. O&P**
1 Labor Foreman (outside)	$18.10	$144.80	$26.55	$212.40	$17.03	$24.85
4 Building Laborers	16.10	515.20	23.65	756.80		
1 Cement Finisher	19.70	157.60	27.95	223.60		
2 Gas Engine Vibrators		51.00		56.10	1.06	1.16
48 M.H., Daily Totals		$868.60		$1248.90	$18.09	$26.01
Crew C-7	**Hr.**	**Daily**	**Hr.**	**Daily**	**Bare Costs**	**Incl. O&P**
1 Labor Foreman (outside)	$18.10	$144.80	$26.55	$212.40	$17.38	$25.38
5 Building Laborers	16.10	644.00	23.65	946.00		
1 Cement Finisher	19.70	157.60	27.95	223.60		
1 Equip. Oper. (med.)	20.75	166.00	30.35	242.80		
2 Gas Engine Vibrators		51.00		56.10		
1 Concrete Bucket, 1 C.Y.		15.55		17.10		
1 Hyd. Crane, 55 Ton		556.20		611.80	9.73	10.70
64 M.H., Daily Totals		$1735.15		$2309.80	$27.11	$36.08
Crew C-8	**Hr.**	**Daily**	**Hr.**	**Daily**	**Bare Costs**	**Incl. O&P**
1 Labor Foreman (outside)	$18.10	$144.80	$26.55	$212.40	$18.07	$26.25
3 Building Laborers	16.10	386.40	23.65	567.60		
2 Cement Finishers	19.70	315.20	27.95	447.20		
1 Equip. Oper. (med.)	20.75	166.00	30.35	242.80		
1 Concrete Pump (small)		498.80		548.70	8.90	9.79
56 M.H., Daily Totals		$1511.20		$2018.70	$26.97	$36.04
Crew C-9	**Hr.**	**Daily**	**Hr.**	**Daily**	**Bare Costs**	**Incl. O&P**
1 Cement Finisher	$19.70	$157.60	$27.95	$223.60	$19.70	$27.95
1 Gas Finishing Mach.		26.25		28.90	3.28	3.61
8 M.H., Daily Totals		$183.85		$252.50	$22.98	$31.56
Crew C-10	**Hr.**	**Daily**	**Hr.**	**Daily**	**Bare Costs**	**Incl. O&P**
1 Building Laborer	$16.10	$128.80	$23.65	$189.20	$18.50	$26.51
2 Cement Finishers	19.70	315.20	27.95	447.20		
2 Gas Finishing Mach.		52.50		57.75	2.18	2.40
24 M.H., Daily Totals		$496.50		$694.15	$20.68	$28.91
Crew C-11	**Hr.**	**Daily**	**Hr.**	**Daily**	**Bare Costs**	**Incl. O&P**
1 Struc. Steel Foreman	$24.10	$192.80	$38.50	$308.00	$21.71	$34.10
6 Struc. Steel Workers	22.10	1060.80	35.30	1694.40		
1 Equip. Oper. (crane)	21.20	169.60	31.00	248.00		
1 Equip. Oper. Oiler	17.55	140.40	25.65	205.20		
1 Truck Crane, 150 Ton		1055.00		1160.50	14.65	16.11
72 M.H., Daily Totals		$2618.60		$3616.10	$36.36	$50.21

xvi

Figure 14.10

	3.2 Reinforcing Steel		CREW	DAILY OUTPUT	MAN-HOURS	UNIT	BARE COSTS MAT.	LABOR	EQUIP.	TOTAL	TOTAL INCL O&P	
100	3800	Subgrade chairs, (SCSGC) ½" diameter, 3-½" high				C	210			210	230	100
	3850	12" high					615			615	675	
	3900	¾" diameter, 3-½" high					280			280	310	
	3950	12" high					640			640	705	
	4200	Subgrade stakes (SCSGS) ½" diameter, 16" long					170			170	185	
	4250	24" long					240			240	265	
	4300	¾" diameter, 16" long					180			180	200	
	4350	28" long				↓	280			280	310	
	4500	Tie wire, 16 ga. annealed steel, under 500 lbs.				Cwt.	67			67	74	
	4520	2,000 to 4,000 lbs.				"	60			60	66	
	4550	Tie wire holder, plastic case				Ea.	28			28	31	
	4600	Aluminum case				"	30			30	33	
150	0010	**COATED REINFORCING** Add to material										150
	0100	Epoxy coated				Cwt.	43			43	47	
	0150	Galvanized, #3					19.80			19.80	22	
	0200	#4					18.75			18.75	21	
	0250	#5					16.50			16.50	18.15	
	0300	#6 or over					15.35			15.35	16.90	
	1000	For over 20 tons, #6 or larger, minimum					14.25			14.25	15.70	
	1500	Maximum				↓	16			16	17.60	
200	0010	**PRESTRESSING STEEL** Post-tensioned in field										200
	0020											
	0100	Grouted strand, 50' span, 100 kip (40)	C-3	1,200	.053	Lb.	1.45	1.10	.07	2.62	3.36	
	0150	300 kip		2,700	.024		1.17	.49	.03	1.69	2.07	
	0300	100' span, grouted, 100 kip		1,700	.038		.95	.77	.05	1.77	2.29	
	0350	300 kip		3,200	.020		.85	.41	.03	1.29	1.60	
	0500	200' span, grouted, 100 kip		2,700	.024		.88	.49	.03	1.40	1.75	
	0550	300 kip		3,500	.018		.82	.38	.02	1.22	1.50	
	0800	Grouted bars, 50' span, 42 kip		2,600	.025		.95	.51	.03	1.49	1.86	
	0850	143 kip		3,200	.020		.84	.41	.03	1.28	1.58	
	1000	75' span, grouted, 42 kip		3,200	.020		.79	.41	.03	1.23	1.53	
	1050	143 kip	↓	4,200	.015		.73	.31	.02	1.06	1.30	
	1200	Ungrouted strand, 50' span, 100 kip	C-4	1,275	.025		.92	.57	.02	1.51	1.93	
	1250	200 kip		1,475	.022		.92	.49	.02	1.43	1.80	
	1400	100' span, ungrouted, 100 kip		1,500	.021		.79	.48	.02	1.29	1.65	
	1450	200 kip		1,650	.019		.79	.44	.02	1.25	1.58	
	1600	200' span, ungrouted, 100 kip		1,500	.021		.73	.48	.02	1.23	1.58	
	1650	200 kip		1,700	.019		.73	.43	.02	1.18	1.49	
	1800	Ungrouted bars, 50' span, 42 kip		1,400	.023		.84	.52	.02	1.38	1.75	
	1850	143 kip		1,700	.019		.79	.43	.02	1.24	1.56	
	2000	75' span, ungrouted, 42 kip		1,800	.018		.79	.40	.02	1.21	1.52	
	2050	143 kip		2,200	.015		.73	.33	.01	1.07	1.33	
	2220	Ungrouted single strand, 100' slab, 25 kip		1,200	.027		.84	.60	.03	1.47	1.89	
	2250	35 kip	↓	1,475	.022	↓	.79	.49	.02	1.30	1.66	
300	0010	**REINFORCING IN PLACE** A615 Grade 60										300
	0100	Beams & Girders, #3 to #7	4 Rodm	1.60	20	Ton	525	440		965	1,275	
	0150	#8 to #14		2.70	11.850		515	260		775	975	
	0200	Columns, #3 to #7		1.50	21.330		525	470		995	1,325	
	0250	#8 to #14 (38)		2.30	13.910		515	305		820	1,050	
	0300	Spirals, hot rolled, 8" to 15" diameter		2.20	14.550		1,105	320		1,425	1,725	
	0320	15" to 24" diameter		2.20	14.550		1,055	320		1,375	1,675	
	0330	24" to 36" diameter		2.30	13.910		1,030	305		1,335	1,625	
	0340	36" to 48" diameter		2.40	13.330		1,055	295		1,350	1,625	
	0360	48" to 64" diameter		2.50	12.800		1,105	285		1,390	1,650	
	0380	64" to 84" diameter		2.60	12.310		1,185	270		1,455	1,725	
	0390	84" to 96" diameter		2.70	11.850		1,185	260		1,445	1,725	
	0400	Elevated slabs, #4 to #7		2.90	11.030		525	245		770	960	
	0500	Footings, #4 to #7		2.10	15.240		515	335		850	1,100	
	0550	#8 to #14		3.60	8.890		505	195		700	865	
	0600	Slab on grade, #3 to #7	↓	2.30	13.910	↓	520	305		825	1,050	

For expanded coverage of these items see *Means Concrete Cost Data 1987*

71

Figure 14.11

		3.3 Cast in Place Concrete	CREW	DAILY OUTPUT	MAN-HOURS	UNIT	BARE COSTS MAT.	LABOR	EQUIP.	TOTAL	TOTAL INCL O&P
380	0400	Columns, square or round, 12" thick, pumped	C-20	45	1.420	C.Y.		25	12.20	37.20	50
	0450	With crane and bucket	C-7	40	1.600			28	15.55	43.55	58
	0600	18" thick, pumped	C-20	60	1.070			18.55	9.15	27.70	37
	0650	With crane and bucket	C-7	55	1.160			20	11.30	31.30	42
	0800	24" thick, pumped	C-20	92	.696			12.10	6	18.10	24
	0850	With crane and bucket	C-7	70	.914			15.90	8.90	24.80	33
	1000	36" thick, pumped	C-20	140	.457			7.95	3.93	11.88	15.95
	1050	With crane and bucket	C-7	100	.640			11.10	6.25	17.35	23
	1400	Elevated slabs, less than 6" thick, pumped	C-20	110	.582			10.10	5	15.10	20
	1450	With crane and bucket	C-7	95	.674			11.70	6.55	18.25	24
	1500	6" to 10" thick, pumped	C-20	130	.492			8.55	4.23	12.78	17.15
	1550	With crane and bucket	C-7	110	.582			10.10	5.65	15.75	21
	1600	Slabs over 10" thick, pumped	C-20	150	.427			7.40	3.67	11.07	14.85
	1650	With crane and bucket	C-7	130	.492			8.55	4.79	13.34	17.75
	1900	Footings, continuous, shallow, direct chute	C-6	120	.400			6.80	.43	7.23	10.40
	1950	Pumped	C-20	100	.640			11.10	5.50	16.60	22
	2000	With crane and bucket	C-7	90	.711			12.35	6.90	19.25	26
	2100	Deep continuous footings, direct chute	C-6	155	.310			5.25	.33	5.58	8.05
	2150	Pumped	C-20	120	.533			9.25	4.58	13.83	18.60
	2200	With crane and bucket	C-7	110	.582			10.10	5.65	15.75	21
	2400	Footings, spread, under 1 C.Y., direct chute	C-6	55	.873			14.85	.93	15.78	23
	2450	Pumped	C-20	50	1.280			22	11	33	45
	2500	With crane and bucket	C-7	45	1.420			25	13.85	38.85	51
	2600	Spread footings, over 5 C.Y., direct chute	C-6	110	.436			7.45	.46	7.91	11.35
	2650	Pumped	C-20	105	.610			10.60	5.25	15.85	21
	2700	With crane and bucket	C-7	100	.640			11.10	6.25	17.35	23
	2900	Foundation mats, over 20 C.Y., direct chute	C-6	350	.137			2.34	.15	2.49	3.57
	2950	Pumped	C-20	325	.197			3.42	1.69	5.11	6.85
	3000	With crane and bucket	C-7	300	.213			3.71	2.08	5.79	7.70
	3200	Grade beams, direct chute	C-6	150	.320			5.45	.34	5.79	8.35
	3250	Pumped	C-20	130	.492			8.55	4.23	12.78	17.15
	3300	With crane and bucket	C-7	120	.533			9.25	5.20	14.45	19.25
	3500	High rise, for more than 5 stories, pumped, add per story	C-20	2,100	.030			.53	.26	.79	1.06
	3510	With crane and bucket, add per story	C-7	2,100	.030			.53	.30	.83	1.10
	3700	Pile caps, under 5 C.Y., direct chute	C-6	90	.533			9.10	.57	9.67	13.90
	3750	Pumped	C-20	85	.753			13.10	6.45	19.55	26
	3800	With crane and bucket	C-7	80	.800			13.90	7.80	21.70	29
	3850	Pile cap, 5 C.Y. to 10 C.Y., direct chute	C-6	175	.274			4.67	.29	4.96	7.15
	3900	Pumped	C-20	160	.400			6.95	3.44	10.39	13.95
	3950	With crane and bucket	C-7	150	.427			7.40	4.15	11.55	15.40
	4000	Pile cap, over 10 C.Y., direct chute	C-6	215	.223			3.80	.24	4.04	5.80
	4050	Pumped	C-20	195	.328			5.70	2.82	8.52	11.45
	4100	With crane and bucket	C-7	185	.346			6	3.37	9.37	12.50
	4300	Slab on grade, 4" thick, direct chute	C-6	110	.436			7.45	.46	7.91	11.35
	4350	Pumped	C-20	120	.533			9.25	4.58	13.83	18.60
	4400	With crane and bucket	C-7	110	.582			10.10	5.65	15.75	21
	4600	Slab over 6" thick, direct chute	C-6	165	.291			4.96	.31	5.27	7.55
	4650	Pumped	C-20	165	.388			6.75	3.33	10.08	13.50
	4700	With crane and bucket	C-7	145	.441			7.65	4.29	11.94	15.95
	4900	Walls, 8" thick, direct chute	C-6	90	.533			9.10	.57	9.67	13.90
	4950	Pumped	C-20	85	.753			13.10	6.45	19.55	26
	5000	With crane and bucket	C-7	80	.800			13.90	7.80	21.70	29
	5050	12" thick, direct chute	C-6	100	.480			8.20	.51	8.71	12.50
	5100	Pumped	C-20	95	.674			11.70	5.80	17.50	23
	5200	With crane and bucket	C-7	90	.711			12.35	6.90	19.25	26
	5300	15" thick, direct chute	C-6	105	.457			7.80	.49	8.29	11.90
	5350	Pumped	C-20	100	.640			11.10	5.50	16.60	22
	5400	With crane and bucket	C-7	95	.674			11.70	6.55	18.25	24
	5600	Wheeled concrete dumping, add to placing costs above (46)									
	5610	Walking cart, 50' haul, add	C-18	32	.281	C.Y.		4.59	1.09	5.68	7.95

For expanded coverage of these items see *Means Concrete Cost Data 1987*

79

Figure 14.12

Walls

Formwork: Line No. 3.1 650 7860 (see Figure 14.13):
Forms in Place – Walls, Modular Prefabricated Plywood to 8′ high (4 uses per month).

Daily Output: 970 S.F. C-2 Crew (see Figure 14.10)

$$\frac{6880 \text{ S.F.}}{970 \text{ S.F./day}} = 7.09 \text{ days}$$

Reinforcing: Line No. 3.2 300 0700 (see Figure 14.14)
Reinforcing in Place Walls, #3 to #7.

Daily Output: 3 Tons/Day Using 4 Rodmen.

$$\frac{2.4 \text{ Tons}}{3 \text{ Tons/day}} = .80 \text{ days}$$

Placing Concrete: Line No. 3.3 380 5050 (see Figure 14.12)
Placing Concrete and Vibrating, including Labor and Equipment.

Walls 12″ thick, Direct Chute.

Daily Output: 100 C.Y./day C-6 Crew (see Figure 14.10)

$$\frac{127 \text{ C.Y.}}{100 \text{ C.Y./day}} = 1.27 \text{ days}$$

Based on the preceding information, the total time required to form, reinforce, and place concrete in 860 L.F. of concrete footing and wall would be:

Form continuous footing	3.55 days
Reinforce continuous footing	2.38 days
Place concrete continuous footing	.53 days
Form wall	7.09 days
Reinforce wall	.80 days
Place concrete	1.27 days
Total time to place footing and wall	**15.62 days**

$$\frac{860 \text{ L.F.}}{15.62 \text{ days}} = 55 \text{ L.F./day}$$

For a faster method of developing duration times, use the concrete-in-place section of *Building Construction Cost Data*, 1987, as follows:

Line No. 3.3 140 3950 (see Figure 14.15)
Concrete in Place, Footings Strip 36″ x 12″ reinforced.
Daily Output 49.07 C.Y.
Using Composite Crew C-17B (see Figure 14.16)

$$\frac{64 \text{ C.Y.}}{49.07 \text{ C.Y./day}} = 1.30 \text{ days}$$

Line No. 3.3 140 4260 (see Figure 14.15)
Concrete in Place, Grade Walls 12″ thick, 8′ high.
Daily Output 13.5 C.Y.
Using Composite Crew C-17A (see Figure 14.16)

$$\frac{127 \text{ C.Y.}}{13.5 \text{ C.Y./day}} = 9.4 \text{ days}$$

Total time to place footing and wall:

Strip Footing:	1.3 days
Wall	9.4 days
	10.7 days

		3.1 Formwork	CREW	DAILY OUTPUT	MAN-HOURS	UNIT	BARE COSTS MAT.	BARE COSTS LABOR	BARE COSTS EQUIP.	BARE COSTS TOTAL	TOTAL INCL O&P	
650	7800	Modular prefabricated plywood, to 8′ high, 1 use per month	C-2	910	.053	SFCA	.74	1.06	.05	1.85	2.42	650
	7820	2 use per month		930	.052		.42	1.04	.05	1.51	2.04	
	7840	3 use per month (34)		950	.051		.31	1.02	.04	1.37	1.88	
	7860	4 use per month		970	.049		.25	1	.04	1.29	1.79	
	8000	To 16′ high, 1 use per month		550	.087		.96	1.76	.08	2.80	3.72	
	8020	2 use per month		570	.084		.54	1.70	.07	2.31	3.16	
	8040	3 use per month		590	.081		.41	1.64	.07	2.12	2.93	
	8060	4 use per month		610	.079		.36	1.58	.07	2.01	2.80	
	8600	Pilasters, 1 use		270	.178		1.51	3.58	.16	5.25	7.10	
	8620	2 use		330	.145		.91	2.93	.13	3.97	5.45	
	8640	3 use		370	.130		.73	2.61	.11	3.45	4.76	
	8660	4 use		385	.125		.65	2.51	.11	3.27	4.53	
	9000	Steel framed plywood, to 8′ high, 1 use per month (34)		600	.080		1.10	1.61	.07	2.78	3.65	
	9020	2 use per month		640	.075		.57	1.51	.07	2.15	2.92	
	9040	3 use per month		655	.073		.38	1.48	.06	1.92	2.66	
	9060	4 use per month		665	.072		.34	1.45	.06	1.85	2.57	
	9200	Over 8′ to 16′ high, 1 use per month		455	.105		1.23	2.12	.09	3.44	4.57	
	9220	2 use per month		505	.095		.65	1.91	.08	2.64	3.62	
	9240	3 use per month		525	.091		.44	1.84	.08	2.36	3.27	
	9260	4 use per month		530	.091		.38	1.82	.08	2.28	3.18	
	9400	Over 16′ to 20′ high, 1 use per month		425	.113		1.30	2.27	.10	3.67	4.88	
	9420	2 use per month		435	.110		.70	2.22	.10	3.02	4.14	
	9440	3 use per month		455	.105		.49	2.12	.09	2.70	3.76	
	9460	4 use per month		465	.103		.43	2.08	.09	2.60	3.62	
700	0010	**GAS STATION FORMS** Curb fascia, with template,										700
	0050	12 ga. steel, left in place, 9″ high	1 Carp	50	.160	L.F.	5	3.29		8.29	10.35	
	1000	Sign or light bases, 18″ diameter, 9″ high		9	.889	Ea.	28	18.25		46.25	58	
	1050	30″ diameter, 13″ high		8	1		50	21		71	85	
	2000	Island forms, 10′ long, 9″ high, 3′- 6″ wide	C-1	10	3.200		120	62	3.18	185.18	225	
	2050	4′ wide		9	3.560		130	69	3.53	202.53	250	
	2500	20′ long, 9″ high, 4′ wide		6	5.330		220	105	5.30	330.30	400	
	2550	5′ wide		5	6.400		250	125	6.35	381.35	465	
750	0010	**SCAFFOLDING** See division 1.1-440										750
800	0010	**SHORES** Erect and strip, by hand, horizontal members										800
	0500	Aluminum joists and stringers	2 Carp	60	.267	Ea.		5.50		5.50	8.05	
	0600	Steel, adjustable beams		45	.356			7.30		7.30	10.70	
	0700	Wood joists		50	.320			6.60		6.60	9.65	
	0800	Wood stringers		30	.533			10.95		10.95	16.10	
	1000	Vertical members to 10′ high		55	.291			6		6	8.75	
	1050	To 13′ high		50	.320			6.60		6.60	9.65	
	1100	To 16′ high		45	.356			7.30		7.30	10.70	
	1500	Reshoring		1,400	.011	S.F.	.14	.23		.37	.50	
	1600	Flying truss system	C-17D	9,600	.009	SFCA		.19	.05	.24	.33	
	1760	Horizontal, aluminum joists, 6′ to 30′ spans, buy				L.F.	8			8	8.80	
	1770	Aluminum stringers, 12′ & 16′ spans				"	12			12	13.20	
	1810	Horizontal, steel beam, adjustable, 4′ to 7′ span				Ea.	85			85	94	
	1830	6′ to 10′ span					110			110	120	
	1920	9′ to 15′ span					200			200	220	
	1940	12′ to 20′ span					235			235	260	
	1970	Steel stringer, 6′ to 15′ span				L.F.	6.30			6.30	6.95	
	2000											
	3000	Rent for job duration, aluminum, first month				SF Flr.	.16			.16	.18	
	3050	Steel				"	.13			.13	.14	
	3500	Vertical, adjustable steel, 5′-7″ to 9′-6″ high, 10,000# cap., buy				Ea.	45			45	50	
	3550	7′-3″ to 12′-10″ high, 7800# capacity					55			55	61	
	3600	8′-10″ to 12′-4″ high, 10,000# capacity					60			60	66	
	3650	8′-10″ to 16′-1″ high, 3800# capacity					65			65	72	
	4000	Frame shoring systems, aluminum, 10,000# per leg,										
	4050	6′ wide, 5′ & 6′ high				Ea.	180			180	200	

For expanded coverage of these items see *Means Concrete Cost Data 1987*

67

3

Figure 14.13

		3.2 Reinforcing Steel	CREW	DAILY OUTPUT	MAN-HOURS	UNIT	BARE COSTS MAT.	LABOR	EQUIP.	TOTAL	TOTAL INCL O&P	
300	0700	Walls, #3 to #7	4 Rodm	3	10.670	Ton	520	235		755	940	300
	0750	#8 to #14		4	8		510	175		685	840	
	1000	Typical in place, 10 ton lots, average		1.70	18.820		525	415		940	1,225	
	1100	Over 50 ton lots, average		2.30	13.910		505	305		810	1,025	
	1200	High strength steel, Grade 75, #14 bars only, add (39)					40			40	44	
	2000	Unloading & sorting, add to above	C-5	100	.560			12.10	4.08	16.18	23	
	2200	Crane cost for handling, add to above, minimum		135	.415			8.95	3.02	11.97	17.10	
	2210	Average		92	.609			13.15	4.43	17.58	25	
	2220	Maximum		35	1.600			35	11.65	46.65	66	
	2400	Dowels, 2 feet long, deformed, #3 bar	2 Rodm	140	.114	Ea.	.75	2.53		3.28	4.78	
	2410	#4 bar		125	.128		.90	2.83		3.73	5.40	
	2420	#5 bar		110	.145		1.08	3.21		4.29	6.20	
	2430	#6 bar		105	.152		1.40	3.37		4.77	6.80	
	2450	Longer and heavier dowels		450	.036	Lb.	.35	.79		1.14	1.61	
	2500	Smooth dowels, 12" long, ¼" or ⅜" diameter		140	.114	Ea.	.57	2.53		3.10	4.58	
	2520	⅝" diameter		125	.128		.99	2.83		3.82	5.50	
	2530	¾" diameter		110	.145		1.19	3.21		4.40	6.35	
	2550											
	2700	Dowel caps, 5" long, ½" to ¾" diameter	2 Rodm	800	.020	Ea.	.06	.44		.50	.76	
	2720	1-¼" diameter	"	750	.021	"	.08	.47		.55	.83	
350	0010	**SPLICING REINFORCING BARS** Incl. holding bars in										350
	0020	place while splicing										
	0100	Butt weld columns #4 bars	C-5	190	.295	Ea.	.72	6.35	2.15	9.22	12.95	
	0110	#6 bars		150	.373		1.08	8.05	2.72	11.85	16.60	
	0130	#10 bars		95	.589		1.44	12.75	4.29	18.48	26	
	0150	#14 bars		65	.862		1.85	18.60	6.25	26.70	38	
	0280	Column splice clamps, sleeve & wedge, or end bearing										
	0300	#7 or #8 bars	C-5	190	.295	Ea.	2.37	6.35	2.15	10.87	14.75	
	0310	#9 or #10 bars		170	.329		2.42	7.10	2.40	11.92	16.25	
	0320	#11 bars		160	.350		3.30	7.55	2.55	13.40	18.05	
	0330	#14 bars		150	.373		3.80	8.05	2.72	14.57	19.60	
	0340	#18 bars		140	.400		5.95	8.65	2.91	17.51	23	
	0500	Reducer inserts for above, #14 to #18 bar					2.35			2.35	2.59	
	0520	#14 to #11 bar					2			2	2.20	
	0550	#10 to #9 bar					.62			.62	.68	
	0560	#9 to #8 bar					.62			.62	.68	
	0580	#8 to #7 bar					.62			.62	.68	
	0600	For bolted speed sleeve type, deduct						15%				
	0700											
	0800	Mechanical butt splice, sleeve type with filler metal, compression										
	0810	only, all grades, columns only #11 bars	C-5	68	.824	Ea.	9.80	17.80	6	33.60	45	
	0900	#14 bars		62	.903		10	19.50	6.55	36.05	48	
	0920	#18 bars		62	.903		12.30	19.50	6.55	38.35	51	
	1000	125% yield point, grade 60, columns only, #6 bars		68	.824		13.50	17.80	6	37.30	49	
	1020	#7 or #8 bars		68	.824		11.90	17.80	6	35.70	47	
	1030	#9 bars		68	.824		11.60	17.80	6	35.40	47	
	1040	#10 bars		68	.824		12.60	17.80	6	36.40	48	
	1050	#11 bars		68	.824		15.70	17.80	6	39.50	51	
	1060	#14 bars		62	.903		19.85	19.50	6.55	45.90	59	
	1070	#18 bars		62	.903		29	19.50	6.55	55.05	69	
	1080											
	1200	Full tension, grade 60 steel, columns,										
	1220	slabs or beams, #6, #7, #8 bars	C-5	68	.824	Ea.	11.50	17.80	6	35.30	47	
	1230	#9 bars		68	.824		13.10	17.80	6	36.90	48	
	1240	#10 bars		68	.824		14.35	17.80	6	38.15	50	
	1250	#11 bars		68	.824		17.05	17.80	6	40.85	53	
	1260	#14 bars		62	.903		23	19.50	6.55	49.05	63	
	1270	#18 bars		62	.903		37	19.50	6.55	63.05	78	
	1400	If equipment handling not required, deduct						50%				

3

72

For expanded coverage of these items see *Means Concrete Cost Data 1987*

Figure 14.14

3.3 Cast in Place Concrete

140		CREW	DAILY OUTPUT	MAN-HOURS	UNIT	BARE COSTS MAT.	LABOR	EQUIP.	TOTAL	TOTAL INCL O&P
1700	Curbs, formed in place, 6" x 18", straight,	C-15	400	.180	L.F.	2.95	3.47	.12	6.54	8.50
1750	Curb and gutter	"	170	.424	"	4.70	8.15	.28	13.13	17.45
1900	Elevated slabs, flat slab, 125 psf Sup. Load, 20' span	C-17A	13.36	6.060	C.Y.	115	130	9.40	254.40	325
1950	30' span	C-17B	18.25	4.490		105	95	13.20	213.20	270
2100	Flat plate, 125 psf Sup. Load, 15' span	C-17A	10.28	7.880		115	165	12.20	292.20	385
2150	25' span	C-17B	17.01	4.820		99	100	14.20	213.20	275
2300	Waffle const., 30" domes, 125 psf Sup. Load, 20' span		14.10	5.820		125	125	17.10	267.10	340
2350	30' span		17.02	4.820		115	100	14.15	229.15	295
2500	One way joists, 30" pans, 125 psf Sup. Load, 15' span	C-17A	11.07	7.320		110	155	11.35	276.35	360
2550	25' span		11.04	7.340		125	155	11.35	291.35	380
2700	One way beam & slab, 125 psf Sup. Load, 15' span		7.49	10.810		130	230	16.75	376.75	500
2750	25' span		10.15	7.980		125	170	12.35	307.35	400
2900	Two way beam & slab, 125 psf Sup. Load, 15' span		8.22	9.850		125	210	15.25	350.25	465
2950	25' span	C-17B	12.23	6.700		110	140	19.70	269.70	350
3100	Elevated slabs including finish, not									
3110	including forms or reinforcing									
3150	Regular concrete, 4" slab	C-8	2,685	.021	S.F.	.66	.38	.19	1.23	1.48
3200	6" slab		2,585	.022		1.04	.39	.19	1.62	1.92
3250	2-½" thick floor fill		2,685	.021		.46	.38	.19	1.03	1.26
3300	Lightweight, 110# per C.F., 2-½" thick floor fill		2,585	.022		.58	.39	.19	1.16	1.42
3400	Cellular concrete, 1-⅝" fill, under 5000 S.F.		2,000	.028		.26	.51	.25	1.02	1.30
3450	Over 10,000 S.F.		2,200	.025		.22	.46	.23	.91	1.16
3500	Add per floor for 3 to 6 stories high		31,800	.002			.03	.02	.05	.06
3520	For 7 to 20 stories high		21,200	.003			.05	.02	.07	.10
3800	Footings, spread under 1 C.Y.	C-17B	31.82	2.580	C.Y.	74	55	7.60	136.60	170
3850	Over 5 C.Y.	C-17C	70.45	1.180		70	25	5.20	100.20	120
3900	Footings, strip, 18" x 9", plain	C-17B	34.22	2.400		63	51	7.05	121.05	150
3950	36" x 12", reinforced		49.07	1.670		69	35	4.91	108.91	135
4000	Foundation mat, under 10 C.Y.		32.32	2.540		115	54	7.45	176.45	215
4050	Over 20 C.Y.		47.37	1.730		105	37	5.10	147.10	175
4200	Grade walls, 8" thick, 8' high	C-17A	10.16	7.970		105	170	12.35	287.35	380
4250	14' high	C-20	7.30	8.770		155	150	75	380	475
4260	12" thick, 8' high	C-17A	13.50	6		125	125	9.30	259.30	335
4270	14' high	C-20	11.60	5.520		115	96	47	258	320
4300	15" thick, 8' high	C-17B	20.01	4.100		85	87	12.05	184.05	235
4350	12' high	C-20	14.80	4.320		99	75	37	211	260
4500	18' high	"	12	5.330		115	93	46	254	310
4510										
4650	Ground slab, not including finish, 4" thick	C-17C	75.28	1.100	C.Y.	63	23	4.87	90.87	110
4700	6" thick	"	113.47	.731	"	60	15.50	3.23	78.73	92
4750	Ground slab, incl. troweled finish, not incl. forms									
4760	or reinforcing, over 10,000 S.F., 4" thick slab	C-8	3,520	.016	S.F.	.73	.29	.14	1.16	1.37
4820	6" thick slab		3,610	.016		1.10	.28	.14	1.52	1.77
4840	8" thick slab		3,275	.017		1.48	.31	.15	1.94	2.25
4900	12" thick slab		2,875	.019		2.20	.35	.17	2.72	3.12
4950	15" thick slab		2,560	.022		2.75	.40	.19	3.34	3.82
5200	Lift slab in place above the foundation, incl. forms, (48)									
5210	reinforcing, concrete and columns, minimum	C-17E	745	.107	S.F.	2.05	2.28	.06	4.39	5.70
5250	Average		675	.119		3.30	2.51	.07	5.88	7.40
5300	Maximum		430	.186		3.55	3.94	.11	7.60	9.85
5500	Lightweight, ready mix, including screed finish only, (54)									
5510	not including forms or reinforcing									
5550	1:4 for structural roof decks (43)	C-8	80	.700	C.Y.	71.30	12.65	6.25	90.20	105
5600	1:6 for ground slab with radiant heat		90	.622		69	11.25	5.55	85.80	98
5650	1:3:2 with sand aggregate, roof deck		80	.700		71	12.65	6.25	89.90	105
5700	Ground slab		105	.533		71	9.65	4.75	85.40	97
5900	Pile caps, incl. forms and reinf., sq. or rect., under 5 C.Y.	C-17C	47.26	1.760		72	37	7.75	116.75	145
5950	Over 10 C.Y.		76.47	1.090		71	23	4.79	98.79	115
6000	Triangular or hexagonal, under 5 C.Y.		47.95	1.730		70	37	7.65	114.65	140
6050	Over 10 C.Y.		77.85	1.070		72	23	4.71	99.71	120

For expanded coverage of these items see *Means Concrete Cost Data 1987*

Figure 14.15

CREWS

Crew No.	Bare Costs Hr.	Bare Costs Daily	Incl. Subs O & P Hr.	Incl. Subs O & P Daily	Cost Per Man-hour Bare Costs	Cost Per Man-hour Incl. O&P
Crew C-12	**Hr.**	**Daily**	**Hr.**	**Daily**	**Bare Costs**	**Incl. O&P**
1 Carpenter Foreman (out)	$22.55	$180.40	$33.10	$264.80	$20.25	$29.70
3 Carpenters	20.55	493.20	30.15	723.60		
1 Building Laborer	16.10	128.80	23.65	189.20		
1 Equip. Oper. (crane)	21.20	169.60	31.00	248.00		
1 Hyd. Crane, 12 Ton		243.00		267.30	5.06	5.56
48 M.H., Daily Totals		$1215.00		$1692.90	$25.31	$35.26
Crew C-13	**Hr.**	**Daily**	**Hr.**	**Daily**	**Bare Costs**	**Incl. O&P**
1 Struc. Steel Worker	$22.10	$176.80	$35.30	$282.40	$21.58	$33.58
1 Welder	22.10	176.80	35.30	282.40		
1 Carpenter	20.55	164.40	30.15	241.20		
1 Gas Welding Machine		54.75		60.25	2.28	2.51
24 M.H., Daily Totals		$572.75		$866.25	$23.86	$36.09
Crew C-14	**Hr.**	**Daily**	**Hr.**	**Daily**	**Bare Costs**	**Incl. O&P**
1 Carpenter Foreman (out)	$22.55	$180.40	$33.10	$264.80	$19.79	$29.41
5 Carpenters	20.55	822.00	30.15	1206.00		
4 Building Laborers	16.10	515.20	23.65	756.80		
4 Rodmen (reinf.)	22.10	707.20	34.60	1107.20		
2 Cement Finishers	19.70	315.20	27.95	447.20		
1 Equip. Oper. (crane)	21.20	169.60	31.00	248.00		
1 Equip. Oper. Oiler	17.55	140.40	25.65	205.20		
1 Crane, 80 Ton, & Tools		927.80		1020.60		
Power Tools		31.80		35.00		
2 Gas Finishing Mach.		52.50		57.75	7.02	7.73
144 M.H., Daily Totals		$3862.10		$5348.55	$26.81	$37.14
Crew C-15	**Hr.**	**Daily**	**Hr.**	**Daily**	**Bare Costs**	**Incl. O&P**
1 Carpenter Foreman (out)	$22.55	$180.40	$33.10	$264.80	$19.27	$28.31
2 Carpenters	20.55	328.80	30.15	482.40		
3 Building Laborers	16.10	386.40	23.65	567.60		
2 Cement Finishers	19.70	315.20	27.95	447.20		
1 Rodman (reinf.)	22.10	176.80	34.60	276.80		
Power Tools		21.20		23.30		
1 Gas Finishing Mach.		26.25		28.90	.65	.72
72 M.H., Daily Totals		$1435.05		$2091.00	$19.92	$29.03
Crew C-16	**Hr.**	**Daily**	**Hr.**	**Daily**	**Bare Costs**	**Incl. O&P**
1 Labor Foreman (outside)	$18.10	$144.80	$26.55	$212.40	$18.97	$28.10
3 Building Laborers	16.10	386.40	23.65	567.60		
2 Cement Finishers	19.70	315.20	27.95	447.20		
1 Equip. Oper. (med.)	20.75	166.00	30.35	242.80		
2 Rodmen (reinf.)	22.10	353.60	34.60	553.60		
1 Concrete Pump (small)		498.80		548.70	6.92	7.62
72 M.H., Daily Totals		$1864.80		$2572.30	$25.89	$35.72
Crew C-17	**Hr.**	**Daily**	**Hr.**	**Daily**	**Bare Costs**	**Incl. O&P**
2 Skilled Worker Foremen	$22.80	$364.80	$33.65	$538.40	$21.20	$31.29
8 Skilled Workers	20.80	1331.20	30.70	1964.80		
80 M.H., Daily Totals		$1696.00		$2503.20	$21.20	$31.29
Crew C-17A	**Hr.**	**Daily**	**Hr.**	**Daily**	**Bare Costs**	**Incl. O&P**
2 Skilled Worker Foremen	$22.80	$364.80	$33.65	$538.40	$21.20	$31.29
8 Skilled Workers	20.80	1331.20	30.70	1964.80		
.125 Equip. Oper. (crane)	21.20	21.20	31.00	31.00		
.125 Crane, 80 Ton, & Tools		120.60		132.70		
.125 Hand Held Pwr. Tools		1.40		1.50		
.125 Walk Power Tools		3.40		3.75	1.54	1.70
81 M.H., Daily Totals		$1842.60		$2672.15	$22.74	$32.99
Crew C-17B	**Hr.**	**Daily**	**Hr.**	**Daily**	**Bare Costs**	**Incl. O&P**
2 Skilled Worker Foremen	$22.80	$364.80	$33.65	$538.40	$21.20	$31.29
8 Skilled Workers	20.80	1331.20	30.70	1964.80		
.25 Equip. Oper. (crane)	21.20	42.40	31.00	62.00		
.25 Crane, 80 Ton, & Tools		231.95		255.15		
.25 Hand Held Power Tools		2.65		2.90		
.25 Power Tools		6.55		7.20	2.94	3.23
82 M.H., Daily Totals		$1979.55		$2830.45	$24.14	$34.52
Crew C-17C	**Hr.**	**Daily**	**Hr.**	**Daily**	**Bare Costs**	**Incl. O&P**
2 Skilled Worker Foremen	$22.80	$364.80	$33.65	$538.40	$21.20	$31.29
8 Skilled Workers	20.80	1331.20	30.70	1964.80		
.375 Equip. Oper. (crane)	21.20	63.60	31.00	93.00		
.375 Crane, 80 Ton & Tools		352.55		387.80		
.375 Hand Held Power Tools		4.05		4.45		
.375 Power Tools		10.00		10.95	4.41	4.85
83 M.H., Daily Totals		$2126.20		$2999.40	$25.61	$36.14
Crew C-17D	**Hr.**	**Daily**	**Hr.**	**Daily**	**Bare Costs**	**Incl. O&P**
2 Skilled Worker Foremen	$22.80	$364.80	$33.65	$538.40	$21.20	$31.29
8 Skilled Workers	20.80	1331.20	30.70	1964.80		
.5 Equip. Oper. (crane)	21.20	84.80	31.00	124.00		
.5 Crane, 80 Ton & Tools		463.90		510.30		
.5 Hand Held Power Tools		5.30		5.85		
.5 Power Tools		13.15		14.45	5.74	6.31
84 M.H., Daily Totals		$2263.15		$3157.80	$26.94	$37.60
Crew C-17E	**Hr.**	**Daily**	**Hr.**	**Daily**	**Bare Costs**	**Incl. O&P**
2 Skilled Worker Foremen	$22.80	$364.80	$33.65	$538.40	$21.20	$31.29
8 Skilled Workers	20.80	1331.20	30.70	1964.80		
1 Hyd. Jack with Rods		47.65		52.40	.59	.65
80 M.H., Daily Totals		$1743.65		$2555.60	$21.79	$31.94
Crew C-18	**Hr.**	**Daily**	**Hr.**	**Daily**	**Bare Costs**	**Incl. O&P**
.125 Labor Foreman (out)	$18.10	$18.10	$26.55	$26.55	$16.32	$23.97
1 Building Laborer	16.10	128.80	23.65	189.20		
1 Concrete Cart, 10 C.F.		35.00		38.50	3.88	4.27
9 M.H., Daily Totals		$181.90		$254.25	$20.20	$28.24
Crew C-19	**Hr.**	**Daily**	**Hr.**	**Daily**	**Bare Costs**	**Incl. O&P**
.125 Labor Foreman (out)	$18.10	$18.10	$26.55	$26.55	$16.32	$23.97
1 Building Laborer	16.10	128.80	23.65	189.20		
1 Concrete Cart, 18 C.F.		54.30		59.75	6.03	6.63
9 M.H., Daily Totals		$201.20		$275.50	$22.35	$30.60
Crew C-20	**Hr.**	**Daily**	**Hr.**	**Daily**	**Bare Costs**	**Incl. O&P**
1 Labor Foreman (outside)	$18.10	$144.80	$26.55	$212.40	$17.38	$25.38
5 Building Laborers	16.10	644.00	23.65	946.00		
1 Cement Finisher	19.70	157.60	27.95	223.60		
1 Equip. Oper. (med.)	20.75	166.00	30.35	242.80		
2 Gas Engine Vibrators		51.00		56.10		
1 Concrete Pump (small)		498.80		548.70	8.59	9.45
64 M.H., Daily Totals		$1662.20		$2229.60	$25.97	$34.83

xvii

Figure 14.16

As can be seen from the two duration times established, the duration depends on the number of workers used to complete the activity. If a number of workers has been previously established, it may be advisable to use the man-hour per unit data to determine the duration time.

Line No. 3.3 140 3950 (see Figure 14.15)
Concrete in Place Footings, Strip 36″ x12″ reinforced
Man-hours: 1.67/C.Y. Using 6-worker crew.

$$\frac{64 \text{ C.Y. x } 1.67 \text{ hours/C.Y.}}{8 \text{ hours x } 6 \text{ workers}} = 2.23 \text{ days}$$

Line No. 3.3 140 4260 (see Figure 14.15)
Concrete in Place, Grade Walls 12″ thick, 8′ high.
Man-hours: 6/C.Y. Using 6-worker crew.

$$\frac{127 \text{ C.Y. x } 6 \text{ Man-hours/C.Y.}}{8 \text{ Man-hours/day x } 6 \text{ Workers}} = 15.88 \text{ days}$$

Total time to place footing and wall using a composite 6-worker crew:

Strip footing	2.23 days
Wall	15.88 days
	18.11 days

$$\frac{860 \text{ L.F. Wall}}{18.11 \text{ days}} = 47.49 \text{ L.F./day}$$
using a 6-worker crew.

The same logic may be used with field-reported information to develop activity durations.

The schedule in Figure 14.17 shows a continuous operation without constraints, such as available formwork curing time, etc. Assume a schedule using four pours with a three day during time required before working on the concrete surface or removing forms for the 860 L.F. of footing and wall. A bar chart and C.P.M. schedule are shown in Figures 14.6 and 14.7. The logic may be followed easier on the C.P.M. schedule.

Both schedules show no lost time for weather or other conditions that might cause delay. Time should be added to compensate for anticipated delays. The schedules as shown require a complete set of formwork. If the wall forms are removed from (1) and reused on (3), and one day is allowed to strip the forms, the scheduled time would be increased by one day.

Owner Decision Points

In the course of the process – from concept to completion – the owner has to make a contribution in furnishing information, performing certain functions, and making timely decisions.

Most of the information gathering is accomplished prior to the presentation. Nevertheless, clarifications of needed items and more details are still required. While this information should ideally be resolved all at once, at the outset of the design process, things do not usually happen that quickly. The information tends to flow as the need requires. For example, data on storage racks, floor loading, storage rack layouts, sprinkler head installation and density of spray, aisle lighting, delivery and installation times of items may not be known from the start. The reason this information comes later is that the owner/client has yet to negotiate for this work. All that is known is that the height of the loaded racks will not exceed a certain elevation. This latter piece of information is critical prior to the presentation since it contributes to the determination of the building's height. The management schedule must indicate the date when this storage rack information is to be furnished.

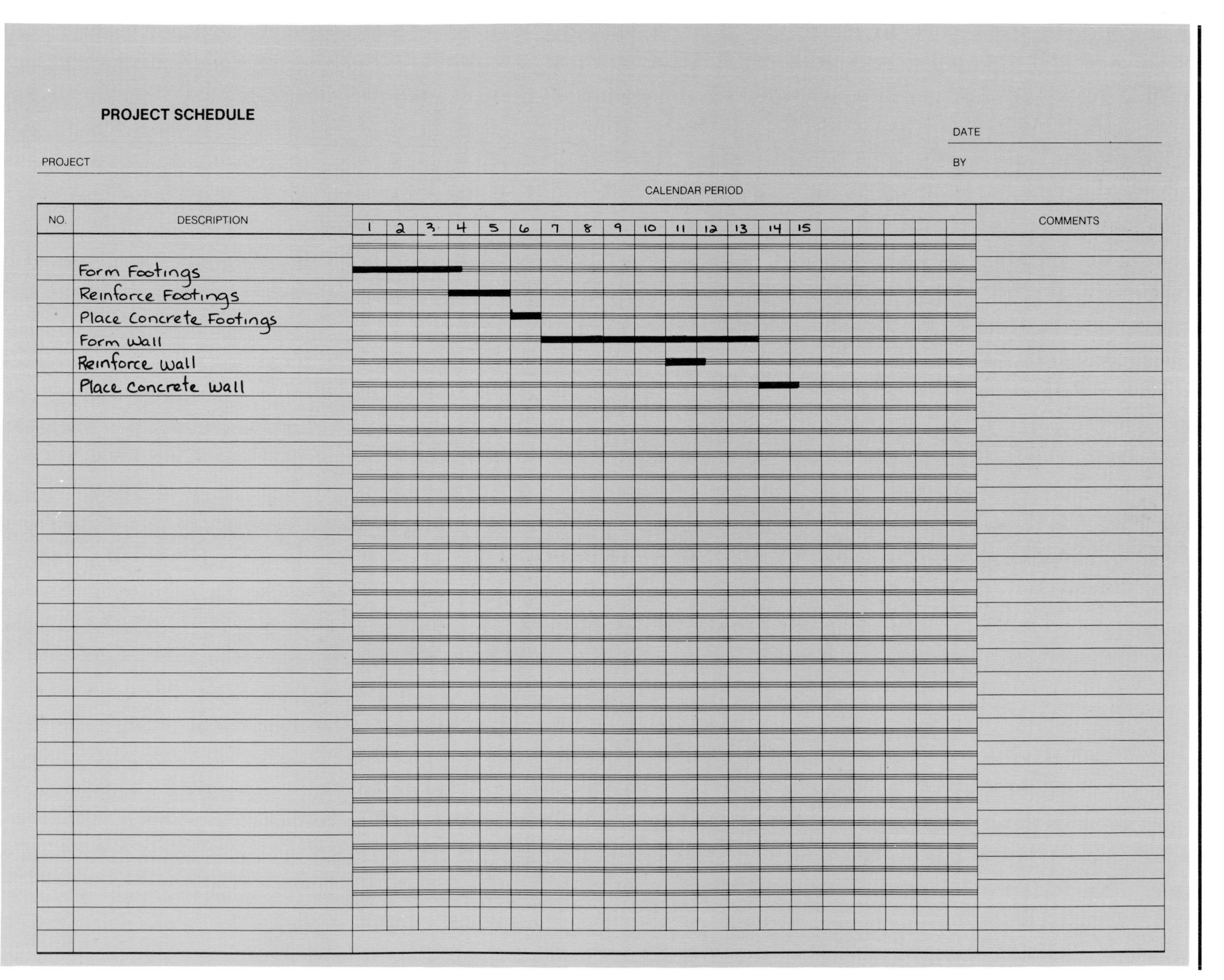

PROJECT SCHEDULE

PROJECT

DATE

BY

CALENDAR PERIOD

NO.	DESCRIPTION	1	2	3	4	5	6	7	8	9	10	11	12	13	14	15	COMMENTS
	Form Footings																
	Reinforce Footings																
	Place Concrete Footings																
	Form Wall																
	Reinforce Wall																
	Place Concrete Wall																

Figure 14.17

The electrical load requirements for all of the equipment should be given at the presentation. If this work is (at the time of presentation) carried as an allowance, then more definitive information on the electrical loads is needed and the dates required to receive that data should be shown on the schedule. The same is true for any mechanical ventilation needs, and additional heating and cooling loads to accommodate the owner's equipment.

Decision dates on the balance of other allowance items should be shown. Design completion sign-off meetings are scheduled before proceeding with the particular work.

A sample list of the owner's decision points is as follows:

- Provide structural loads for owner equipment
- Furnish electrical, mechanical loads
- Resolve allowance items
- Finish schedule resolution
- Select colors for all products and paint
- Provide office layouts
- Approve design for special signage and graphics
- Sign off on design plans
- Instruction periods of building systems
- Conduct final inspection and close out

Issues such as land suitability and securing of financing are not factors in the management schedule. Work does not usually start until the owner has secured buildable land (land that has been cleared of any environmental and zoning incumbencies). Financing also must be in place prior to the start of any work, including design. An exception to this rule is if an interim agreement is made in order to guarantee the payment of costs for prescribed work.

Milestone Dates

Milestone dates are those that serve as indicators of the project's performance. The milestone dates have a direct effect on the completion of the project. They are broadly interdependent on a number of other activities. Sometimes, milestone dates are synonymous with critical path dates. Although certain critical items of work may be listed as milestone dates, the reverse is not true. Typical milestone dates are as follows:

- Contract signing
- Ordering structural steel
- Selection of exterior finish color
- Ordering of elevators
- Summation of design electric loads
- Summation of design mechanical loads
- Securing financing
- Securing building permit
- Start construction
- Start structural steel
- Tenant information, if a commercial project
- Office layout, if one-user project
- Job completion

Contract Signing

The contract signing triggers all of the activities. Lacking a formal contract, a specific letter of intent from the client may be sufficient to proceed.

Ordering Structural Steel

The ordering of structural steel is on the critical path. However, the final design may require input from the owner regarding structural loading of the building. If the milestone date for furnishing that information is missed, the structural design completion would be delayed, the structural steel procurement would be delayed, and the steel not ordered. In addition, certain architectural items may be affected, such as the detailing of the members to fit in a given space. This delay, or missed milestone date, would delay the architectural design as well.

Selection of Exterior Finish Color

This item could also be on the critical path if the delay in installing the exterior finish resulted in the closing of the building for the winter. This finish selection process might not be a milestone date if it did not have an effect on other areas.

Ordering the Elevators

This item is almost always on both the critical path and the milestone lists because it has a long lead time and will affect the completion of the building.

Summation of Design Electrical Loads

Summation of the electrical loads due to owner's equipment affects the architectural design of the building, the electrical design and the furnishing of the electric service to the building by the utility company. The utility company requires total load calculations prior to computing electrical service capacity.

Summation of Design Mechanical Loads

The mechanical loads affect the mechanical design, the structural design and the architectural design. The effects on the mechanical design are obvious; the structural design is affected because the mechanical systems rest on or are suspended from the structure and the weight and location of each major piece of equipment is important. The architectural design is affected because the mechanical systems occupy large amounts of space in the building. The utility companys' service and the electrical design are also affected due to electrical needs of the mechanical equipment.

Securing Financing

Prior to the start of any work (including design work), the design/build company must be assured that the owner has adequate financing. Without the assurance that the work can be paid for, the start of the project will be put off.

Securing the Building Permit

The building permit represents the official approval to commence with the work. Securing this permit at a designated milestone date assures that all of the work can be performed as scheduled. A delay in securing the permit may be the result of an incomplete design. As a result, the entire project schedule must be pushed back.

Start Construction

The construction start date is both logistical and psychological. If the project begins on schedule, it usually follows that it will end on schedule. There is an old axiom in construction which states, "a bad job never gets good and a good job only gets better"

Start Structural Steel

The start of structural steel work affects all of the work that comes after – which is everything. This item is always a milestone date and most times, a critical path date.

Tenant Information/Office Layout

This milestone-date item is a convenience to the client. Prompt information will allow this work to be accomplished during the course of the building construction. Delay would mean that the spaces would not be finished with the building. Calling back workers to complete this work would involve extra expense for the client. If design coordination issues are involved, then the milestone date becomes more critical and is not just an accommodation.

Job Completion

This may appear an obvious item, but the completion of the project is not necessarily the date of occupancy. The owner may be occupying the space and the building in advance of completion as provided by the usual terms of the contract. Issues of unfinished work and "punch list" items (items of correction and repair) may still be outstanding. It takes a great deal of effort and detail scheduling to complete a project. Materials needed may not have been delivered. It may be difficult to get workers who have left the job to return once they have other commitments, especially when the owner's occupancy of the building makes it difficult to get to the work areas at will or at all. This work is not the most preferred, and it tends to be a low priority.

The milestone dates should not be distractions in the schedule, but rather an aid to help in achieving the successful completion of the project. The strict observance of those dates provides the format for a successful project.

Chapter Fifteen

PERMITS

Most state or regional building codes are based on standards such as the Southern Building Code or the one prepared by Building Officials and Code Administrators (BOCA). These standards may or may not be modified, according to the specific requirements of a state or municipality. The statute that regulates particular building code and permit requirements varies from state to state; interpretation and application of the statutes often vary from community to community.

What is a Building Permit?

Most codes state that it is unlawful to construct, reconstruct, alter, repair, remove, or demolish a structure without first filing a written application with the building official and obtaining the required permit. The permit is filed in the name of the owner, but may require the signature of a licensed builder and "engineer of record" as part of the application. The term "engineer of record" is generic and refers to a licensed architect or engineer.

The building permit is issued by the building department. The precise names of local building departments vary and could be termed *Department of Inspectional Services, Engineering Department*, or the *Department of Building Safety and Conformance,* to name a few. Plans are usually submitted to the fire marshal's office at the same time. Approval for the permit must be obtained from both departments. The approval process starts with the submission of the application and required documentation. The review usually requires, as a minimum, the following documentation:

1. General description of the proposed work, its location, and the use and occupancy of all parts of the building.
2. The accurate locations and dimensions of all fire exits and an occupancy schedule of persons for all occupiable space.
3. Drawings and specifications with sufficient clarity and detail dimensions to show the nature and character of the work.
4. The method and amount of ventilation and sanitation.
5. Engineering details.

6. The methods of firestopping as required by the code.
7. Schedule and details indicating compliance of interior trim and finish to code requirements. (Reference 1986 BOCA and Section 111.0.)

In addition to the above, all of the plans submitted are to bear the state-conferred seal of registration of a qualified professional engineer or architect. Explanatory calculations are submitted, as necessary, to validate the design of the building systems and to confirm the structural design. Calculations used to develop the heating, cooling and electrical loads (and any other pertinent data), are also submitted.

Permit Fees

The permit cost will vary from state to state and by community. The requirements might vary from one permit for all work on a project to individual permits for the various aspects of a project. The cost of the permits can also vary from a fixed amount for each permit to a formula of fees computed on a sliding scale and based on the cost of the work or on the area of the building. The fee is meant to cover the expense of the building department's activity for the project. These activities involve:

- Plan review
  - Architectural, Structural, Civil
  - Mechanical, Electrical, Plumbing, Sprinkler
- Assessor's expenses
- Inspection Services

Plan Review

The purpose of the plan review by the building official is to verify conformance to the Building Code and to assure public safety. For example, he examines the travel distance for egress from the building, the size of safe exit-ways, and the types of finishes within the space for fire considerations, all in conformance to code. ("Egress" is the term used to define the posted exit-ways from a building.) The building official examines these and other elements of the structure to ensure conformance to code for the type of building occupancy, the location in fire districts, and the type of building construction selected. The code allows for different types of construction, such as all wood, wood and masonry, all masonry, and structural steel, to name a few; each type has unique requirements.

The building official reviews the proposed plumbing to verify adequate sanitary facilities, number of fixtures based upon projected occupancy, waste water disposal, and potable water distribution and treatment.

The electrical review is based on the building electrical code, but is concerned mostly with the conformance of the electrical design to NEC (National Electric Code).

The mechanical system is reviewed in terms of the conditioned space. The systems are examined for adequate air movement and exchange, smoke evacuation, and energy conservation.

The sprinkler system and other life safety items such as fire alarms, fire extinguishers, smoke control devices, fire compartmentation and fireman access are reviewed by the fire marshal.

The elevator system usually involves a state review and approval function. Specification for elevator use and function (as specified by the local fire department) are reviewed by the fire marshal as well.

Property Assessment

Improvements to an existing structure or the construction of a new structure require an updating of the assessor's records and an assessment of the new value of the property. The update is done in order to determine the amount of property tax to be levied. The work required in this department involves revisions to official records, plot plans, and (if necessary) street maps.

Inspection Services

The building permit is to be posted on the project site and should be available for the use of the inspection officials. The inspection official(s) may be the same individual(s) as the reviewing official(s). The design/build contractor must also maintain, on the site, the stamped (approved) set of permit drawings for the inspector's use. The inspector refers to the stamped drawings when conducting inspections. He is obliged to sign off on the permit at different stages of the project.

One of the key inspections can be that of foundations. The inspector examines the installation of reinforcing steel and the condition of the forms prior to placement of the concrete. When satisfied, he gives his release for the work progress, initialing the permit to indicate his approval for that stage of the work. Inspection of walls includes an examination of plumbing piping and electrical wiring in the walls (along with other items) prior to their being closed over. This phase, known as "rough-in", is also initialed as approved.

When the work has been completed and all of the inspecting agencies (including the state elevator inspector) have indicated their approval of the systems, the building department issues a document called a Certificate of Use and Occupancy, more commonly known as a "C of O". This document allows the building to be occupied for its intended use.

Types of Permits

A building permit is not the only type of permit that may be required for a project. There are a number of other types of permits that may have to be filed depending on the building use and location of the project. Urban projects tend to require a larger number of permits because of the public impact and ramifications of a project in a city. These conditions are generally not found in rural or suburban locales. Some additional permits that may be necessary are as follows:

1. *Street Opening Permits.* This permit would be necessary to do excavation in an existing, active street. These permits are required of whomever is doing the work, such as the excavation subcontractor or the installer of the line.
2. *Street Closing Permit.* This permit sounds as if it should be the opposite of a Street Opening Permit, but it is not. A Street Closing Permit is required in the instance that an active street needs to be barricaded to traffic in order to perform certain operations which may be short-term or, conceivably, for the duration of a project. This permit is usually required in urban centers and is issued to the design/build contractor.
3. *Sidewalk Bridges.* To protect pedestrians from falling debris from an urban building under construction, temporary overhead shelters are erected, often over existing sidewalks. The design/build contractor is obliged to secure this permit.

4. *Street-Stored Trash Container Permit*. In an urban setting, large trash containers can usually only be located in the street. Permission allowing this practice must be obtained in the form of a permit issued to the design/build contractor.
5. *Police Detail*. Police are needed to control traffic and pedestrians in both urban and suburban sites. Payment is usually by contract. This arrangement is not necessarily in the form of a true permit. The responsibility for this municipal service is with the contractor or subcontractor who needs the service.
6. *Traffic Control*. Traffic control is different from police detail because it also involves detours and traffic light controls. The design/build contractor is usually responsible for this permit.
7. *Water Taps*. A permit is required for tapping into an existing water main. This tap permit or assessment is required in addition to the Street Opening Permit. This permit could be issued separately, or it may be included in the Plumbing Permit. (See Number 18.)
8. *Sewer/Drain Opening Permits*. These are similar to the Water Tap Permit. The excavation subcontractor or the installer is responsible for obtaining this permit.
9. *Carbon Monoxide Exhaust Review/Permit*. This permit is two-fold and involves a review of enclosed parking structure fumes, sensing equipment, and exhaust equipment capabilities, as well as permission to install the same. This is an owner's requirement as part of the design and construction process.
10. *Boiler Review/Permit*. This permit is usually involved with the replacement of a boiler rather than new construction or large scale retrofit construction. In the latter instances, the Plumbing Permit covers this work.
11. *Elevator Permits/Inspection*. This is normally a municipal or state function. The permit is required of the elevator subcontractor.
12. *Fossil Fuel Utilization Review/Permit*. Fossil fuel includes petroleum derivative products, natural gas, or coal. These products are reviewed for use, type of fuel, pollutants, and energy conservation. This type of permit is granted to the owner based on the use of those products designated for the project. (See Nos. 14 and 15.)
13. *Curb Cut Permit*. Most communities require a permit in order to construct a curb depression onto a roadway to allow permanent access to the site. State issuance is needed on state roadways. This is an owner's requirement.
14. *Gasoline Storage Permit*. The storage of gasoline for private or commercial use requires a permit. This is an owner's requirement.
15. *Diesel Fuel Storage Permit*. The storage of diesel fuel has the same requirements as the gasoline storage. Although storing gasoline in commercial buildings is a rare occurrence, diesel fuel is commonly stored for use in emergency generators and fire pumps.
16. *Electrical Permit*. The electrical subcontractor is obliged to file for a separate permit to do the electrical work, apart from the building permit. The issuance of the electrical permit is, however, contingent upon the issuance of the building permit.
17. *Temporary Electric Permit*. The electrical service to a building under construction (or a building whose renovation first stipulates removal of all of the existing electrical service) requires a temporary service permit. The temporary electrical service design must meet the same NEC (National Electric Code) standards as the permanent service in terms of capacity, code compliance, and

protection of the equipment (which may be remote from the building in an exposed locale). The electrical subcontractor is responsible for obtaining this permit.

18. *Plumbing Permit*. The plumbing subcontractor is responsible for securing a separate plumbing permit. (The requirements are similar to those of the electrical subcontractor. See No. 16.)
19. *Sprinkler Permit*. The sprinkler subcontractor is also responsible for securing a separate permit (as required by the electrical and the plumbing subcontractors).
20. *Bottled Oxygen, Propane, etc*. Tanks of gas used for welding and for temporary heating on a site require a plan for storage, handling, and use. The users are obliged to secure permits.
21. *Over the Road Haulage*. The transportation (using public roadways) of earth, fill, stone, and exceptionally large, wide, or heavy items for the building requires a separate permit. This permit may be secured by the design/build contractor or by the transporter.
22. *Temporary Construction Signs*. The use of temporary signs identifying the project and the contractors requires a permit, which should be obtained by the design/build contractor.
23. *Permanent Building Signs and/or Marquees*. Permanent signs require permits secured by the owner.
24. *Egress Review*. The reason for this last step is to reaffirm that the code conformance reviewed in the original plans has not changed with successive revisions and (in the matter of commercial ventures) due to tenant improvements. This permit is an owner requirement.

A number of other permits and licenses may be required as a consequence of the building use such as:

- Building Cleaning
- Outside Window Washing (using scaffolding)
- Innkeeper's License (Hotel)
- Liquor License
- Common Victualler's Licenses (Restaurants)
- Navigation Lights (for tall buildings in airport flight paths)

All of the permits designated as *required by owner* can be secured, or the process can be aided by the design/build company. This is true in the design phase and also for the construction phase. The fee for the building permit is included as part of the GMP (Guaranteed Maximum Price). Miscellaneous permit fees are usually reimbursable at cost and not included in the GMP.

Part Four

CONSTRUCTION

Chapter Sixteen

PURCHASING

Purchasing is the construction activity that procures the manpower and materials necessary to build the project. Work is bid, negotiated, or simply awarded to vendors and subcontractors. The subcontractors provide field labor and products and are responsible for the installation and performance of those products. The vendors supply services and materials to the project. The materials are usually delivered to the site to be installed by others. Examples of vendor items are doors and frames, hardware, concrete, and fill materials for earthwork. The services include inspection and testing of the work by independent laboratories, base line and final certified plot plan survey work, and work by the design professionals. The procedure for purchasing involves many activities and requires attention to detail. The first item to be addressed is the contractual documentation needed for the procurement.

There are two basic forms of agreement used in purchasing – the subcontract agreement and the purchase order.

Subcontract Agreement

The contract for subcontractors may be a standard form such as AIA Document A401, *Standard Form of Agreement Between Contractor and Subcontractor*, or it may be independently drafted. The sections and content of the independently drafted contract should be the same as those drafted for the design/build owner contract. The content of the clauses in the subcontract agreement is to be the same as the contract with the client. The reason for maintaining this continuity is the risk factor associated with contracting. The normal procedure is to share the risk burden for a project with all of the participants. Whether it is a standard form or specially drafted, the contract should particularly define the following areas:

- Performance criteria
- Time of completion
- Incentives
- Payments
- Liability

Performance Criteria

Performance of a project includes the quality and quantity areas of the work. The quality of the materials must be assured and there must be a standard established to which the materials furnished must conform. Once standards are established, changes are not permitted, without approval. Approval usually comes from the design/build contractor and the owner, with the review and concurrence of the architect. If adequate clauses to stipulate this condition of the agreement are not included, the subcontractor would have the latitude to make changes at will. Since the design/build company is obligated to provide the building at a certain level of quality (stipulated in the Design/build/Client Agreement), an omission of proper clauses would place the design/build company at a disadvantage. The quantity factor involves furnishing all of the required product, based on the plans. This quantity factor could be a problem if the content or extent of the product is in question. For example, if a reflected ceiling drawing is not complete, the omitted portions could be construed to indicate unfinished work. As a rule, this interpretation is considered only by the most unscrupulous subcontractors.

Performance of a project also includes warranties and guarantees for the work completed. The satisfactory operation of the building is the design/build company's responsibility (the subcontractors are responsible to the design/build company for their work). The design/build company's reputation and ability to get future work is based upon the satisfaction of the clients. The usual time period of the warranties and guarantees is one year, unless extended by the manufacturer or requested by the client. Many clients do request a two-year warranty period, which can be provided at additional cost from the subcontractors. The contract terms must be clear on the points of performance.

Time of Completion

The time required to complete a project is tied to the construction and management schedule. A standard clause in most contracts cautions that "time is of the essence" and that the work is to be accomplished "in due haste". These standard statements are meant to imply to the contractor (and through him to the subcontractor) the necessity of constructing the building as rapidly as possible. For the design/build contractor, the ability to build quickly means more than a standard clause; it is the essence of the sale. The ability to produce the building faster with design/build than with any other method is one of the key selling points and it is a principle that must be upheld. The design/build company must include as part of the subcontract documents the dates of milestone events in the progress of the job and, as necessary, the management schedule. Clauses should also be included allowing the design/build company the right to insist on subcontractors performing overtime work if they fall behind in their progress. The contract clauses must tie the schedule into the other clauses, such as an incentive clauses. A word of caution, however, on the matter of including the schedule: the obligation to have all of the trades meet the schedule is with the design/build company. If some of the work is performed by design/build company personnel, then the obligations for these workers are the same as for the subcontractors. The failure of one subcontractor to perform can affect all of the other subcontractors. Recourse against that subcontractor may not be enough to match the recourse demanded by the remaining subcontractors. The contract will give the design/build company the authority to take certain action in order to pick up a lagging schedule, but good project management and adherence to the schedule is the preferred approach.

Incentives

Associated with the schedules and performance are contract clauses that provide for incentives. There are two basic types of incentives with many variations for each.

Profit incentives are shared by all of the participants. The subcontractor is motivated by the fact that he is working with a lump sum quotation and stands to gain by doing the work more quickly and efficiently. The profit incentive for the design/build company is more defined and deals with the GMP (Guaranteed Maximum Price) savings provisions of the contract, and the circumstances under which the profit might have to be expended. The latter condition occurs only in the event that the cost of the work exceeds the GMP cost guaranteed.

Schedule incentive can be both a bonus and a penalty. Most public bid work includes a liquidated damages clause in the standard form contract. This clause refers to the penalty, usually in dollars per day, to be assessed in the event that the completion of the project exceeds the stipulated time span. Such a condition may then lead to a situation where time extensions are requested for all manner of real or imagined delay. In design/build, the penalty may or may not be included as an "incentive". Usually when a liquidated damages clause is included, a bonus clause is included to define the real incentive. The client may have the building pre-leased and will realize a certain amount of income (by occupying the building early) that he is willing to share – if the project can be completed as quickly as possible. He also may want recourse for any lost revenue due to the delay in completion of the project. This is a normal relationship when a schedule incentive is included. Even if the owner is not losing money by a delayed occupancy of the building, he may be highly motivated to get the project done in order to convert from a construction loan to a permanent mortgage, at a lower rate. No matter how great the client's motivation to complete the project quickly, the design/build company has its own strong incentive: establishing and maintaining its reputation for successful and timely work.

Payments

The payment clauses stipulate the manner of payment. The four points of information regarding payment are: the progress payment, retainage, payment for stored material, and final payment.

The *progress payment* clause will explain the procedure for submittal, approval and payment, including all of the intervening time periods. The *retainage*, or money withheld, is usually ten percent of the total amount billed; it is always the same as the formula imposed in the design/build-client contract. *Payment for stored material* pertains to the ability of the subcontractor to be paid for material stored at the job site, but not installed. Sometimes material stored off-site is paid in advance of installation. *Final payment* is defined in terms of substantial completion. Final payment occurs at a certain time span, usually 30 days after substantial completion of the work, less uncompleted work and the value of punch list work remaining. The final payment clause may also stipulate beneficial occupancy as a condition for final payment. The owner's ability to use and occupy the building in advance of substantial completion may not be a contradiction. This clause needs to be stated clearly. If such clauses are not to be included in the body of the subcontract agreement, they should be included in the General Conditions.

Liability

Liability extends to insurance, indemnification, and bonding. The subcontractors are required to provide liability insurance coverage to the extent that coverage is required of the design/build contractor. The limits are set by the owner's liability company. The insurance covers Worker's Compensation, comprehensive automobile for bodily injury and property damage liability, and comprehensive coverage for the work to include bodily injury and property damage liability. The indemnification to be included in the subcontract agreement should be the same as that included in the design/build-client contract. The furnishing of labor and material and performance bonds provisions for the subcontractor may be included whether or not they are required of the design/build company. This one-way provision may be the owner's condition to waive the bonding for the design/build company.

Purchase Order

A purchase order is a short form document used for vendor procurement. The terms and conditions for this document need not be as stringent as those for the subcontract agreement. Furnishing of products to the site focuses more on the product and its performance, rather than the performance of the tradesman. Consequently, the purchase order does not require as many clauses as the subcontract agreement, and those that are included can be much simpler. Examining the five major clauses listed as part of the subcontract agreement, we find only one particularly referenced and three partially referenced in the purchase order.

Payments

The terms of payment for products clause is particularly important. A purchase order usually does not contain a provision for retainage. Thus, payment will be concerned with the *time* of payment, such as 30 days after receipt of material, or the discount statement, such as 2% net ten days.

Liability

Liability coverage is sometimes stipulated for the trucks making deliveries to the job. Delivery of materials from established vendors (such as ready-mix concrete suppliers) would include blanket policies that would be effective for all of the jobs and therefore would not require individual documentation. Indemnification and bonding are not covered on the purchase order nor are they generally required of the vendors.

The purchase order might also refer to the quality and quantity of the product and the time of delivery. These are, however, conditions of the price and not "hard boiler plate" items in the agreement.

Purchase Contract

While the purchase order is a simple, one-page document for limited use in securing material, some design/build contractors have introduced a more involved document, called a *Purchase Contract*. The purchase contract is used for large material purchases such as structural steel, reinforcing steel, cabinetry, case work, and any other large dollar value items to be furnished. The purchase contract agreement contains many of the subcontract agreement clauses, especially those pertaining to time of completion, incentive, payments (including retainage), and liability. The liability clauses include automobile insurance for the delivery trucks, performance bonds (if required), and indemnification.

The agreements used to secure services can be any of the three described (subcontract agreements, purchase order, or purchase contract) depending on the the terms and conditions of the work. Testing, laboratory, and inspection work could be documented on a subcontract agreement form since it involves furnishing labor to the site, or it could be done on a simple purchase order with a liability clause added. Design services could be secured through the use of the subcontract agreement or purchase contract agreement with certain appended clauses due to field labor and liability. The design services could also be contracted through the standard AIA Document B901, *Standard Form of Agreements between Design/Builder and Architect* (see Appendix).

It is important to make decisions pertaining to the form of the contract agreements early in the process of a design/build construction project. The form and initial draft of all documents – either between the design/build company and the owner or between the design/build company and a subcontractor or vendor – are initiated by the design/build company.

Bid Packaging

Bid packaging is simply the gathering of all of the bid documents in preparation for issuance to bidders. This term is also used in the fast track method. The procedure is exactly the same in both cases. Preparation of the bid package in purchasing should include the *Request for Proposal*. In addition to the request for proposal of a fixed price lump sum subcontract, the bid package should contain all of the information needed to prepare the proposal, including the following additional items:

- Invitation or request to bid
- Proposal form
- Form of subcontractor agreement
- General conditions
- Special conditions
- Technical specifications
- Working Drawings

Invitation to Bid

The Invitation to Bid contains the following information: type of work, bid due date, bid location, and meeting dates.

Proposal Form

The proposal form is the format in which the bid is to be returned. This format contains: the lump sum price, price breakdowns of the lump sum price, unit prices, alternate prices, acknowledgments of additional documentation such as addenda, bond information, and other data relevant to the bid, as well as the signature and title of the signer.

Subcontract Agreement

A sample subcontract agreement is included in the request for proposal in order to allow the bidders the opportunity to review in advance the subcontract agreement that they will be obliged to sign. This preview saves time when awarding the project. Questions and objections can be made early in the process. It is conceivable that some subcontractors will be discouraged from bidding due to the requirements of the contract's "boiler plate". The boiler plate of a contract consists of the preprinted articles that constitute the fixed portions of the agreement. Those subcontractors who have done work with the design/build company in the past would be familiar with their standard agreement. The agreement is included to inform the new bidders and to confirm for the old bidders the current edition.

General Conditions

The General Conditions define the conditions by which the work is to be performed. They also define specifics of the relationship between the design/build company and the owner, and between the design/build company and the subcontractor. The General Conditions can be drafted for a particular project, or the design/build company can use a standard form such as AIA Document A201, *General Conditions of the Contract for Construction*.

Special Conditions

Special conditions include the scoping of the work, conditions of the work and site, any alternative to the work, and deletion or addition to the General Conditions or Specifications.

Technical Specification

The technical specification defines the quality of the work and the standards with which the product is to comply. Each section is intended to complement the drawings which define the content of the work. The technical specification is not the same as the outline specification prepared for the proposal. The outline specification consists of statements on quantity, quality and standards, and acts as a complement to the definition of work. The technical specification is in the same form and content used for lump sum bids. A technical specification is generic and could fit a number of projects. The outline specification is specific for one particular project.

Although all of the specifications are included as documents in the subcontract agreement, not all of them would be issued to the bidders. Only those specifications needed to define the particular work are issued. The remaining specifications are made available for viewing at certain prescribed locations.

Working Drawings

The working drawings presented to bidders are 100% complete. In the case of fast track, they are 100% complete for a particular trade, while further design for other trades may yet be included on the same documents. The drawings provide sufficient information so that the bidders will be able to submit a competitive and complete quotation. As with the specifications, not all of the drawings would be included in the bid packages, but only those needed to define the work. The successful subcontractor (who wins the bid) would, however, be responsible for all of the drawings as part of his contract. The complete drawings would also be available during the bidding phase for viewing at a defined location. The MEPS (mechanical, electrical, plumbing, and sprinkler) are exceptions to the rule in that they customarily receive a complete set of documents. The structural steel subcontractor would receive a complete set of architectural drawings along with the structural drawings.

Figures 16.1a through 16.1g are examples of a request for proposal for precast concrete that contains all of the elements stated. In this example, the documents are included in a separate listing. This is the same procedure used for the fast track bid format. In the fast track method, the same documents will be subject to successive revisions; it is important that the subcontractors, secured as part of the bid package period, and the design/build contractor know on which revisions the documents that particular subcontract agreement is based. Ultimately, all of the subcontractors will be issued the latest and final revised set of documents. The adjusted cost of each agreement will be based upon those latest drawings.

This example uses the preprinted AIA forms for the Subcontract Agreement and the General Conditions. The form of the contract will be a subcontract agreement because the bidders are being asked to erect their product. If the request for proposal was only to furnish the precast panels, then a purchase contract agreement or purchase order could have been used. The Special Conditions in this example lists the scope of work, conditions of site, and the timetable for erection.

Figure 16.2 shows another Request for Proposal, this time for furnishing ready-mix concrete. This is a vendor request for material delivered to the site. For ready-mix concrete, the amount of information is reduced and the conditions of the bid are simpler.

As usual, there is a statement for invitation to bid. The proposal form is a fill-in-the-blank type for the material required. The special conditions describe the scope of work. The nature of this product is such that no drawings are required; it will be a unit price contract and the cost will be for the amount and type of concrete used. The vendor does not need to know the source of the quantities or to verify their accuracy. What the vendor does need to know is when the work will be done and the approximate quantities. That information allows the bidders to structure their unit prices, which are fixed, for the periods indicated, reflecting the costs of raw material and labor.

The relevant documents are part of the standard specification and the applicable section, in this case 03300 is indicated. This vendor would not be responsible for all of the remaining specifications and drawings as part of the purchase order agreement that will be signed for this service. The purchase order will be a simple document reflecting the unit price arrangement and methods of payment. The request for proposal for ready-mix concrete needs no more information than what is furnished in the example in order to receive competitive and complete quotations.

Figures 16.3a and 16.3b show a Request for Proposal for a personnel/material hoist. The hoists are exterior elevators used to vertically transport men and material to various points in the building while it is under construction. These are temporary pieces of equipment and will be removed when the project is finished or when the permanent elevators are made available. This request for proposal contains only two of the categories: special conditions scope of the work, and the proposal form. As a rule, there is neither a specification section nor a drawing available to define its location. The work to be performed is defined by the design/build contractor for bidding. The information needed to secure this service is contained in the request for proposal as written. Any further information or contractual ties to the plans and specifications have no meaning in this context and are not included.

PRECAST CONCRETE

REQUEST FOR PROPOSALS

1. Invitation to Bid

   a. Submit on or before 3:00 p.m., November 16, 1987, at the office of Design/Build Construction Co., 80 Build St., Boston, MA, a written proposal for the furnishing and erection of the precast concrete work. All proposals shall be written on the company's letterhead.

2. Proposal Form

   a. Written quotation shall be lump sum for the precast panels, including brick work and all precast coping.

   b. Precast concrete contractor shall submit a breakdown of the lump sum price of all work as follows:

      1) Copings-(type), linear footage, and unit price for a total cost in place for each type of coping.

      2) Panels-(type), square footage, and unit price for a total cost in place for each type of panel.

      3) Brickwork-square footage and unit price for the total cost in place for "infill brick."

      4) Engineering costs for anchorage design.

   c. Alternates

      A1 - Iron clad cement (grey), red sand and Kingston red granite.

      A2 - Iron clad cement (grey), sand with red dye and Kingston red granite.

      A3 - Grey cement, Farmington gravel (non-bleeding) and Manchester sand.

      A4 - Deductive price to furnish only copies to the job site to be installed by others.

Figure 16.1a

PRECAST CONCRETE

REQUEST FOR PROPOSALS (Cont'd.)

3. The Contract Agreement

   a. The contract agreement should use AIA Document A401 - Standard Form of Agreement Between Contractor and Subcontractor, latest edition.

4. General Conditions

   a. The General Conditions are to be based on AIA Document A201 - General Conditions of the Contract for Construction.

5. Special Conditions

   a. Scope of Work

      1. The precast concrete contractor shall be responsible for the furnishing and erection of all precast concrete panels and copings. The work shall include all items in conformance with the specifications and the drawings.

      2. This contractor is responsible for all anchoring of his panels inclusive of inserts, loose anchors, bracing, plates, items installed in the field and the design of same subject to Architect's approval.

      3. This contractor shall coordinate, through Design/Build Construction Co., with other contractors to insure correct fit and interfacing in order to have a complete wall system.

      4. This contractor shall be responsible for all in-fill brick work using the allowance for brick material indicated in the specifications. All brick anchorages shall be cast into panel as required.

      5. The base bid precast product is to contain white cement, red sand and Kingston red granite in conformance with the material specification. (See Item 2c, Alternates).

      6. The work shall include, but not be limited to, all galvanized reinforcing bars, galvanized attachments for panel support, galvanized dovetail brick anchors, galvanized regrets, all lifting attachments as required, and all engineering work including shop drawings.

Figure 16.1b

PRECAST CONCRETE

REQUEST FOR PROPOSALS (Cont'd.)

a. Scope of Work (cont'd.)

7. All work within the panel such as caulking soft joints and flashings as required to be included in the price.

8. The brick allowance of $400/M shall include cost of stretchers only and freight to job site. All taxes and special shapes are excluded.

b. Conditions of Site

1. The project is accessible from three sides. Erection cranes can travel on Main Avenue and on Access Parkway inside of the protection system set by Design/Build Construction Co., and along the west side of the project.

2. There is no storage at the project site. The precast contractor shall arrange for his own storage in order to maintain an orderly flow of material to the project.

3. Precast contractor shall make a site visit to acquaint himself with field conditions. Mr. Director is the Project Superintendent. Telephone number is (617) 555-1212.

4. The Structural steel erection will go column tier by column tier and will proceed from east to west. The tower sections will be erected simultaneously. Erection will begin mid-January.

c. Timetable for Erection

1. It is anticipated that precast panel erection will start on or about March 1, 1987 and the erection duration, exclusive of copings, will be accomplished in 10 to 12 weeks.

6. Documents

a. Specifications as listed on Exhibit "A"
b. Drawings listed in Exhibit "B"

Figure 16.1c

EXHIBIT "A"

SPECIFICATIONS

Specifications prepared by Architects Associates, Inc.

Section Number	Description	Date
SCI	Supplementary Conditions	10/17/87
01010	Summary of Work	10/17/87
01045	Cutting and Patching	10/17/87
01345	Submittals	10/17/87
01410	Testing Laboratories Services	10/17/87
01500	Temporary Facilities	10/17/87
01501	Layout of Work	10/17/87
01600	Material and Equipment	10/17/87
01700	Contract Closeout	10/17/87
01710	Cleaning	10/17/87
01720	Project Record Documents	10/17/87
03450	Precast Concrete	10/17/87

Figure 16.1d

EXHIBIT "B"

All drawings prepared by Design/Build Associates, Inc.

Drawing No.	Description	Original Date	Revision No.	Revision Date
0.0	Title Sheet	10/17/87		
1.1	Site Plan	9/26/87		
2.1	Floor Plans - Basement and Entry Level	10/17/87		
2.2	Floor Plans - Parking and Mezzanine	10/17/87		
2.3	Floor Plans - Third and Fourth Levels	10/17/87		
2.4	Floor Plans - Fifth through Ninth Levels	10/17/87		
2.5	Roof Plan, Upper & Lower Penthouse Floor Plans	10/17/87		
3.1	Elevations	10/17/87		
3.2	Elevations	10/17/87		
3.3	Riverfront & West Wing Part 1/8" Plans & Elevations	10/17/87		
3.4	Building Section at East Wing	10/17/87		
3.5	Building Section at Elevator Core	10/17/87		
3.6	Longitudinal Building Section	10/17/87		
3.7	Partial Sections at North Entry	10/17/87		
3.8	Partial sections at Garage and Loading Dock	10/17/87		
4.1	Wall sections at Base of West Wing	10/17/87		
4.2	Wall Sections at North Side of Building Base	10/17/87		

Figure 16.1e

EXHIBIT "B" (Cont'd.)

All drawings prepared by Design/Build Associates, Inc.

Drawing No.	Description	Original Date	Revision No.	Revision Date
4.3	Wall Sections at East Building Line	10/17/87		
4.4	Wall Sections	10/17/87		
4.5	Wall Sections	10/17/87		
4.6	Wall Sections	10/17/87		
4.7	Interior Garage Wall Sections	10/17/87		
4.8	Spandrel & Pier Detailed Sections	10/17/87		
4.9	Upper & Lower Roof Terrace Wall and Plan Sections	10/17/87		
4.10	Archway Sections & Details	10/17/87		
4.11	Wall Sections at Penthouse	10/17/87		
4.12	Upper & Lower Roof Terrace Wall Sections	10/17/87		

Figure 16.1f

EXHIBIT "B"

Structural drawings as prepared by Structural Associates.

Drawing No.	Description	Original Date	Revision No.	Revision Date
14.1	General Notes and Abbreviations	9/26/87	7	10/17/87
14.2	Foundation Plan Parking Level A/B Details	9/26/87	7	10/17/87
14.3	Pilecap Details	7/16/87	8	10/19/87
14.4	Parking Level C/D First Floor Parking Level E/F Second Floor (Riverfront) Framing Plans	9/26/87	7	10/17/87
14.5	Parking Level G/H Second Floor (Main) Third Floor/ Plaza Framing Plans	9/26/87	7	10/17/87
14.6	Fourth Floor Through Eight Floors Framing Plans	9/26/87	7	10/17/87
14.7	Ninth Floor and Roof Framing Plans	9/26/87	7	10/17/87
14.8	Penthouse Framing Plans and Details - Bracing Details	9/26/87	7	10/17/87
14.9	Bracing Elevations - Sections and Details	9/26/87	7	10/17/87
14.10	Column Schedule	9/26/87	7	10/17/87
14.11	Column Details - Beam Details	12/15/87	7	10/17/87
14.12	Foundation Sections and Details	9/16/87	7	10/17/87
14.13	Foundation Sections and Details	9/26/87	7	10/17/87
14.14	Sections and Details	No Date	7	10/17/87
14.15	Sections and Details	No Date	7	10/17/87
14.16	Sections and Details	No Date	7	10/17/87
14.17	Sections and Details	No Date	7	10/17/87

Figure 16.1g

FURNISH READY-MIX-CONCRETE

REQUEST FOR PROPOSALS

1. Invitation to Bid

   a. Submit on or before 3:00 p.m., November 1, 1987, at the office of Design/Build Construction Co., 80 Build St., Boston, MA 02119, a written proposal for the furnishing of ready-mix concrete. All proposals shall be written on the company's letterhead.

2. Proposal Form

   a. Your quotation is to be submitted in the following manner for price per cubic yard without tax. Indicate discounts, as applicable.

Class	Regular	Pump Mix	Heated For Winter Conditions
1,500 PSI Normal Wt.	________	________	________
3,000 PSI Normal Wt.	________	________	________
4,000 PSI Normal Wt.	________	________	________
5,000 PSI Normal Wt.	________	________	________
3,000 PSI Lt. Wt.	________	________	________

Please quote on the five (5) different classes of concrete for a pumped mix, and for winter conditions.

Figure 16.2a

FURNISH READY-MIX-CONCRETE

REQUEST FOR PROPOSALS (Cont'd.)

3. Special Conditions

a. Approximate schedule for pouring and approximate quantity of the concrete required in c.y.

Item	4,000 PSI Normal	5,000 PSI Normal	3,000 PSI Lt. Wt.	Dates
Pile caps	1,300			Nov.-Jan.
Grade beams	250			Nov.-Jan.
Walls	500			Nov.-Jan.
Slab on grade	1,300			
Parking slabs (structural)		6,000		Mar.-Apr.
Office slab (suspended)			1,500	Apr.-June
Stairs				

We will require approximately 100 c.y. of 1,500 PSI for concrete pile caps during November-January, and approximately 400 c.y. of 3,000 PSI concrete for pits and miscellaneous.

b. Superplasticizers are not in the specifications; however, the architect will consider their use.

4. Documents

a. Preliminary concrete specifications - 03300 dated 10/7/87. All paragraphs marked "NA" are for information only and not part of the contract documents that pertain to supplying ready-mix concrete.

Figure 16.2b

REQUEST FOR PROPOSAL

SCOPE OF WORK

Personnel/Material Hoist

Quote by 10/17/87 to Design/Build Construction Co. on supplying for rental on a monthly basis, one personnel/material hoist based on the following criteria:

1. Single tower to be erected to initial height of 80'± with one cab.

2. One jump approximately two months later shall be required to 120'± with the addition of a second cab at time of jump. Jump to be done on premium time.

3. Rental period for tower and one cab shall be approximately twelve (12) months and the rental period for second cab shall be approximately five (5) months. Both rental periods should have an option to extend for the same rental price.

4. Each cab shall have a minimum capacity of 5,000 pounds, be usable for the material and/or personnel and shall be approximately 12' long x 5' wide x 8' high clear inside.

5. Each cab shall be equipped with electric heater for operator.

6. Each cab shall meet all code requirements.

7. Speed of cab to be 150 fpm.

8. All erection of initial tower and final dismantling shall be done on a straight time basis. Jump of tower shall be on a premium time basis.

9. All scheduled routine maintenance to be done on premium time by mechanics employed by the hoist company. The hoist company shall maintain the equipment to meet all code requirements during the period of rental.

10. Eighteen (18) sets of door interlocks shall be included in the rental price.

11. Double shift operations shall be estimated for an approximate twelve (12) week period.

12. All tests, permits, etc., required by authorities having jurisdiction shall be obtained by and paid for by the hoist supplier.

Figure 16.3a

REQUEST FOR PROPOSAL

SCOPE OF WORK (Cont'd.)

Personnel/Material Hoist (cont'd.)

13. Elevator supplier shall warrant the equipment for a period of 30 days.

14. All taxes to be included.

15. All freight, to and from project site, shall be included in the rental.

16. Electric power on project is 480 V, 3 phase.

Quotation, on company letterhead, shall list and include all costs of the following items:

1. Capacity of cab in pounds__________________.
2. Size of cab: length____________feet x width____________feet x height____________feet.
3. Operating speed____________fpm.
4. Initial cost of tower and cab installation_______________.
5. Cost of tower jump and installation of second cab____________.
6. Removal cost of second cab_______________.
7. Removal cost of all equipment from project_______________.
8. All freight costs to and from project._______________.
9. Rent for 1st cab for approximately 12 months based on 200 hrs per month_______________.
10. Rent for 2nd cab for approximately 5 months based on 200 hrs per month_______________.
11. Scheduled routine maintenance per month for approximately twelve month period_______________.
12. Location of maintenance yard. If maintenance is by local company, state company and location_______________________ ____.
13. State insurance requirements_____________________________.
14. Cost of all tests and permits____________________________.
15. Cost of eighteen door interlocks_________________________.
16. Rental costs/door/month______________________________________.
17. Costs of double shift/car____________________________________.
18. Costs, including fringe, overhead and profit

Elevator mechanic foreman Str. time ____________________.
Elevator mechanic foreman Prem. time ____________________.
Elevator mechanic Str. time ____________________.
Elevator mechanic Prem. time ____________________.

Figure 16.3b

The form in which the contract is drawn up will depend upon the nature of the labor and the operator, and on whose payroll the operator will be included. If included on the hoist rental company payroll, then the form of the agreement will be a subcontract. Otherwise, it would be a simple purchase order including liability clauses to cover the equipment.

The form of the request for proposal (for construction purchasing) is developed only to the degree needed for each case. This is the most successful and cost effective approach. Making the information voluminous and confusing increases the cost of the work. The examples shown in Figures 16.1 through 16.3 demonstrate the different degrees of information necessary for an effective request for proposal. The Appendix includes a Request for Proposal for HVAC work and illustrates the degree of information that is necessary when a specification is confused, the work complicated, and the scope of work in a state of change. Although the documentation has been omitted from this example and the form is not as previously outlined, this request for proposal contains all of the categories needed.

Scoping

The term "scoping" is used in purchasing to define specific work. The much abused *plans and specs* approach is too general a method for construction purchasing and is prone to error and disputes. Another problem with the plans and specs approach is that it obliges the subcontractor to pick his work out from its place among all of the other project documents. The time and trouble involved in such a search could discourage many small trades. It could frustrate competing subcontractors, or force them to add to their price in order to cover any work that they might have overlooked.

The preferred method is scoping, which provides the specifics of the work to bid. To achieve such a detailed record, the plans and specifications must first be reviewed and the individual documents of each trade noted. Next, all of the details and sections must be examined and labeled by trade (see Figure 16.4).

Figure 16.4 shows a detail of the point at which a ceiling meets a wall. The wall would be part of the drywall work, and the painting included with the finish work. The ceiling is part of acoustic tile work. In the corner is a light fixture, which is part of the electrical work. The wall trim (A), cove enclosure (B), and end trim (C) are not so easily categorized. Does the wall trim belong to the acoustic ceiling subcontractor even though it does not touch the ceiling system and is not needed to finish the ceiling work? Is the cove enclosure an electrical item of work, along with the end trim? The point here is that the work may not be claimed by any of the trades who can excuse themselves with one or more valid reasons. The result is that at the end of the job, the work is not done, and to have it completed at that stage is costly and delays the job. Purchasing by plans and specifications often results in such end-of-the-job disputes. The preferred method is to scope the work by designating it to specific trades. The wall trim (A) could be assigned to the acoustic ceiling subcontractor, the cove enclosure (B) to the electrical subcontractor, the end trim (C) to the acoustic ceiling trade, and the finish of the enclosure might be designated as factory-finished or assigned to the painting trade. If the selection is wrong or unworkable, the trades will inform you and the work can be reassigned to other trades. The final decision could be based on coordination and/or costs. If it is more convenient for one trade to do certain work (such as the window subcontractor caulking around window systems), and if they are equally capable (such as the waterproofing subcontractor), then the decision to assign the work to that more convenient trade (the window subcontractor) would not be based on scheduling, but rather on lower cost. To determine the lowest cost, the scope of work for one item can be assigned to more than one trade as an alternate. The trade submit- ting the least cost would be assigned the work. The point is that work priced in this manner, at the beginning of the job, when included as part of the total bid, should be less than its cost as an extra item of work at the end of the job. Masonry work, drywall work, and windows, for example, are separate trade items that can be clearly categorized. However, related installation components may require further definition. The window system requires sealants around the perimeter of the frames as it abuts adjacent trade work surfaces (see Figure 16.5), such as precast and drywall. Is the sealant or caulking to be installed by the window subcontractor or by a waterproofing/dampproofing subcontractor? The design/build contractor makes that choice. These decisions must be made and the bidders notified so that purchasing can be done correctly.

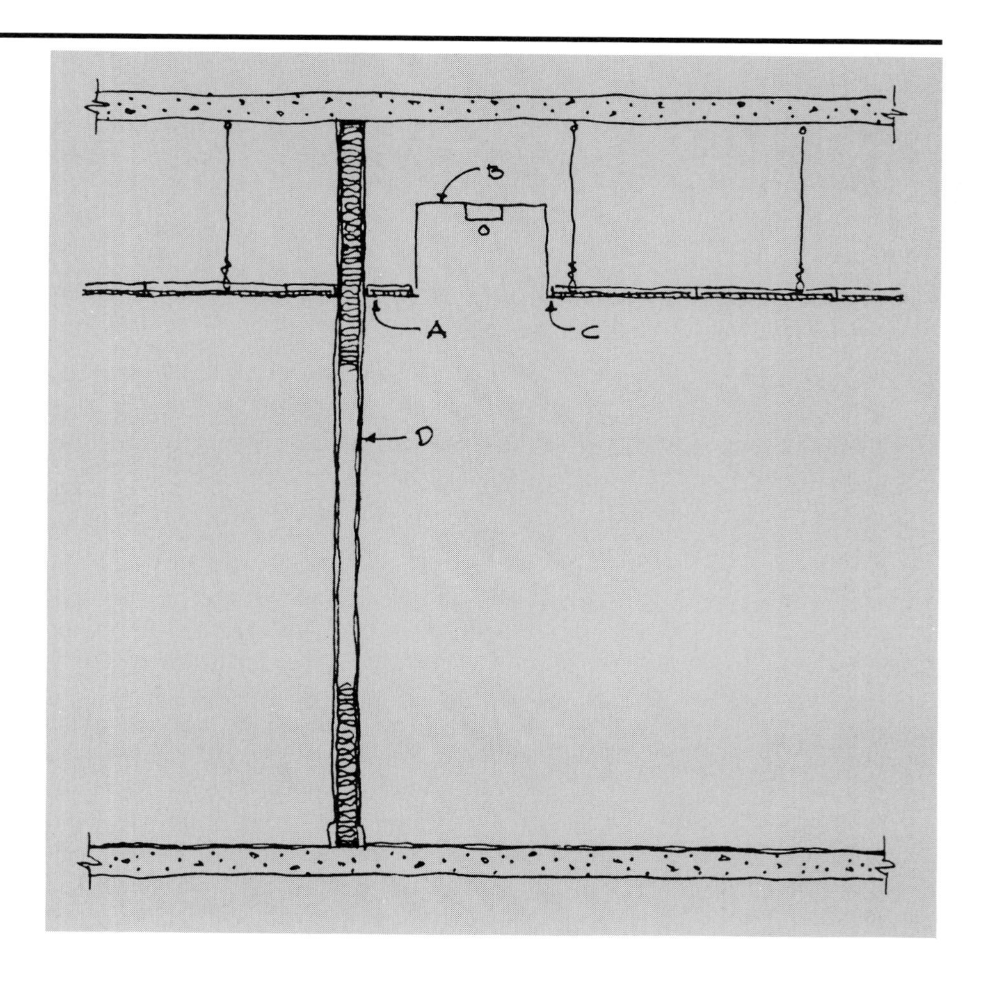

Figure 16.4

Scoping the trades considerably reduces the possibility of an omission. The grey areas of work that might otherwise be prone to end-of-the-job disputes are removed. The goal is having *no surprises and no mistakes*.

The experienced purchasing agent who secures the products and resources for a project starts by examining all of the documents. Then all of the details are labeled so that no item of work is left unassigned. Finally, in the Request for Proposal there are statements specifically assigning the work found in the details.

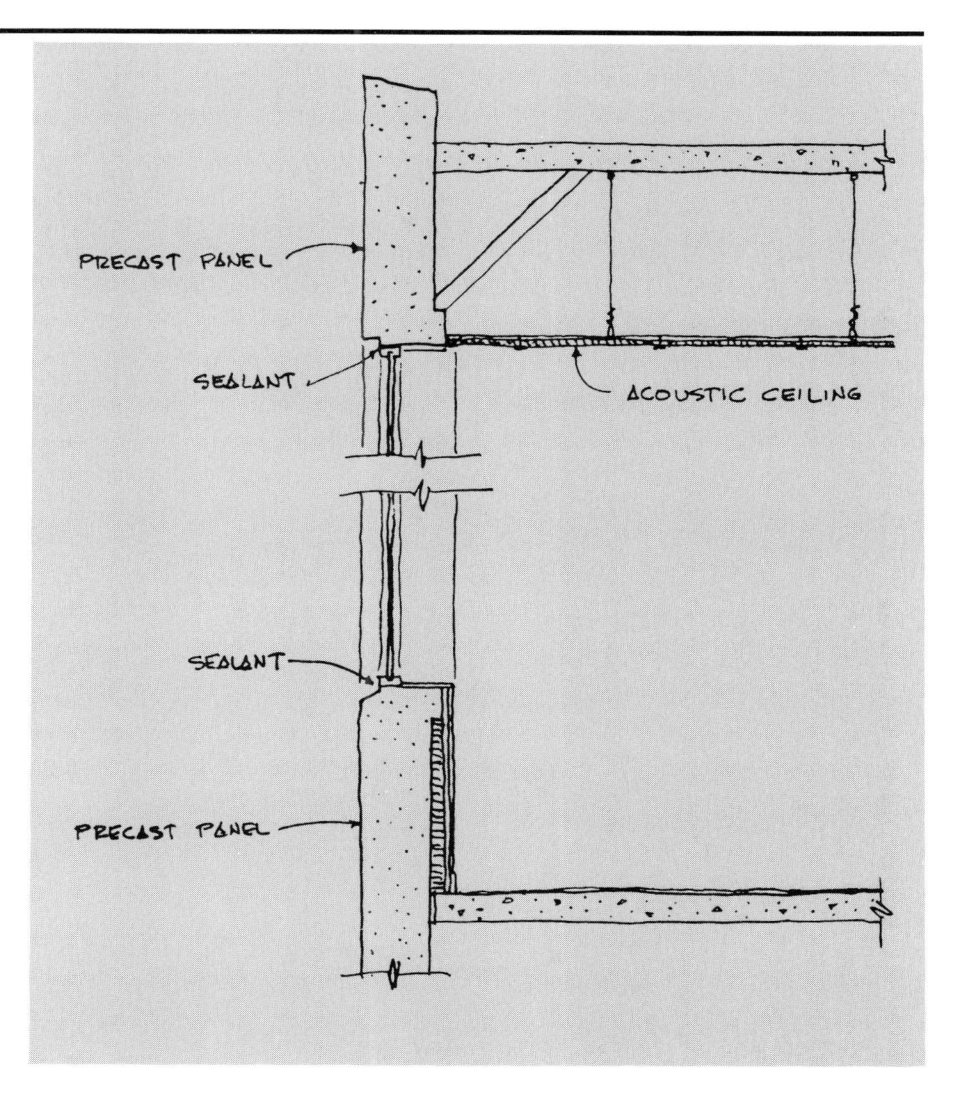

Figure 16.5

Subcontractor Qualification

Qualification of subcontractors and vendors can be accomplished in two ways – pre-qualification and post-qualification. Pre-qualification is the process of screening potential bidders before allowing them to bid work. Post-qualification allows anyone to bid and then provides for screening of the candidates to arrive at the list of those most qualified. From that list is selected the subcontractor who will be awarded the work. The post-qualification method is commonly used by public agencies, and the work is awarded on the basis of the lowest qualified bid.

The design/ build company has the choice of conducting the purchasing using either method of qualification – or both. Most commonly, both the post- and pre-qualification are used. A select bid list is assembled and those candidates asked to bid the work. The request for proposal is then issued to the invited bidders. At the same time, completed sets of documents are sent to the bid rooms. These bid rooms are open to subscribers who can examine the documents (including the request for proposal) and submit quotations for the work. The purchasing agent reviews all of the bids and determines the low bidder from both categories – those who are qualified and those yet to be qualified. Obviously, the process of procurement would take a very long time if needed information was either not forthcoming or could not be checked rapidly. A qualification questionnaire is a good way to expedite the process. Such a form is needed for prequalification and should also be included with unqualified quotations in post-qualification. The form not only simplifies the data gathering process, but it includes sufficient questions to determine the bidders' qualifications. An example is shown in Figure 16.6.

The form is simple and records basic information that can also be used for future bid lists. The *trade categories of work to bid* is included to determine which other trades, if any, the subcontractor performs. For example, a drywall subcontractor may also do acoustic ceiling work and painting. This information is overlooked if it is not asked. *Geographic work regions* represent another information category that should be referenced for future work. It would not do to include a subcontractor on a bid list for work in an area outside of the zones he would work. The *type of structure* category is included for the same reason. If a subcontractor's experience has only been with low-rise buildings, then he may not be interested in or qualified for work on high-rise buildings.

The *company contact* requirement is for the names of the subcontractor's key people who are accessible and have the power to make decisions. The *bonding capacity* and *years in business* are included in order to determine stability.

New companies are not excluded because of a lack of experience; they are selectively included for projects that they are capable of handling. The *union, non-union, minority, and female-owned* category represents information needed to comply with legal requirements. A *financial statement* is requested so that the company stability can be judged against industry standards.

SUBCONTRACTOR QUALIFICATION APPLICATION

NAME: ______________________________

ADDRESS: ______________________________

PHONE NUMBER: ______________________________

TRADE CATEGORIES OF WORK TO BID:
- ______________________________
- ______________________________
- ______________________________

REFERENCES: LAST THREE YEARS

PROJECT	CONTRACTOR	CONTACT	AMOUNT

GEOGRAPHIC WORK REGIONS
- ______________ • ______________
- ______________ • ______________
- ______________ • ______________

TYPE OF STRUCTURE
- ______________ • ______________
- ______________ • ______________

COMPANY CONTACT 1st ______________________
2nd ______________________

BONDING CAPACITY ______________________
BONDING COMPANY ______________________

YEARS IN BUSINESS __________ UNION__________ NON-UNION__________
MINORITY OWNED__________ FEMALE OWNED__________

SUBMIT LATEST FINANCIAL STATEMENT PLUS PREVIOUS 2 YEARS.

Figure 16.6

Financial Statement

Financial statements are a summary of a company's earnings and balance sheet position. The *Earnings Statement* lists the revenue from all sources. From that amount, operating costs, administrative expenses, debt, and income taxes are subtracted. The result is net earnings which are added to the retained earnings. (See Figure 16.7.)

The *balance sheet* lists the assets and liabilities. The assets include cash, accounts receivable, inventories, land, buildings and improvements, machinery, and equipment. The liabilities include the accounts payable and notes payable, and long term debt. Net worth includes stockholders' equity, and retained earnings (see Figure 16.8).

The income statement and the balance sheet are presented at the end of the fiscal year. The form is as prescribed by standard accounting practice. The income statement shown has been prepared using the *completed contract basis*. This basis is one of two accounting methods available to contractors. The other is the cash accrual (cash receipts and disbursements) method. The difference between the two is that the contract basis method includes the total contract value of work completed in a certain period – whether or not it is paid. The cash basis method includes only the amount of the completed contract for which payment has been received.

The income statement includes the contract revenue which is the total received for that fiscal year. Subtracted from this income are the contract costs for all subcontractors, labor, material, equipment, as well as other costs that are charged to the projects.

Next, administrative expenses and taxes are subtracted to arrive at the company's earnings. These earnings can be retained within the company and added to the previous total retained earnings of (in the sample case) $110,250; or they can be partially distributed in the form of a dividend (as shown in the example). What remains is a total of retained earnings: $186,432 remained in the company.

The balance sheet compares the assets with the liabilities and net worth. The balance sheet formula states that the assets are equal to the liabilities plus the net worth.

$$\text{Assets} = \text{Liabilities} + \text{Net Worth}$$

This form of financial statement is common for subcontractors and contractors. Usually, but not always, the statements are audited as required by the bonding company. If there is no bonding company or other need requirement, then the reports are simply reviewed by the accountant. The audit process is more involved and is prescribed by strict accounting rules in the preparation and certification of the financial records.

Financial statements are baffling to many people who have little idea how to read the statement or what to look for. One method of examining a financial statement is through the use of financial ratios.

Earnings Statement	
Contract Revenue	$8,725,370
Contract Cost	8,310,540
Net Operating Revenue	$ 414,830
Administrative Expense	202,661
Net Profit	$ 212,169
Taxes	95,415
Net Earnings	$ 116,754
Retained Earnings:	
Beginning	$ 110,250
Paid Dividends	40,572
Retained Earnings	$ 69,678
New Balance Retained	$ 186,432

Figure 16.7

Balance Sheet

Assets		Liabilities	
Current Assets:		**Current Liabilities:**	
Cash	$ 389,297	Notes Payable	$ 24,309
Accounts Receivable	1,224,254	Accounts Payable	308,614
Inventory	78,340	Due Subcontractors	770,528
Total Current Assets	$1,691,891	Income Tax Provisions	95,415
Property, Plant and Equipment			
Land	52,527		
Buildings & Improvements	128,123		
Equipment	35,520		
Motor Vehicles	45,235		
Office Equipment	12,855	Total Current Liabilities	$1,198,866
Total Capital Assets	274,260	Capital Stock	468,333
Less: Accumulated Depreciation	112,520	Retained Earnings	186,432
Net Capital Assets	161,740	Tangible Net Worth	$ 654,765
TOTAL ASSETS	$1,853,631	**TOTAL LIABILITY AND NET WORTH**	$1,853,631

Figure 16.8

Financial Ratios

Financial ratios are used by a number of organizations. Banks use certain ratios as guidelines for determining issuance of a loan. Bonding companies might use ratios in an initial determination for bond approval and continue to use them as a form of monitoring. Creditors of all types, especially when examining a small company, would also use financial ratios. The size of the figures in the financial statement can be deceiving, but the relationship between the numbers is more revealing. In the U.S. Small Business Administration Publication (series No. 20) *Ratio Analysis for Small Business*, eight commonly used ratios are listed. They are:

1. Current assets to current liabilities
2. Current liabilities to tangible net worth
3. Annual sales to tangible net worth
4. Annual sales to working capital
5. Net earnings to tangible net worth
6. Net earnings to working capital
7. Net earnings to annual sales
8. Fixed assets to tangible net worth

Using the previous example financial statement, the following information was taken:

Current Assets	$1,691,891
Current Liabilities	1,198,866
Total Assets	1,853,631
Total Liabilities	1,198,866
Tangible Net Worth	654,765
Fixed Assets (Property)	274,260
Net Working Capital	493,025
Net earnings	116,754
Annual sales	8,725,370

(current assets – current liability) ($1,691,891 – $1,198,866)

1. *Current Assets to Current Liabilities*. This ratio is known as the current ratio and is a test of solvency. It measures the liquid assets available to meet debts within a year's time.

$$\frac{\text{Current assets}}{\text{Current liabilities}} = \frac{\$1{,}691{,}891}{\$1{,}198{,}866} = 1.41$$

For building construction, the commonly accepted ratio is about 1.5.

2. *Current Liability to Tangible Net Worth*. Like the current ratio, this is another means of evaluating the financial condition of the company. This ratio is expressed as a percentage. This high percentage suggests not having enough resources to pay the debt. Due to the lag in payments to contractors versus costs, the ratio of liability to tangible net worth in the building industry is commonly over 100%.

$$\frac{\text{Current liabilities}}{\text{Tangible net worth}} = \frac{\$1{,}198{,}866}{\$\ 654{,}765} = \times\ 100 = 183\%$$

For building construction, the accepted ratio is about 130%.

3. *Annual Sales to Tangible Net Worth*. This is called the *turnover ratio*. It shows how actively a company's capital is put to work.

$$\frac{\text{Annual Sales}}{\text{Tangible net worth}} = \frac{\$8{,}725{,}370}{\$\ 654{,}765} = 13.3$$

For building construction, the accepted ratio is about 9.0.

4. *Annual Sales to Working Capital*. This ratio is also called the *turnover ratio*. The difference between the two ratios is based on the source of the figures. The ratio using the tangible net worth involves a fixed figure and measures how actively the company capital is put to work. The ratio using the working capital involves a variable figure, the *working capital*. The working capital is the difference between the current assets and the current liabilities – a figure which can change throughout the fiscal year. A financial ratio using the working capital to annual sales is indicative of the margin available to pay for the interim between payment and costs. A high ratio can indicate vulnerability to creditors.

   $$\frac{\text{Annual Sales}}{\text{Working Capital}} = \frac{\$8,725,370}{\$\ 493,025} = 17.7$$

   For building construction, this ratio is about 12.0.

5. *Net Earnings to Tangible Net Worth*. As the measure of return on investment, this is considered a criteria for profitability and management efficiency. This ratio is expressed as a percentage. If this percentage is too low, then the capital involved might be better invested elsewhere.

   $$\frac{\text{Net Earnings}}{\text{Tangible Net Worth}} = \frac{\$116,754}{\$654,765} \times 100 = 17.8\%$$

6. *Net Earnings to Working Capital*. This ratio is also a criteria for determining investment and management efficiency. The use of working capital versus tangible net worth allows for a varying factor for different points in time.

   $$\frac{\text{Net Earnings}}{\text{Working Capital}} = \frac{\$116,754}{\$493,025} \times 100 = 23.7\%$$

7. *Net Earnings to Annual Volume*. This ratio measures the return on sales. The resulting percentage indicates the sale dollar remaining after deduction of all costs and taxes.

   $$\frac{\text{Net Earnings}}{\text{Annual Sales}} = \frac{\$\ 116,754}{\$8,725,370} \times 100 = 1.34\%$$

   For large volume building construction, 1% to 1.25% is normal.

8. *Fixed Assets to Tangible Net Worth*. This ratio shows the relationship between investment in plant and equipment and the capital. The higher this ratio percentage, the less liquid the net worth and the less capital is available for use.

   $$\frac{\text{Fixed Assets}}{\text{Tangible Net Worth}} = \frac{\$274,260}{\$654,765} \times 100 = 41.9\%$$

   For building construction, the accepted ratio is about 26.8%.

Not all of the ratios are used at one time. Information needed for a specific reason will cause the investigator to examine certain ratios. The purchasing agent examines the current ratio, net earnings to annual volume, net earnings to tangible net worth, net earnings to net working capital, and current liabilities to tangible net worth.

Vendor Purchasing

Vendor purchases involve material, equipment, and services. The process of purchasing these items requires drawing up a bill of materials. The purchasing of wood doors, for example, requires the purchaser to list the requirements as shown in Figures 16.9a and 16.9b.

REQUEST FOR PROPOSAL

1. Sealed lump sum proposals are to be furnished no later than 3:00 p.m. on November 23, 1987 at the offices of Design/Build Company, 80 Build Street, Boston, MA 02119.

2. The scope of the work will be in accordance to this document and as follows:

   a. Furnish solid core particle board flush wood doors, factory machined for 1 1/2 pair of butts, undercut and with beveled edges. The face of the doors is to be oak veneer rotary cut stain grade. Factory machining for mortise locks hardware is required. Templates are to be furnished by others.

   b. Door schedule of sizes and quantities is as follows:

No.	Size	Hand
25	3070	R
25	3070	L
125	3670	R
137	3084	R
154	3084	L
15	2668	R
5	2668	L

   c. Delivery to be tailgate F.O.B. at the job site.

   d. All doors are to be in individual cartons adequately padded to prevent scratches and scars during transport, handling, and storage.

   e. Delivery is to be on or about April 23, 1988. Coordinate delivery with the project superintendent.

Figure 16.9a

REQUEST FOR PROPOSAL (Cont'd.)

3. Furnish unit prices and minimum door order quantity, if any, for all of the sizes indicated.

4. The quotation is to include all costs for transportation and taxes.

5. Proof of adequate insurance to the limits stated in the General Conditions, attached, for automobile liability will be required.

6. Design/Build standard form purchase order will be used as the contract document, sample copy attached.

Figure 16.9b

Other vendor purchases, such as hoists, are done in the same manner by stipulation of a bill of materials (similar to that shown in Figure 16.9a for wood doors). Interestingly, a subcontract purchase for *miscellaneous materials* is more effective when it stipulates items and their locations on the plans as the scope of work.

Unit price vendor purchases are also commonly done to allow flexibility in the field for certain items of work. Securing heavy equipment for earthwork can be on a unit price basis for all of the equipment needed. The precise number of hours used for each of the pieces is not necessary. The same type of purchase order is prepared for formwork. If a standard form is rented, all of the sizes available are listed and their unit costs shown. As the quantities are used, they are reported from the field; that report provides a basis for checking invoices. To establish a cost for a unit price purchase order, a method called *not-to-exceed* is used.

The value stated as the not-to-exceed amount of the purchase order is commonly 90% of the budget amount for that item. The 10% discounting is a management technique to control the expenditure. At the 90% level, a change order to the purchase order would be needed if the expenditure is to exceed the allowed amount. The person performing the evaluation then has an opportunity to examine the efficiency of that product's use.

In Figures 16.2a and 16.2b, Request for Proposal for ready-mix concrete, the unit price purchase order written for that vendor would have a not-to-exceed value. The inclusion of a not-to-exceed amount requires a close-out change order for that purchase order to close the books.

Those items which can be quantified (as in the case of the wood doors) will have the purchase order written for the full lump sum. The unit price alternate stated on the purchase order is used for future (door) purchases. The future purchases would require a change order to the lump sum purchase order.

Purchasing services such as survey work would be based on the work description and a lump sum purchase order. Design services are also based on a definition of the work and a lump sum order.

Mass Purchases

Mass purchases involve securing products from vendors for more than one job. The advantage of this method is in the cost reduction for buying quantity. Examples of mass purchases are: ready-mix concrete (for multiple jobs in an approximate common locale), asphalt paving, reinforcing steel, welded wire mesh, and lumber. The method of purchasing is the same through the use of request for proposals, bill of materials, and unit prices. The unit price purchase orders are written for each of the projects as not-to-exceed lump sums.

More sophisticated mass purchases may involve standard items such as light fixtures, unit heaters, ventilation fans, toilet partitions, overhead doors, loading docks, hollow metal doors, and hollow metal frames. For this kind of mass purchase, the type of construction must be fixed and the design/build company must be able to convince the client(s) to allow these items to be used. The advantage of this type of purchase is in the incremental savings which can be passed on to the client in the form of lower cost, while enhancing the design/build company's competitive edge.

Bid Evaluation

When the bids are received, they have to be examined for conformance to the request for proposal and then evaluated to determine the low bidder. This part of the process is a mechanical analysis involving little theoretical input. The bid results report is helpful for this evaluation. See Figure 16.10 for an example of a peripheral form which my be used for this purpose.

The bid results report lists all of the subcontractors or vendors, the total quotation, and any alternates. This report could also be used to extend a unit price contract by estimated quantities in order to find the low bidder.

To summarize all of the quotations, Means Bid Spreadsheet forms can be used. See Figure 16.11 for such a form filled out as an example.

The spreadsheet is used as a report or in the event that many trades are evaluated at one time. When doing an evaluation, certain questions may remain as to the included work – especially if the bidders qualify their quotation with exceptions. It is good practice before awarding the contract to sit down with the apparent low two or three bidders and discuss the scope. This is a good time to review any work found or added after the request for proposal was issued. Final pricing of that added work and the assurance of a true comparison of the work, item for item, will allow for the final selection of the subcontractor or vendor.

Some contractors claim that this final interview is the time for negotiations. Such negotiations would be aimed at reducing the cost by asking the bidders to lower their prices further in order to get the job. The response from the low two or three bidders influences the decision. Still another approach is to poll the low bidders on their lowest price, and then return to each bidder asking them to meet a lower stated price to be offered if they want the job. This method is called *ratcheting*. Neither of the above strategies is recommended and both are considered unprofessional and unethical.

Award

The award of the contract is a time of elation for the successful bidder and disappointment for the losers. The successful subcontractor is usually notified by telephone of the award of the work. A follow-up appointment is made at the same time to review the details. During the bid evaluation, questions concerning the surety bonds and the liability insurance levels should be discussed and resolved. The draft contract terms and form are also discussed during the evaluation and any objections resolved. After the award, the preparation of the documents of agreement, signing of the contract, securing of the surety bonds, and receipt of the certificate of insurance still remain. These steps represent the correct method; time should be allowed in the master schedule to complete all of these functions. All documentation must be in place prior to the subcontractor starting work on the project. If this is not possible, then some expedient other solutions can be used to avoid delay in starting the work.

Means Forms
BID
RESULTS REPORT

DATE 5-1-86
TIME 2:00 P.M.

PROJECT R.S. MEANS OFFICE BUILDING
LOCATION KINGSTON, MA

A/E ESTIMATE

RANK	BIDDER	BID BOND	BASE BID	ALTERNATES				NOTES
				1	2	3	TOTAL W/3 ALT.	
	A/E ESTIMATE	10%	$ 4,600,000	210,000	64,000	11,000	$4,885,000	BUDGET = $4,800,000
4	GENERAL CONSTRUCTION	Y	$ 4,755,000	200,000	72,000	9,900	$ 5,036,900	
2	CONTRACTING CORP	Y	$4,224,000	189,500	67,400	11,200	$4,492,100	
1	OFFICE BUILDING	Y	$4,170,000	226,200	66,400	9,540	$4,472,140	CASHIER CHECK FOR BOND
3	PROFESSIONAL CONTRACTING	Y	$4,676,000	211,400	63,200	12,350	$4,962,950	
–	CONSTRUCTION SERVICES	N	$3,954,000	176,400	62,000	9,450	$4,201,850	DISQ – NO BOND

Figure 16.10

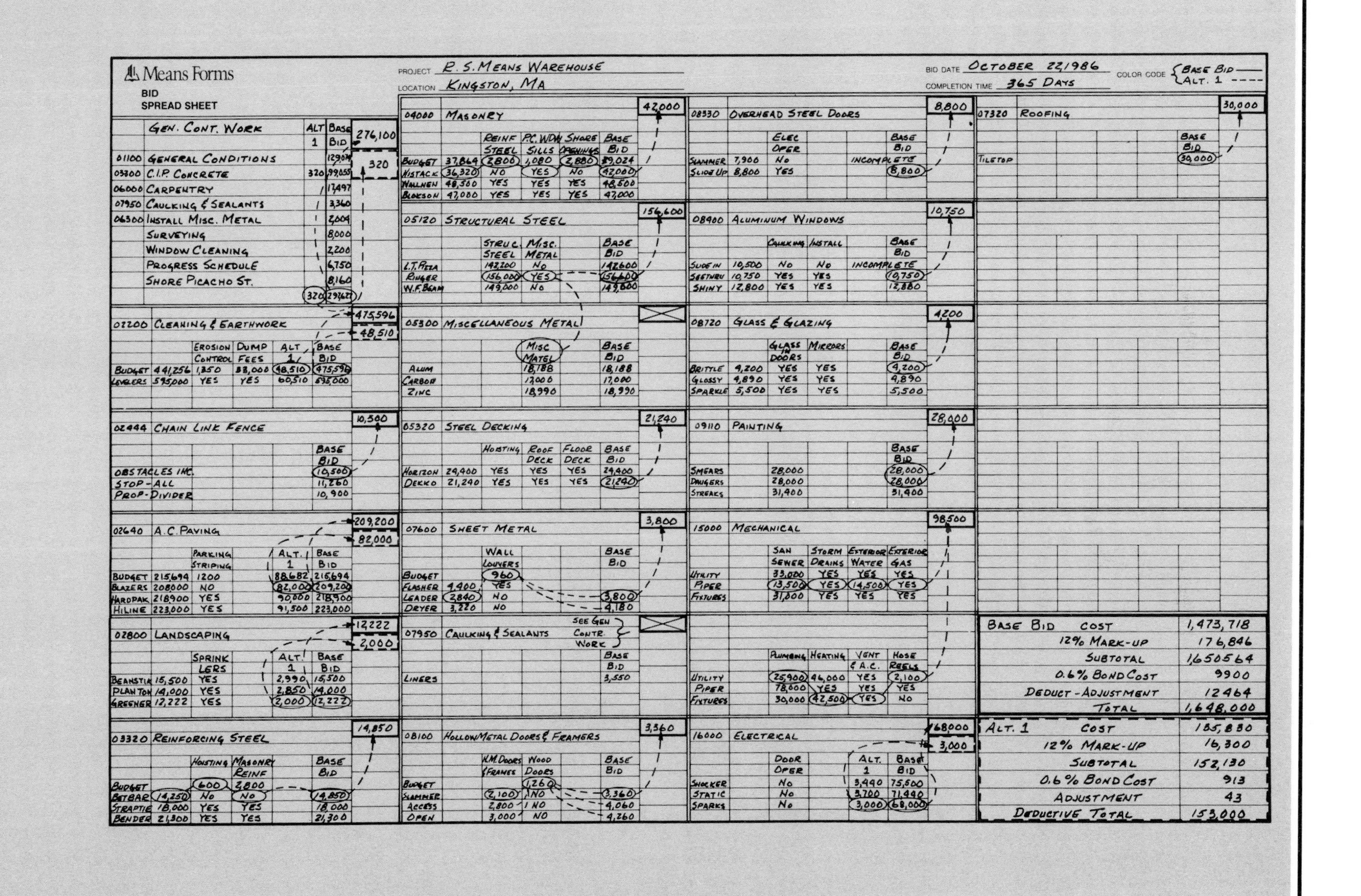

Means Forms
BID
SPREAD SHEET

PROJECT: R. S. MEANS WAREHOUSE
LOCATION: KINGSTON, MA
BID DATE: OCTOBER 22, 1986
COMPLETION TIME: 365 DAYS
COLOR CODE: BASE BID ——— / ALT. 1 - - - -

	GEN. CONT. WORK	ALT 1	BASE BID	276,100 / 320
01100	GENERAL CONDITIONS		129,074	
03300	C.I.P. CONCRETE	320	99,055	
06000	CARPENTRY		17,497	
07950	CAULKING & SEALANTS		3,360	
06300	INSTALL MISC. METAL		2,004	
	SURVEYING		8,000	
	WINDOW CLEANING		2,200	
	PROGRESS SCHEDULE		6,750	
	SHORE PICACHO ST.		8,160	
		320	276,100	

02200 CLEANING & EARTHWORK — 475,596 / 48,510

		EROSION CONTROL	DUMP FEES	ALT 1	BASE BID
BUDGET	441,256	1,350	33,000	48,510	475,596
LEVELERS	595,000	YES	YES	60,510	595,000

02444 CHAIN LINK FENCE — 10,500

	BASE BID
OBSTACLES INC.	10,500
STOP-ALL	11,260
PROP-DIVIDER	10,900

02640 A.C. PAVING — 209,200 / 82,000

		PARKING STRIPING	ALT. 1	BASE BID
BUDGET	215,694	1200	88,682	215,694
BLAZERS	208,000	NO	82,000	209,200
HARDPAK	218,900	YES	90,000	218,900
HILINE	223,000	YES	91,500	223,000

02800 LANDSCAPING — 12,222 / 2,000

		SPRINKLERS	ALT. 1	BASE BID
BEANSTIK	15,500	YES	2,990	15,500
PLANTON	14,000	YES	2,850	14,000
GREENER	12,222	YES	2,000	12,222

03320 REINFORCING STEEL — 14,850

		HOISTING	MASONRY REINF	BASE BID
BUDGET		600	2,800	
BETBAR	14,250	NO	NO	14,850
STRAPTIE	18,000	YES	YES	18,000
BENDER	21,300	YES	YES	21,300

04000 MASONRY — 42,000

		REINF STEEL	P.C. WDW SILLS	SHORE OPENINGS	BASE BID
BUDGET	37,864	2,800	1,080	2,880	39,024
HISTACK	36,320	NO	YES	NO	42,000
WALLNEN	48,500	YES	YES	YES	48,500
BLOKSON	47,000	YES	YES	YES	47,000

05120 STRUCTURAL STEEL — 156,600

	STRUC. STEEL	MISC. METAL	BASE BID
L.T. PIZZA	142,200	NO	142,600
RINGER	156,000	YES	156,600
W.F. BEAM	149,000	NO	149,000

05300 MISCELLANEOUS METAL — ☒

	MISC MATEL	BASE BID
ALUM	18,188	18,188
CARBON	17,000	17,000
ZINC	18,990	18,990

05320 STEEL DECKING — 21,240

		HOISTING	ROOF DECK	FLOOR DECK	BASE BID
HORIZON	24,400	YES	YES	YES	24,400
DEKKO	21,240	YES	YES	YES	21,240

07600 SHEET METAL — 3,800

		WALL LOUVERS	BASE BID
BUDGET		960	
FLASHER	4,400	YES	
LEADER	2,840	NO	3,800
DRYER	3,220	NO	4,180

07950 CAULKING & SEALANTS — SEE GEN CONTR. WORK — ☒

	BASE BID
LINERS	3,550

08100 HOLLOW METAL DOORS & FRAMERS — 3,360

	H.M. DOORS & FRAMES	WOOD DOORS	BASE BID
BUDGET		1,260	
SLAMMER	2,100	NO	3,360
ACCESS	2,800	NO	4,060
OPEN	3,000	NO	4,260

08330 OVERHEAD STEEL DOORS — 8,800

		ELEC OPER		BASE BID
SLAMMER	7,900	NO	INCOMPLETE	
SLIDE UP	8,800	YES		8,800

08900 ALUMINUM WINDOWS — 10,750

		CAULKING	INSTALL		BASE BID
SLIDE IN	10,500	NO	NO	INCOMPLETE	
SEETHRU	10,750	YES	YES		10,750
SHINY	12,800	YES	YES		12,880

08720 GLASS & GLAZING — 4,200

		GLASS IN DOORS	MIRRORS	BASE BID
BRITTLE	4,200	YES	YES	4,200
GLOSSY	4,890	YES	YES	4,890
SPARKLE	5,500	YES	YES	5,500

09110 PAINTING — 28,000

		BASE BID
SMEARS	28,000	28,000
DAUGERS	28,000	28,000
STREAKS	31,400	31,400

15000 MECHANICAL — 98,500

	SAN SEWER	STORM DRAINS	EXTERIOR WATER	EXTERIOR GAS
UTILITY	33,000	YES	YES	YES
PIPER	13,500	YES	14,500	YES
FIXTURES	31,000	YES	YES	YES

	PLUMBING	HEATING	VENT & A.C.	HOSE REELS
UTILITY	25,900	46,000	YES	2,100
PIPER	78,000	YES	YES	YES
FIXTURES	30,000	42,500	YES	NO

16000 ELECTRICAL — 68,000 / 3,000

	DOOR OPER	ALT. 1	BASE BID
SHOCKER	NO	3,490	75,500
STATIC	NO	3,700	71,490
SPARKS	NO	3,000	68,000

07320 ROOFING — 30,000

	BASE BID
TILETOP	30,000

BASE BID	
COST	1,473,718
12% MARK-UP	176,846
SUBTOTAL	1,650,564
0.6% BOND COST	9900
DEDUCT - ADJUSTMENT	12464
TOTAL	1,648,000

ALT. 1	
COST	135,830
12% MARK-UP	16,300
SUBTOTAL	152,130
0.6% BOND COST	913
ADJUSTMENT	43
DEDUCTIVE TOTAL	153,000

Figure 16.11

One of the expedients is the issuance of a *letter of intent*. The letter of intent is an interim document. It may be drawn up when some terms of the subcontract agreement are still unresolved. A letter of intent should contain clear language regarding the scope of work, and the relationship of the parties to the agreement. A time limit should be established for execution of the final contract. If a contractual agreement has not been signed by the end of this time limit, the offer may be withdrawn. Until the contract is executed, the subcontractor is normally not allowed to come onto the site to start work. If it is absolutely essential for him to start work, the signing of an agreement may be deferred temporarily. Nevertheless, the subcontractor must submit the Certificate of Insurance prior to coming on site. The lack of a performance bond for a short period of time at the outset of the work is not important. The signed contract is more important to the subcontractor than to the design/build contractor, because it is required in order for the subcontractor to get paid. Without the proof of insurance, however, the client and the design/build contractor are most likely vulnerable and liable for all damages. Other requirements (in addition to bonds or insurance) which would become contractual upon signing the contract should be reviewed before work is started.

Unsuccessful Bidders

At this time, the unsuccessful bidders should also be notified and thanked for the effort that they have invested. This last step is both professional and courteous, and helps to maintain a good relationship with these subcontractors for future bidding. A telephone call to each is recommended. If the list is very large, then a letter to each may suffice.

It should be remembered that information concerning bids is confidential. Release of such information to others is a betrayal of the low bidder and could compromise his bidding practices.

Chapter Seventeen

The Superintendent

The role of the superintendent has been compared to that of a traffic cop in that he spends so much of his time directing the activities of upwards of 30 to 60 different trades, depending on the type of project. His job involves the smooth transition of one trade to another in the work area. To be effective, the superintendent must avoid congestion and confusion. While the traffic cop has to deal only with the four points of the intersection, the superintendent's responsibility is multiplied by the number of floors and work areas involved with the project. In addition, the superintendent is required to process a great deal of paperwork and act as a liason between management and workers. Clearly, the superintendent plays a key role in the success of the project.

Client Contact

The superintendent may have more contact with the client than any other member of the design/build team. Clients tend to visit the site often as the work progresses. When one has a vested interest, it is difficult to stay away. As a result, the superintendent may have occasion to exchange pleasantries with the owner every day. He would be the first to respond to direct questions from the owner concerning the conduct and progress of the work. The owner should not address nor should the superintendent respond to questions concerning the financial status of the project. These questions should be addressed to the project manager. This is mentioned here because the owner is concerned with two issues: the progress of the job and the finances, especially if there is a savings provision in the contract.

The superintendent is also present when the sales people bring new prospects to the site. His response could influence future work. Client contact therefore becomes part of the job description in the consideration of candidates for employment as superintendents.

Schedules

One of the superintendent's duties is maintaining and updating the construction schedule. The construction schedule is drawn up by the project manager together with the superintendent. This schedule represents the path to follow to assure the successful completion of the project. The superintendent's emphasis should be that each date is fixed and cannot be changed. Each milestone date must be met. He should maintain that the update of a construction schedule should only reflect the improvement of the construction duration.

Superintendents may not always consider the developed schedule as cast in stone. They may insist that the job be done at its own pace regardless of the predetermined program. When the schedule is not followed, the work goes along without control. The updates then reflect the areas where there has been progress. This is not the most effective use of a progress schedule. That is why a superintendent should be involved – from the start – in development of the schedule. The superintendent is, after all, responsible for job progress, and the construction schedule is one of his tools.

Subcontractors

The superintendent is responsible for subcontractor performance. He must make sure that the work is performed in accordance with the plans and specifications and he must be vigilant of the quality of the work. He must motivate the subcontractors and develop the tempo of work in order to optimize the flow of workmen. The superintendent must treat the good subcontractor producers in such a way that they can continue to work efficiently. He must also have ways to handle the delinquent subcontractor, getting this person involved and working as part of the team.

Safety

The superintendent is responsible for the safety aspects of a project in conformance to the current safety rules and OSHA (Occupational, Safety and Health Act). He holds the weekly safety meetings and sees to it that all of the safety provisions are followed. These provisions may involve installation of railings, stairs, toe boards and railings around openings, and safety cable along the edges of open multi-story buildings. The superintendent is responsible for fire safety and proper lighting in the spaces. First aid, the wearing of hard hats, and authorization of visitors on the site come under his responsibility.

Site

It is said that a well kept, neat and clean site is indicative of a healthy, well run project. The construction site is the exclusive domain of the superintendent. He assigns the location of the offices, trailers, sheds, toilets, and subcontractors' trailers and equipment. He determines the access onto the site and the traffic flow in and out of the site. The superintendent controls the use of common equipment such as hoists and the priority of stored material on the site.

Today's superintendent has a complex role. Fulfilling and completing the contract within the requirements of cost, quality, and time requires considerable skills. The means to this end are efficient planning, attention to detail, realistic scheduling, a cooperative and helpful attitude, and just plain job sense. The successful design/build company selects their superintendents with care.

Chapter Eighteen

PROJECT MANAGEMENT

Project Management is a title used to describe the duties of a manager in the construction industry. Management of a project may be interpreted in a variety of ways depending on the corporate make-up of the construction company. A project manager can be the overall responsible entity for a project, or may be a member of the staff of a construction or development company. A construction company may have a construction manager responsible for the entire project and a project manager (who is part of the staff) responsible for another segment of the work. Both architects and developers give their construction coordinators the title of *project manager*. Yet the duties of these project managers are different. Clearly, the job cannot be defined by title alone; describing the tasks for each position is more meaningful.

For the design/build company, a project manager is a middle manager responsible for the successful execution of the project. The project management tasks explained in the following paragraphs can be performed by a single person or several persons, depending on the size and complexity of the project. The project manager becomes an active member of the project team after the award. His first tasks are to become familiar with the project, the proposal documents, and the contract.

Start Up

The project manager is responsible for organizing the project. The tasks to be accomplished are as follows:

- Contracts with the design professionals
- Conversion of the conceptual estimate to the working estimate
- Design review and coordination
- Preparation of the management schedule
- Procurement of labor and materials
- Mobilization and start-up of the work

Contracts With Design Professionals

The selection of the architect and the design engineers is made in advance of the proposal. A letter of understanding may be drafted, but a formal agreement may not be signed until the project has been awarded by the client. At that time, the design contracts are prepared and executed. Since all of the terms, conditions, and duties are known at this point, the contract is a matter of formality. In some instances, the architect may have been the only one whose services were secured. In this case, the project manager is responsible for negotiating for the services of the other designers. This activity is done by selection rather than bid. Sometimes a contract has been concluded in advance of the proposal. In such cases, the architect may require a minimum fee for conceptual planning with stipulations for the full contract value and terms at the time of the award. The project manager then only requires a signature or the issuance of a contract modification to reflect the final agreement. The responsibility for this task, regardless of the variations, would be with the project manager.

Cost Accounting Conversion

Some of the costs in the conceptual estimate are a composite of a number of associated costs. The unit thus derived is in a convenient form for doing that type of estimating. Once the project becomes a reality, the composite costs must be redistributed to their individual items. The computer is a great help in this process. It accomplishes this task using a cost code assigned to each item in the composite and to the other items in the conceptual estimate. The project manager is responsible for directing the conversion and reviewing the resultant working estimate with the estimator. Both the project manager and the estimator have the authority to adjust the estimate as necessary, to properly reflect the items of work, their proper costs, and method of purchasing. The individual work items should simply be reasonable for the job. This is not a re-estimate, only a chance to make adjustments.

Design Review and Coordination

In Chapter 11, it was stated that the design is reviewed by the design/build company. The design/build company contracts with the design professionals separately and is therefore responsible for the review and the design coordination. The project manager is the one who must follow up with this task.

Management Schedule

A management schedule is initially prepared as part of the proposal. This schedule usually is not of sufficient detail for the construction of the building, but is created to show critical events and to demonstrate the necessary control factors for the project. The project manager is responsible for the preparation of a new management schedule which should include all of the tasks necessary to build the project.

Procurement

If this task is not assigned to someone else, the project manager may be responsible for purchasing labor and material items for the project. If another person is responsible, then the project manager's role in purchasing would be one of support. This support includes defining the scope of the work, managing information bid meetings, and sitting in on the bid evaluation and award.

Mobilization and Start-up

The project manager must arrange for the move and temporary construction facilities and for the installation of temporary utilities. The temporary electrical work might require a separate electrical contractor in advance of the bid for this work. The project manager is responsible for securing the electrician to perform this work. Any other preliminary work such as surveys of the land and the setting of control points must be expedited. Securing the building permit is another of the project manager's duties.

Client Contact

As construction gets under way, the project manager replaces the salesman and the estimator in the role of client contact. The operational decisions, information exchange, and approvals required of a client/owner are done through the project manager. He is the one to advise the client of milestone dates which require the client's input and of the consequences of delay. The project manager also presents the client with the design for approval and explains what it is that the owner is reviewing.

As the project manager takes on the role of client contact, he must practice the same approach that his predecessors used to maintain continuous client confidence. The client should not be made to feel that he is being passed to others, especially to others of lesser stature in the company. Some companies assign their project managers a title such as vice president in order to defuse that potential. The professionalism of the project manager will also allay the owner's concerns. He should be made aware that all who have participated in the process thus far are still available to him at any time. The project manager is, however, his day-to-day contact. The design/build company has assured and sold the client on the fact that a service is to be performed in the construction of his building. That service is based on trust and it is up to the project manager to uphold that trust. His role in the project will have the longest duration, as close contact is maintained for a period spanning several months or even a few years. The role of the initial participants spans only a few months, and even then, the contact is infrequent. For these reasons, the project manager becomes the most visible representative of the design/build company.

Subcontract Relationship

The project manager is not only responsible for client contact, but also for the subcontractor relationship. Since the project manager will have done the purchasing, the subcontractors will consider him as having a strong position in the project. If he was not the purchasing agent, the project manager must establish his authority on the project in other ways – chiefly by his ability and character. Such a relationship is important to the success of the project. The superintendent's performance relates directly to the ability of the project manager to back him up in the event of a problem with the subcontractors.

Subcontractors who are actively bidding work may overextend themselves if they are too successful. This is a recurring phenomenon, especially in an active construction market. When this situation does occur, the subcontractor's resources of men and equipment are taxed. New foremen and labor as well as rented equipment may not perform as well as the regular crew. The job might suffer because of the foreman's lack of experience or unfamiliarity with the workings of the company. The subcontractor's financial resources to sustain work on a project can be a problem. This kind of information may not be available at the time of award. Thus, the ability of the project manager to elicit favorable treatment from a subcontractor is a big asset.

Follow Up

The project manager must make sure that what needs to be done gets done. This requires persistence and consistency. It is not enough to assume that something is done just because it is said to be done. To make sure, the project manager must follow up. He must telephone once again the person who has not returned the previous one or more calls. He must follow up with a new letter or phone call if his previous letter received no response. No one is as concerned about the success of the project as the project manager. The project manager has a sense of urgency greater than the manufacturer of the windows who is behind in shipping, and greater than the architect who has yet to complete a detail for the job.

All of these concerns are the project manager's job, and follow-up is his method of attending to them. The follow-up process is enhanced by a good memory; but there are a number of techniques that can also be applied to this important goal.

One of the most common and successful techniques is the use of the "tickler" file. A tickler file consists of folders labeled by the days of the week and/or by weeks. Items needing to be addressed at some time in the future are placed in the appropriate chronological folder. On the appointed date, the folder is perused and the action indicated for each item is performed. Unresolved items are placed in folders with successively later dates for additional follow-up. Examples of tickler file items are unanswered letters, prearranged or reminder phone calls, and the return or delivery of items.

Another follow-up technique is the use of visual aids which can work quite effectively. Examples are color folders used to designate different types of contracts or contract terms, color tabs, labels, and plastic color clips which can be used for similar identification purposes.

Expediting

Expediting is the process of ensuring that the supplies of construction are provided to the job site on time. All of the skills used for scheduling will be in vain if the activities are not accomplished at the appointed date. If the building is to be constructed as promised, someone must see to it that all materials are on hand when needed. This task – expediting – is another of the project manager's responsibilities. Control of the design, a good relationship with subcontractors, and follow-up insure success. One could argue that all of the participants are obliged by contract to perform and that there should therefore be no need for this effort. Knowing that these participants could be sued if they do not perform does not build buildings. Only attention to detail, dedication to the schedule, and making it happen gets it done.

Contract Administration

In addition to the operational tasks, the project manager is responsible for the contract administration. The contract administration involves:

- changes to the project
- time extensions
- logs, forms, and reports

Changes to the Project

It has been explained that the Guaranteed Maximum Price (GMP) for design/build is given early in the design process. If the building design is completed to the agreed building scope, and constructed by those plans, then the guaranteed price is the most the client will pay. The cost could be less, based on savings provisions which are normally part of the contract. The likelihood is great that there will be some changes in the course of the work. Some of these changes will be minor (such as wallpaper instead of paint in the owner's office) and will not constitute any appreciable cost differential. Other changes are bound to be more significant and involve additional cost.

Changes to the contract can be handled in a number of ways. Minor revisions can be ignored as long as the savings in the project can cover the added cost. If the savings are to be shared between the design/build company and the client, then the design/build company will be contributing to the cost of that extra work. This is not a problem, if minor, and can be allocated to goodwill. If the changes are major, then a change to the contract must be executed. The payment could be deferred and also paid from the owner's share of the savings. This arrangement may, however, also present certain problems. For example, if the change is initiated early in the project, the amount of the savings, if any, may not be known at that point in the project. The best approach is to treat all changes – minor or major – in a formal manner. The final accounting would then confirm the availability of savings and their use.

Changes to a contract initiated by the client and relating to completely new work are easy to identify as changes. Since the design was not complete at the time of the GMP, the owner might question whether this work that is extra to the contract should originally have been included in the contract. The project manager must do the job of design review and owner sign-off well in order to avoid this type of problem.

Not all changes involve revisions to the contract. Changes due to value engineering, for example, must be considered. When does a value change require a change to the contract and when is it savings?

The most obvious saving is the one that comes from the procurement. The ability to secure subcontractors, service, and materials for less money than the budget allowed is pure savings. The efficient use of the design/build company field labor and the ability to complete the work ahead of schedule reduces the general conditions costs, and is also savings. If the change of a work item involves only the substitution of one manufacturer for another, it is not a change and it is a savings. Examples of this type of savings are as follows: membrane roofing instead of tar and gravel, or a different window manufacturer instead of the original without change to the window system.

The common requisites for changes to the contract are work deleted or added, a building system concept that is changed, or the direct substitution of an element of work.

Whatever the change, the resulting cost to the work must be accounted for. Changes in a wall system, for example, would change the blocking requirements and finish material. When calculating these changes, the costs for the associated work must be included. An example of this type of analysis is found in Figure 18.1.

Logs and Forms

A great deal of correspondence is generated in the process of constructing a building. An effective method for maintaining control of this correspondence is through the use of logs and standard forms. The forms fall into two categories, reports and control. Among the report forms are: Daily Superintendent Reports, Daily Time Sheets, Weekly Time Sheets, Job Progress Reports (examples from *Means Forms for Building Construction Professionals* are shown in Figures 18.2 through 18.5). These standard forms are one alternative. Some companies might want to devise their own versions.

Control forms include: Shop Drawing Submittal Log, Material Status Report, Change Order Log, and Purchase Order Log.

Shop Drawing Submittal Log: The Shop Drawing Submittal Log is columnar across the page and includes the following headings:

1. Specification Division
2. Specification Section
3. Specification Page
4. Item
5. Drawing/Sample/Cut Number
6. Submittal Date
7. Submitted to the Architect
8. Date Received
9. Disposition:
   A (Approved)
   AAN (Approved as noted)
   RR (revise and resubmit)
   R (rejected)
10. Returned to subcontractor

If the material requires re-submission, then the log needs to be extended by adding more columns for headings 6 through 10.

Material Status Report: The Material Status Report is also columnar, and has the following headings:

1. Specification Division
2. Specification Section
3. Material Item
4. Page Number
5. Promised Delivery Date
6. Shop Drawing Submittal Date: anticipated ______ actual ______
7. Shop Drawing Approval Date: anticipated ______ actual ______
8. Actual Delivery Date

CHANGES IN WALL SYSTEM

Change wall system from metal panel to brick faced light gauge metal framing system.

ADD

Brick include brick ties
Light gauge framing
Relieving angle to support brick
Framing structural support
Batt insulation
Exterior board
Felt paper covering
Interior gypsum board
Interior finish
Blocking
Fire stop material

DEDUCT

Composite insulated metal panel
Panel structural support
Channel girt (intermediate support)
Furring channels
Rigid insulation
Interior gypsum board
Interior finish
Fire stop material

Figure 18.1

Means Forms

DAILY
CONSTRUCTION REPORT

JOB NO. ____

DATE ____

PROJECT ____ SUBMITTED BY ____

ARCHITECT ____ WEATHER ____ TEMPERATURE ____ AM ____ PM ____

CODE NO.	WORK CLASSIFICATION	FOREMEN	MECHANICS	LABORERS	SUB-CONTR'S	TOTAL HOURS	DESCRIPTION OF WORK
	General Conditions						
	Site Work: Demolition						
	Excavation & Dewatering						
	Caissons & Piling						
	Drainage & Utilities						
	Roads, Walks & Landscaping						
	Concrete: Formwork						
	Reinforcing						
	Placing						
	Precast						
	Masonry: Brickwork & Stonework						
	Block & Tile						
	Metals: Structural						
	Decks						
	Miscellaneous & Ornamental						
	Carpentry: Rough						
	Finish						
	Moisture Protection: Waterproofing						
	Insulation						
	Roofing & Siding						
	Doors & Windows						
	Glass & Glazing						
	Finishes: Lath, Plaster & Stucco						
	Drywall						
	Tile & Terrazzo						
	Acoustical Ceilings						
	Floor Covering						
	Painting & Wallcovering						
	Specialties						
	Equipment						
	Furnishings						
	Special Construction						
	Conveying Systems						
	Mechanical: Plumbing						
	HVAC						
	Electrical						

Page 1 of 2

Figure 18.2a

Means Forms

EQUIPMENT ON PROJECT	NUMBER	DESCRIPTION OF OPERATION	TOTAL HOURS

EQUIPMENT RENTAL - ITEM	TIME IN	TIME OUT	SUPPLIER	REMARKS

MATERIAL RECEIVED	QUANTITY	DELIVERY SLIP NO.	SUPPLIER	USE

CHANGE ORDERS, BACKCHARGES AND/OR EXTRA WORK

VERBAL DISCUSSIONS AND/OR INSTRUCTIONS

VISITORS TO SITE

JOB REQUIREMENTS

Figure 18.2b

Means Forms

DAILY TIME SHEET

PROJECT

FOREMAN

WEATHER CONDITIONS

TEMPERATURE

DATE

SHEET NO.

NO.	NAME		DESCRIPTION OF WORK									TOTALS REG-ULAR	TOTALS OVER-TIME	RATES REG-ULAR	RATES OVER-TIME	OUTPUT	
		HOURS															
		UNITS															
		HOURS															
		UNITS															
		HOURS															
		UNITS															
		HOURS															
		UNITS															
		HOURS															
		UNITS															
		HOURS															
		UNITS															
		HOURS															
		UNITS															
		HOURS															
		UNITS															
		HOURS															
		UNITS															
		HOURS															
		UNITS															
		HOURS															
		UNITS															
		HOURS															
		UNITS															
		HOURS															
		UNITS															
		HOURS															
		UNITS															
		HOURS															
		UNITS															
	TOTALS	HOURS															
	EQUIPMENT	UNITS															

Figure 18.3

WEEKLY TIME SHEET

EMPLOYEE EMPLOYEE NUMBER WEEK ENDING

FOREMAN HOURLY RATE PIECE WORK RATE

PROJECT	NO.	DESCRIPTION OF WORK	DAY								TOTAL	HOURS		RATE	TOTAL COST	UNIT COST
			HOURS									REG.				
			UNITS									O.T.				
			HOURS									REG.				
			UNITS									O.T.				
			HOURS									REG.				
			UNITS									O.T.				
			HOURS									REG.				
			UNITS									O.T.				
			HOURS									REG.				
			UNITS									O.T.				
			HOURS									REG.				
			UNITS									O.T.				
			HOURS									REG.				
			UNITS									O.T.				
			HOURS									REG.				
			UNITS									O.T.				
			HOURS									REG.				
			UNITS									O.T.				
			HOURS									REG.				
			UNITS									O.T.				
			HOURS									REG.				
			UNITS									O.T.				
			HOURS									REG.				
			UNITS									O.T.				
			HOURS									REG.				
			UNITS									O.T.				
			HOURS									REG.				
			UNITS									O.T.				
			HOURS									REG.				
			UNITS									O.T.				
			HOURS									REG.				
			UNITS									O.T.				
			HOURS									REG.				
			UNITS									O.T.				
			HOURS									REG.				
			UNITS									O.T.				
			HOURS									REG.				
			UNITS									O.T.				
			HOURS									REG.				
			UNITS									O.T.				
		TOTAL	HOURS									REG.				
			UNITS									O.T.				

PAYROLL DEDUCTIONS	F.I.C.A.	INCOME TAXES FEDERAL	INCOME TAXES STATE	INCOME TAXES CITY/COUNTY	HOSPITALIZATION			TOTAL DEDUCTIONS		LESS DEDUCTIONS
	$	$	$	$	$	$	$	$		NET PAY

Figure 18.4

Means Forms

JOB PROGRESS REPORT

SHEET ______ OF ______

PROJECT ______ JOB NO. ______

LOCATION ______ YEARS ______

WORK ITEM	QUANTITY, $, OR %	BEGINNING BALANCE	MONTHS												ENDING BALANCE
	DATE														
	YEAR														

Figure 18.5

Change Order Log: The Change Order Log is columnar and contains the following sections:

1. Change Number
2. Description of Change
3. Date Received
4. Approximate Estimate Amount
5. Sent to Subcontractors
6. Received from Subcontractors
7. Actual Estimated Amount
8. Submitted to Owner
9. Received from Owner
10. Disposition
11. Change Order Number

If the change is returned for revision, the log must be extended by adding more columns for headings 5 through 10.

Purchase Order Log: The Purchase Order Log is also columnar and contains the following headings:

1. Cost Code
2. Item of Work
3. Subcontractor/Vendor Name
4. Early Buy Date
5. Late Buy Date
6. Purchased Date
7. Estimated Amount
8. Buy Amount
9. Variance Amount

A Purchase Order log can also be set up on a Means Standard Job Progress form, adapting the headings as necessary.

Meetings

The project manager acts as the moderator for meetings. He creates the agenda and keeps the minutes. There are two types of meetings. One is the job meeting with the subcontractors, architect, and occasionally the owner. The other is the owner's meeting.

The job site meeting held once a week involves a review of the items discussed the previous week and a discussion of new issues. The typical agenda for job meetings is as follows:

- New Business
  - listing of trades in the CSI format
- Shop drawing submittal and approval
- Old Business
  - unresolved issues from the previous week

The owner's meeting involves the financial aspects of the project and any problems or changes. This meeting is usually held once every two weeks to once a month. Changes to the financial status of a project are not meaningful in shorter durations.

The project manager needs to be part businessman, part salesman, and part mediator. The success or failure of the project may be directly attributed to the project manager's capabilities. Along with the superintendent, this middle manager is a key member of the design/build operational team.

Chapter Nineteen

PROJECT ENGINEERING

The role of project engineer can be taken on by the project manager or someone on his staff. The project engineer's duties are mostly technical, but do not involve design. A detailed knowledge of construction is a requisite for someone in this position. The duties of a project engineer vary but can include:

- Shop drawing checking
- Maintaining logs
- Processing change orders
- Quality assurance inspection
- Job site coordination
- Value analysis

Shop Drawing Checking

The processing of the shop drawings requires knowledge of construction and the ability to read and interpret plans. Since the design professionals are hired separately, the design coordination becomes the responsibility of the design/build company. This task could be avoided by the design/build company by allowing one of the design group, usually the architect, to assume that responsibility. The shop drawings require the same attention.

Under a bid method of construction, the design and shop drawing coordination is by the architect. The general contractor is required to review the shop drawings for conformance to the plans and specifications before sending them to the architect. The contract agreement and/or the standard general conditions state, "The contractor shall review, approve and submit . . . . . . all shop drawings . . . ." The approval statement is inconsistent, in this instance, with the responsibility of the work. Only the design professional can approve adherence to design. The common interpretation used in the construction industry is *reviewed for conformance.*

Shop drawings, equipment manufacturer performance data, standard product criteria, and standards called *cuts* and *samples* are submitted.

This information is intended to show specific detail of work items – information that is not available from the design drawings, but which demonstrates conformance to the design. The involvement of the submittals is dictated by custom, industry practice, and specific reference in the specifications.

The structural design shows the arrangement of the members and their sizes. The connection details are left to the fabricator. The structural shop drawings are consequently involved and numerous. Included are the erection drawings that show the arrangement and method of fastening each piece. In support of the erection drawings are the piece drawings. These drawings show every aspect of the piece that allows it to be converted from a uniform member to one that can fit into the overall structure.

Highly detailed shop drawings are created for precast fascia panels, and for structural precast beams and columns. Roofing shop drawings are simply cuts taken from catalogues and submitted. Samples (when required by specification) include brick, drywall, acoustic ceiling panels and supports, wall covering, flooring, and paint colors. Any item that involves conformance to color, texture, and quality usually requires samples to be submitted.

When the design/build company receives the shop drawing, sample, or cut, they review it for conformance to the plans and specifications. The submittals are then stamped with a statement confirming the review and sent to the designer responsible for the item. Submittals may also be sent to the architect for further information and coordination, if those tasks are part of his contract. It is the project engineer's responsibility to follow up on the return of the shop drawing from the designer and its transmittal to the subcontractor or vendor.

The project engineer keeps track of the submittal process by maintaining all project logs. He receives and records all of the change orders and shop drawings, among other items. The project engineer sees to it that the necessary copies are received and the timely exchange of data is maintained.

Processing Change Orders

The change order process should be followed closely for maximum efficiency and timely disposition. When a change has been initiated, the nature of the change and any related documentation is noted in the change order log. The data is examined and sent to the subcontractors who are affected by the change, so that it can be priced. In the interim, the project engineer also prices the work as a check against the returning quotation. He then must maintain contact with the subcontractor to insure the timely return of the quotations.

Upon receiving the quotes, the project engineer checks the scope of work and the cost against his own estimate. Once satisfied that all is in order and that the estimate is a fair representation of the value of the work, he prepares a change request for the owner to approve. Upon approval, the cost of the change is then added to the contract amount by virtue of a change order. As a matter of policy, the project manager may require a review of these changes prior to submission to the owner. It is also the project engineer's responsibility to keep subcontractors informed on the disposition of a change and to process the subcontractor's change order once the owner's change order is approved.

Quality Assurance Inspection

The design/build company is responsible for the quality control of the work. This responsibility is carried out by the project superintendent, who is supported by the project engineer. The project engineer deals with punch list work. The punch list is a list comprised of work deficiencies and work not completed. The superintendent examines the quality during the course of the project. The purpose of this continued scrutiny is to catch deficiencies as they occur so that they are not overlooked and then buried within finished areas. Especially vulnerable to unseen defects are the structural steel and work items in wall cavities. At the end of the job, an unofficial punch list is compiled by the project engineer along with the superintendent, prior to the architect's inspection. Their critical examination of the work accelerates the closeout process and emphasizes the completion and the quality of the work. These are important factors to any good construction company, but even more so for the design/build company. The reputation of the design/build company for speed and quality construction is extremely important for future referrals.

Job Site Coordination

Chapter 12 covers design coordination and states that this aspect of the project is the responsibility of the design/build company. In addition to design coordination, there is also the process of job site coordination of the trades.

The areas that will be covered or closed in on a project represent limited space that needs to be managed. This space is usually the area above the acoustic ceiling and below the floor of the next higher story. In this space may be congregated the mechanical duct work, plumbing piping, electrical conduit, and sprinkler piping. All of the trades working in that space must have clear pathways to go from point to point without interference. To provide these pathways, the scaled size and location of each of the items located must be drawn on one plan.

Once the pathways for all the work are located and dimensioned, that work can proceed. Each trade can work in the assigned areas, confident that their work will fit and not be interfered with. Without this coordination process, it would be a difficult proposition trying to fit all of the items of work into the ceiling cavity. The responsibility for the meetings and the timely processing and production of drawings can be with the superintendent or with the project engineer, or both. If the work areas are exposed to view, as in an airport, the architect must approve the arrangements to conform with the aesthetic design requirements.

Value Analysis

The process of reducing the cost of a project through cost effective choices at the design stage describes *value analysis*. This process involves two steps. The first step is the suggestion and the second is the proof. The suggestion can involve any product or design element of the building. The proof examines the suggestion for quality, design conformance, design compatibility, and cost effectiveness. Many, but not all of the suggestions require both engineering and design analysis. A suggestion regarding quality may not require design input, as in the case of material such as: hardware manufacturers, types of door veneers, and use of paint or wallpaper.

Design/build construction also involves the process of value analysis, but to a lesser degree than other types of construction. The basic design concept and quality level are fixed at the GMP (Guaranteed Maximum

Price) contract price. Any changes usually involve the level of quality of a particular item. This process is ongoing and is between the design/build company and the owner. If for some reason the owner identifies a concept change or a change in use after the GMP and before the design drawings, a value analysis can be made.

After examination for quality, design conformance, and design compatibility, the primary focus of most value analysis decisions is cost. There are two opposing cost factors that have to be examined. The first is the cost to furnish and install the item (also known as the *construction cost*); the second is the long term cost, based on the item's useful life, its energy use cost, and maintenance requirements. Often, the item that has a low first cost has a higher long term cost, and conversely, the lower the long term cost, the higher the first, or construction cost may be. The owner's decision is based on his own available assets, priorities, and plans for the building.

The value analysis process can begin with any item of the work. The most productive approach is to concentrate on the most costly items and work down to the least costly.

Examining the major categories of work for new construction, the proportions of costs are approximately as follows:

Structure:	15% to 25%
Roof:	1% to 3%
Exterior Skin:	9% to 15%
Architectural Features:	15% to 25%
MEPS (Mechanical, Electrical, Plumbing, & Sprinklers):	25% to 35%
General Conditions/Fee:	10% to 15%

Of course, the percentages can vary somewhat due to particular needs of a given project, but the relative relationships remain the same. Changes in one area usually affect others.

Forty-nine to seventy-five percent of the cost of the work involves the compatible elements of MEPS (Mechanical, Electrical, Plumbing, and Sprinklers), exterior skin, and the structure. Since the architectural features only involve 15% to 25% of the cost, it is best not to start with these items. It is more prudent to begin with the items that will produce more return for the effort.

The mechanical/structure/exterior relationship is key to the need for mechanical spaces between stories, and below the building floor and roof framing. The use of an air system would necessitate more mechanical space above the ceiling than the use of another system that does not require large duct to provide conditioned air. The choice of systems varies the space requirements and thereby the structural heights, sizes of the structural members, and the surface area of the exterior skin. Figures 19.1 through 19.3 show some of the variations.

Figure 19.1 shows a variable air volume (VAV) system with a steel beam and bar joist structure. The space (above the ceiling and below the steel structure) required for the ductwork develops a floor-to-floor height of 14'-6".

Figure 19.2 shows the same VAV system, but with a reinforced flat plate structural design. The clear space above the ceiling remains the same, but the floor-to-floor height can now be reduced to 13'-10".

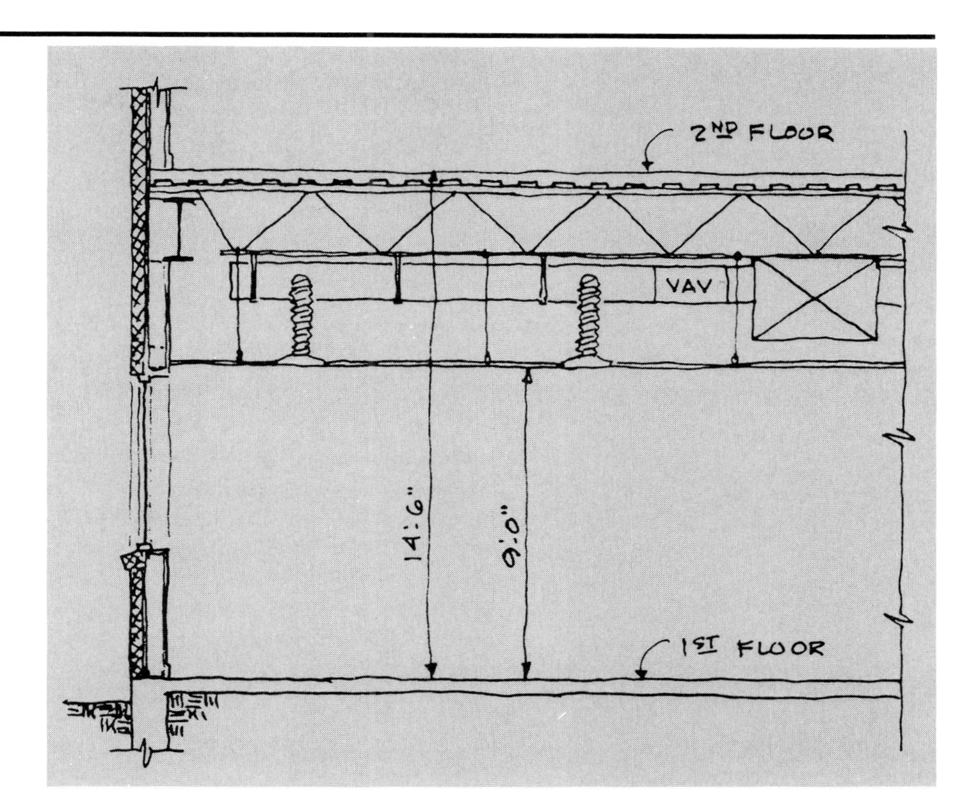

Figure 19.1

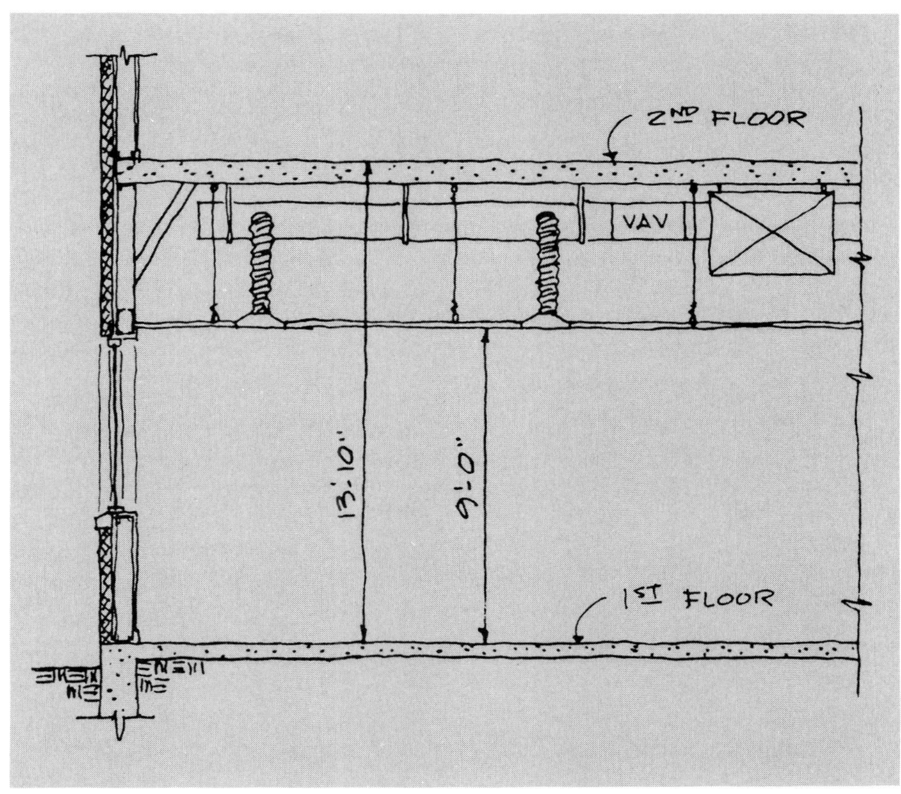

Figure 19.2

Figure 19.3 shows a heat pump system and a beam and bar joist structural system having the heat pumps distributed around the floor. The duct work remaining is only the distribution from the heat pump to the registers. The floor-to-floor height is 12′-8″. If the flat plate reinforced concrete structure is used, then the floor-to-floor height can be further reduced to 12′-0″.

The figures may imply that the obvious, economic solution is for all buildings to be designed using the latter method of flat plate structure and heat pumps. If that was the case, all commercial buildings would be designed this way. While many buildings do use those elements, it is not a solution without a price. The following chart is a comparison of the four systems illustrated.

System	First Cost	Maint.	Life Span	Long Term Cost	Comfort Level	Flexibility for Change
VAV/Steel	HI	Low	Bldg.	Low	HI	Easy
VAV/Plate	Higher	Low	Bldg.	Low	HI	Easy
Pump/Steel	Lower	HI	10 Yrs.	HI	Moderate	Moderate
Pump/Plate	HI	HI	10 Yrs.	HI	Moderate	Moderate

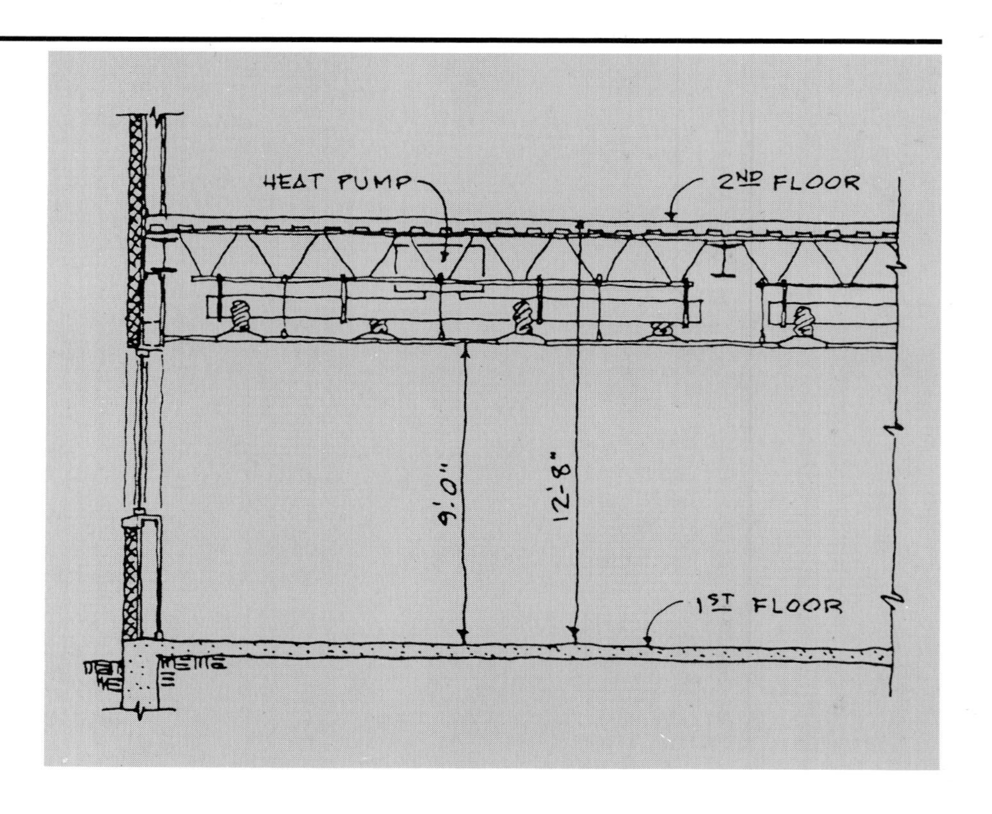

Figure 19.3

The decision rests with the owner. His interest in first cost and long term costs needs to be defined. His interest in flexibility and long term economies as they relate to income will have a bearing on the selection of certain systems. The suggestions can be studied to indicate all of the benefits and their respective costs. When the costs (based on the engineered study) are presented, a decision can be made. The studies for each system are factored, quantified, and estimated. The items to be examined are as follows:

Direct Costs	**Indirect Costs**
HVAC	Floor Efficiency
Structural	Noise Factor
Exterior Skin	Ease in Tenant Conversion
Energy Use	
Maintenance	
Replacement Cost	

The direct costs are self explanatory. The indirect costs are other factors to be considered.

The floor efficiency is affected by the mechanical space that is needed for machinery and not available for tenant use. The VAV system requires a room for the Air Handling Unit (AHU) on each floor. The noise factor affects the quality level of the rented space if the building is to be used for a commercial venture such as an office building. The ease of tenant conversion is another factor for a commercial venture. If there is some difficulty in modifying the selected system for use by different tenants, that difficulty can be estimated and will be another factor in the decision.

As the HVAC suggestion is resolved, the electrical system can also be examined. The cost saving suggestions might involve different quality or numbers of lighting fixtures, and the use of aluminum wire. The cost of the plumbing system may be reduced through the use of different pipe routing schemes and the quality level of the plumbing fixtures.

In addition to a reduction in cost based on the height of the building, the type of exterior skin can be examined and alternatives suggested. Finally, the interior of the building offers several possibilities for changes in design and quality.

In all cases, the process of suggesting alternatives and obtaining approval for the use of particular systems or materials is a collaborative effort. The creator of the building concept works together with the owner, the appropriate designer, and the design/build company to arrive at these decisions. Once changes are selected and the designs modified to reflect those changes, a change order, based on the firm prices received, is prepared and submitted to the owner for approval.

The project engineer maintains a key role in the value analysis process. It is the project engineer who expedites the design schemes and transmits them to the appropriate subcontractors for comment and pricing. It is also his responsibility to receive the returned data and assemble it in a presentation format. Along with his other tasks, he is the responsible member of the design/build team whose role is completely operational. He has little business contact with the owner or subcontractors. His job is directly linked to the conditions of the contract and the design of the building.

Chapter Twenty

Project Accounting

Several techniques may be used to keep an accounting of the project, starting with the first costs that are estimated at the conceptual stage. Costs that are gathered for the conceptual estimate may be reorganized according to the CSI format. Performance and allocation of men and material can then be measured against this standard. The health of the project can be monitored by comparing the actual to the estimated performance. This comparison is, of course, only as valid as the data is accurate.

There are a number of computer programs available for cost accrual and reporting. Selecting the one best suited for a particular company will depend on the company's needs and corporate make-up. The base program should have the capability to provide different *sorts* of the same information. The program should also serve other departments within the company.

Reports

There are two reporting functions in the construction business – accounting and management. The accounting reports deal with the prescribed accounting procedures using ledgers and balance sheets. The accounting reports would carry all of the company activities in one of two ways: lumped together or separately by job, depending on the type of tax reporting used. The management reports are usually on the basis of individual job reports. The minimum reports necessary for a design/build company are as follows:

- Project Management Report
- Job Cost Report
- Vendor Status Report
- Purchase Report

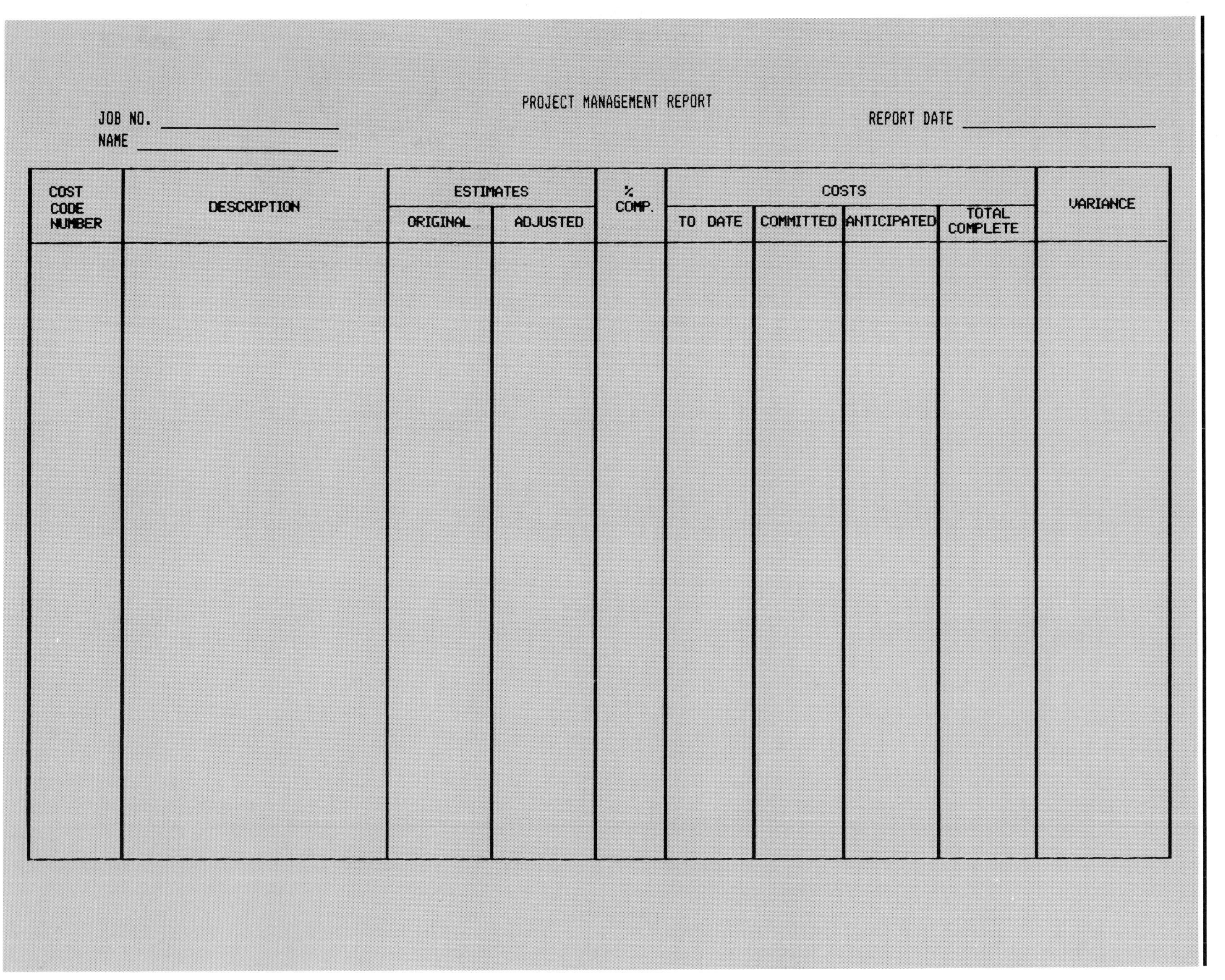

PROJECT MANAGEMENT REPORT

JOB NO. ______________

NAME ______________

REPORT DATE ______________

COST CODE NUMBER	DESCRIPTION	ESTIMATES		% COMP.	COSTS				VARIANCE
		ORIGINAL	ADJUSTED		TO DATE	COMMITTED	ANTICIPATED	TOTAL COMPLETE	

Figure 20.1

Project Management Report

The project management report is a summary of data from all other reports. See Figure 20.1 for an example project management report form.

Cost Code Number

Whether the system is computer based or manual, a method is needed for keeping track of the work items. Assigning a unique cost code number to each item accomplishes this purpose. The cost code number is also used to help organize the conceptual estimate costs according to the CSI format. If all of the work is subcontracted, then the headings relate to trade categories of work. If some of the work is to be done by the design/build company, the following items might be listed: form foundation wall, pour foundation wall, form slab on grade, place and finish slab on grade, etc.

Description

In addition to the number identification of a work item, the description is also helpful. In this column, one might also specify whether the item is labor, material, equipment, or subcontractor.

Original Estimate

The original estimate is the contract amount derived from the conceptual estimate. Once established, this estimate is never changed.

Adjusted Estimate

The adjusted estimate is used for the purpose of including change orders that increase or decrease the estimate. This column is also used for adjustments to the estimate such as the re-allocation of costs from one category of work to another. The re-allocation of cost is necessary to reflect the purchased or anticipated scope of work. For example, if the exterior caulking around the windows had been estimated and collected with the waterproofing work, but now will be assigned to the window installer, that cost would have to be re-allocated to the different trade. Staging work originally estimated with masonry is now charged as a general conditions cost item. All of these adjustments can be made in the adjusted estimate column where there is a clear tract back to the original estimate.

Cost to Date

The cost to date is collected in this column based on data from the other reports. These costs can be entered upon receipt of the bill or when paid. Subcontractor information is usually 30 days out of phase and vendor information about two weeks out of phase.

Percentage Complete

The percentage complete is a mechanical feature of a program. This feature takes the cost to date and divides it by the committed cost times 100.

$$\frac{\text{cost to date}}{\text{committed cost}} \times 100 = \text{percentage completed}$$

Committed Cost

The committed cost is the cost for which the item is contracted. The value of the subcontract is included in this column. Expenditures against this committed cost are included in the *cost-to-date* column. Comparison of the cost to date and the committed cost gives a percentage of completion for the contract.

Anticipated Cost

This is a manually input column used by the project manager to designate the costs yet to be expended to complete the work. The skill and judgment of the project manager and his knowledge of the job are reflected in this column. As these costs are incurred and assigned, they are transferred to the *committed* column.

Total to Complete

The total to complete is the sum of the committed and anticipated cost columns. The amount stated may be more or less than the adjusted budget.

Variance

The variance is the positive or negative difference between the adjusted estimate and the total to complete. The resulting total should represent a savings if the project is going well. The report items are totalled by CSI division and a grand summary of all divisions is made at the end. The variance report is the management tool needed to forecast and monitor the cost progress of the job.

Job Cost Report

The job cost report (Figure 20.2) is needed if the company performs certain work with its own labor force. If all of the work is subcontracted, the only information to report would be on general conditions labor and material, such as the superintendent, survey crews, clean-up crews, and supplies.

Cost Code

The cost code in this report is the same as that reported in the project management report.

Description

The description column is the same as in the project management report. The items of work, as coded, are the same.

Activity

The activity column is used to designate labor (L), material (M), equipment (E), and miscellaneous (X) cost items.

Estimate

The estimated quantity, unit price, and amount are listed in separate sub-columns for this category.

Actual

The actual quality of work performed and the amount of the work is recorded. The computer then calculates the cost per unit. The comparison of the actual cost per unit to the original estimated unit cost is what is monitored.

Totals

The actual totals are listed in the project management report in the *cost to date* column. All of the information contained in the actual report is entered into the project management report. The job cost report is useful to the project superintendent as an indicator of the performance of his men.

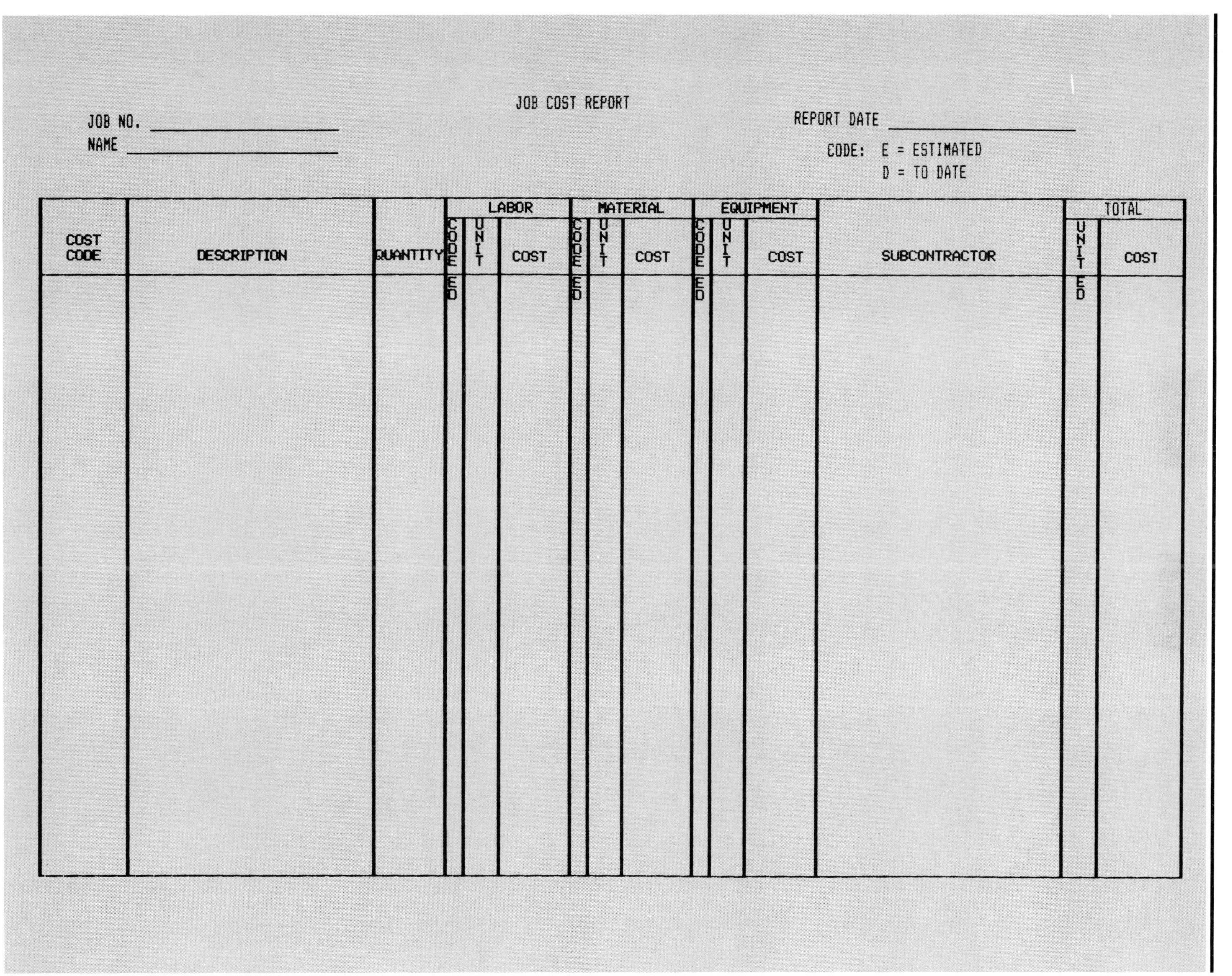

JOB COST REPORT

JOB NO. ____________

NAME ____________

REPORT DATE ____________

CODE: E = ESTIMATED
D = TO DATE

COST CODE	DESCRIPTION	QUANTITY	LABOR			MATERIAL			EQUIPMENT			SUBCONTRACTOR	TOTAL	
			CODE	UNIT	COST	CODE	UNIT	COST	CODE	UNIT	COST		UNIT	COST
			E D			E D			E D				E D	

Figure 20.2

Vendor Status Report

The vendor status report is used to list the status of accounts payable. Although this is an accounting function report, it is useful as a management tool to see in more detail the status of the subcontractors and the material/equipment supplies.

This report is usually listed by vendor number or alphabetically. The vendor number sort is most commonly used. The format is as shown in Figure 20.3. The vendor's number is listed in the *vendor number* location, followed by the vendor's name.

Purchase Report

The purchase report information is not needed for the management report, but it is used in conjunction with the management schedule. This report Figure 20.4 sorts by job, by cost code, and by early start.

The purpose of this report is to provide an as-needed purchasing schedule if all of the projects are to be purchased by one person or department. Sorting by early start dates shows the intensity of the buy from the most critical to the least for all of the jobs. The schedule helps insure that an item of work will be available at a particular job when needed, with no job being sacrificed as in the "buy-each-job-out-completely" method.

Cost Accrual

The gathering of costs is called *cost accrual*. The information is collected in a number of forms. The subcontractor presents a bill called an *application for payment*. These invoices are usually submitted at the end of each month.

The vendors' invoices for material purchases are on their own invoice forms. These bills come in at no particular intervals, but shortly after the material is picked up or delivered.

Field labor costs are expended weekly as payroll payments. The records of the number of hours expended are gathered by a foreman or a superintendent and turned in once a day or once a week. If the project is labor intensive, a job accountant might be assigned to the project for the purpose of keeping the time, sending in the reports, and collecting and submitting the receipts for material. These receipts are used to check against the invoice for the material.

Forms can be devised to record labor information and come in as many different formats as there are companies. A form that can be used to record labor costs is shown in Figure 20.5.

VENDOR REPORT

REPORT DATE ____________________

VENDOR NO.____________________ NAME____________________

JOB NO.____________________ JOB NO.____________________

ITEM OF WORK____________________ ITEM OF WORK____________________

AMOUNT OF CONTRACT____________________ AMOUNT OF CONTRACT____________________

PAYMENT AMOUNT____________________ PAYMENT AMOUNT____________________

RETAINAGE____________________ RETAINAGE____________________

PAID TO DATE ____________________ PAID TO DATE____________________

PAY DATE____________________ PAY DATE____________________

CHECK NO.____________________ CHECK NO.____________________

VENDOR NO.____________________ NAME____________________

JOB NO.____________________ JOB NO.____________________

ITEM OF WORK____________________ ITEM OF WORK____________________

AMOUNT OF CONTRACT____________________ AMOUNT OF CONTRACT____________________

PAYMENT AMOUNT____________________ PAYMENT AMOUNT____________________

RETAINAGE____________________ RETAINAGE____________________

PAID TO DATE____________________ PAID TO DATE____________________

PAY DATE____________________ PAY DATE____________________

CHECK NO.____________________ CHECK NO.____________________

Figure 20.3

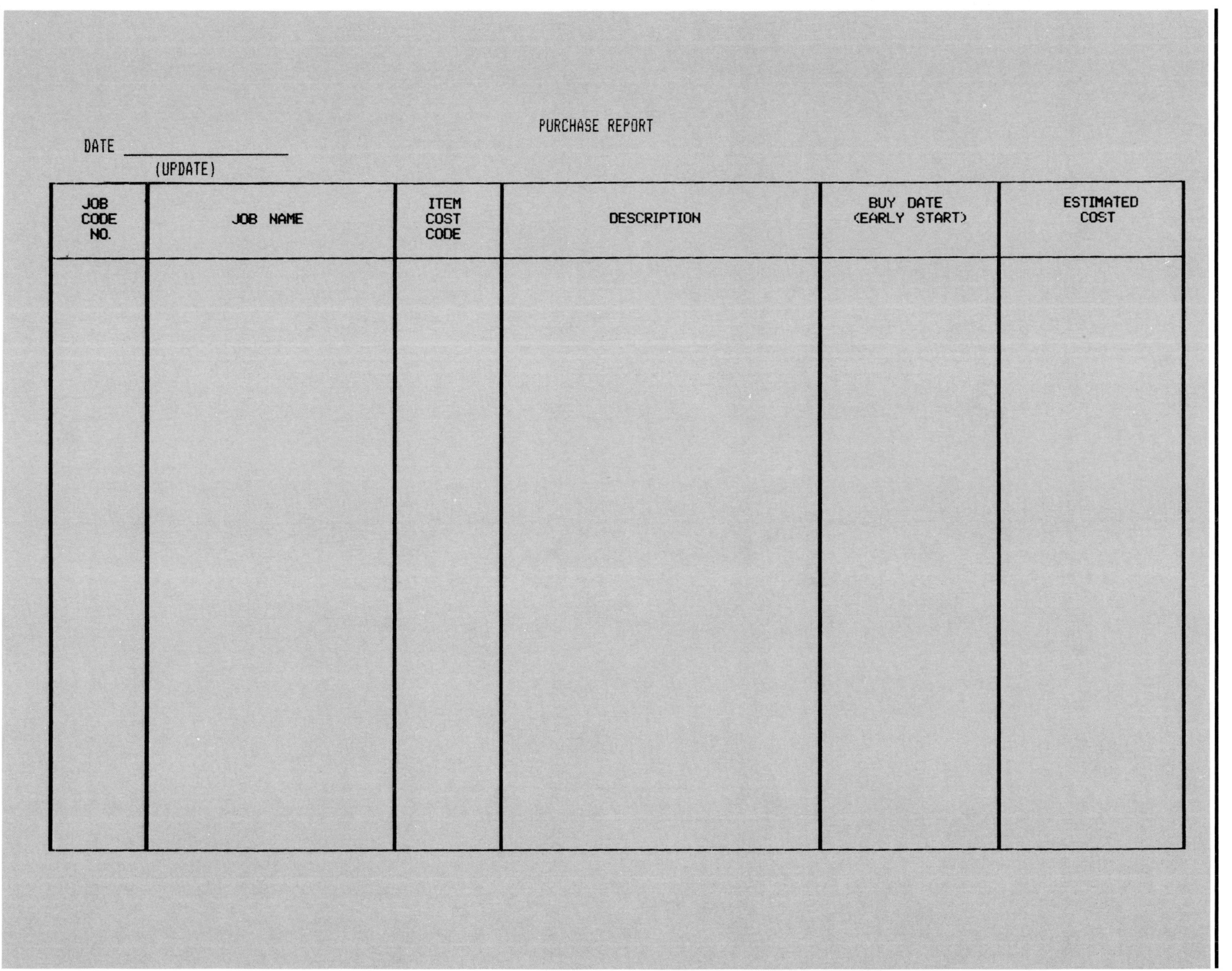

PURCHASE REPORT

DATE ______________ (UPDATE)

JOB CODE NO.	JOB NAME	ITEM COST CODE	DESCRIPTION	BUY DATE (EARLY START)	ESTIMATED COST

Figure 20.4

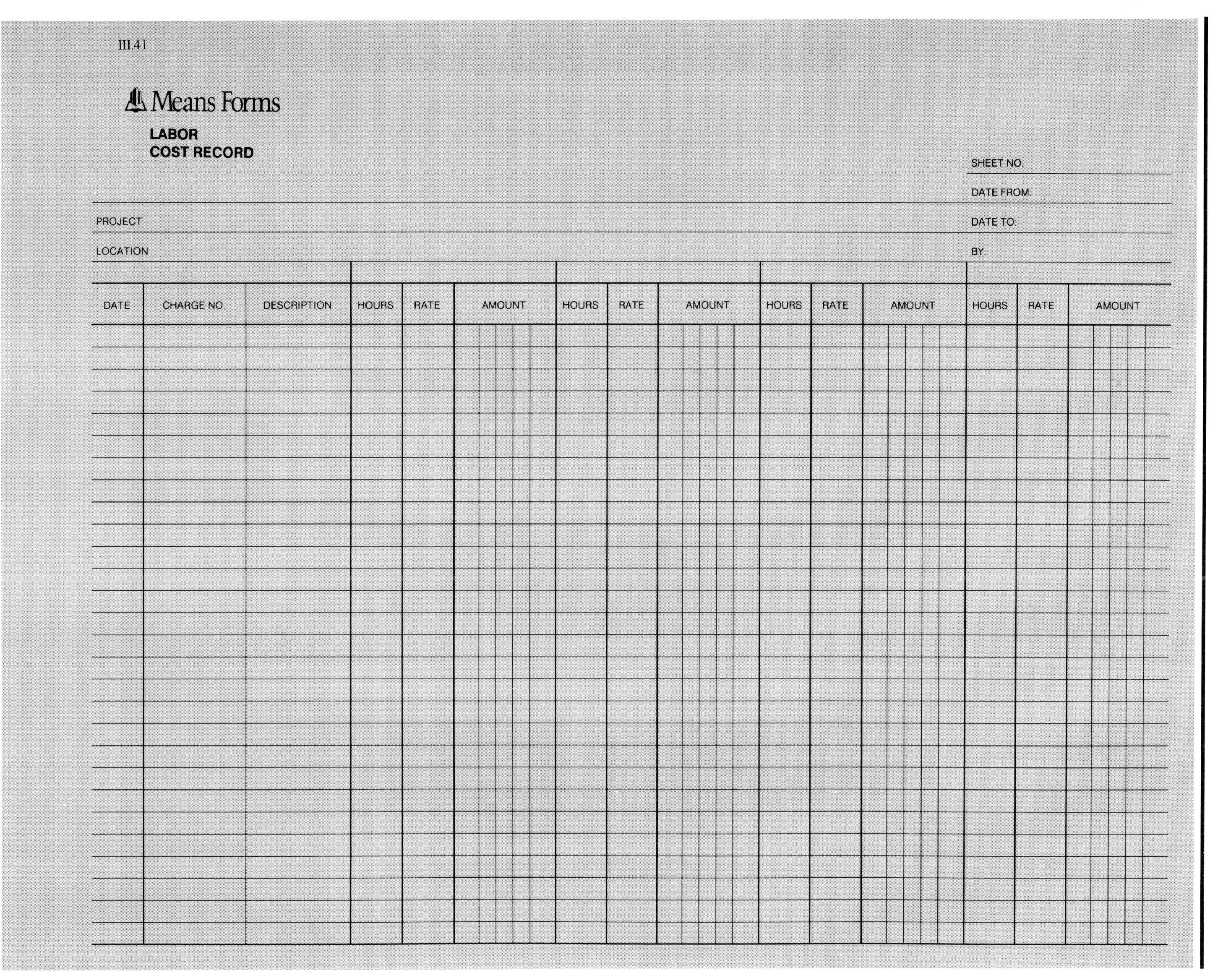

III.41

Means Forms

LABOR
COST RECORD

SHEET NO.

DATE FROM:

PROJECT

DATE TO:

LOCATION

BY:

DATE	CHARGE NO.	DESCRIPTION	HOURS	RATE	AMOUNT	HOURS	RATE	AMOUNT	HOURS	RATE	AMOUNT	HOURS	RATE	AMOUNT

Figure 20.5

Application for Payment

The application for payment is a partial payment submitted monthly as the project progresses. The form of the application can be the standard AIA Document G702 or a custom form used by the design/build company. The application for payment is set up in the same format as the cost reports. At the end of the month, all of the billings from the subcontractors are examined and listed in the appropriate location on the form. The work by the design/build company is judged for its completeness as a percentage of the total estimate for the work. This work is included in the application. The completed application is then submitted to the owner for review, and once approved, is sent on to the bank for payment. Many times, the approval of the application is by a representative of the bank who will inspect the job.

Payment is usually between 15 to 30 days after submittal. The amount paid is the total cost of the work to date less retainage (commonly five to ten percent) less previous payments.

When payment is made to the design/build company, the subcontractors are then paid based on the same conditions as the design/build company. Vendors are usually paid in 30 days of receipt of the invoice and not tied to the subcontractor/design/build contractor schedule.

Stored Material

One item of payment that may cause some problems is the issue of stored material. Material needed for the job can arrive earlier than anticipated or the delivery may be accepted earlier than needed. In such cases, the material can be stored at the site or off site. Problems arise when the subcontractor wants to be paid the material cost, but owners and banks are reluctant because of the liability for damage and pilfering. Material stored on the site constitutes a lesser risk in the eyes of an owner because it can be observed daily. Material stored off site is at greater risk because it is not visible.

The standard method for handling stored material is to require the subcontractor to secure separate insurance for the material in question, provide secure covering, and have all of the items tagged for identification. Material off site must meet the same criteria and also be stored in a secure or bonded location. Payment for stored material is by invoice, at cost less the usual retainage. Inventory records must be maintained as the material is used and is charged at the full value of the work.

Project accounting is the means by which the cost of a project is maintained. The forms and presentation of data can either be computerized or created and maintained manually. Good cost accounting is a contractual necessity, due to the nature of the design/build-client relationship. It is a practical requirement based on the principles of good business.

Chapter Twenty-One

CLOSE-OUT

Close-out is a term used in construction to indicate the tasks needed to complete the project. Close-out also involves the final documentation of the building.

As-Built Drawings

The as-built drawings are exactly what the term states. They are modifications to the construction design drawings, new drawings, or shop drawings that show the actual location (different from the design location) of elements of the work. The areas *most* affected are site work, mechanical, plumbing, sprinkler, and electrical. At the beginning of the job, a set of drawings to be used for as-builts (clean printed drawings or reproducible drawings) is issued to all subcontractors who require them.

For certain types of work, the job site coordination layout drawings are sufficient for as-built drawings. An example is the sprinkler piping and sprinkler head layout. The locations of valves, control devices, and equipment are shown on these layout drawings.

Plumbing routing of water and waste lines (along with the location of valves) are shown as modifications to the design construction drawings. Drawings of electrical routing of conduit and wire are not always required. There must, however, be a clear indication of the circuitry of the electrical distribution panels and the labeling of all electrical devices.

Site work involving different grading patterns and the relocation of drains and other site structures is noted on the design site drawings.

Recording information on as-built drawings aids the building's maintenance staff in locating the valves or switches in the future, after the contractor has left and the workmen have forgotten about the job. It is the responsibility of the superintendent to maintain the as-built drawings for the work of his own labor force and to monitor and insist, if necessary, that the other trades maintain them as well. It is the responsibility of the project engineer to receive and review the drawings, verify them by field observation, and deliver them to the owner as part of the close-out documentation. In addition, the owner is presented with one copy of all of the shop drawings, cuts, and samples submitted and approved.

Warranties and Guarantees

The construction of a building, like the manufacture of products, requires a period of guarantee, usually a year. The conditions of this guarantee are stipulated in the contract terms and conditions. Some design/build companies offer longer warranty periods, for an additional fee, much like the agreements offered for automobiles and appliances. In addition to the blanket one year guarantee which is required of the design/build company and all the trades, there are certain specific additional guarantees and warranties that may be required.

The roof system design, for example, requires an extended warranty from the manufacturer, provided through the installer. This warranty is issued for a certain period of time – usually 10, 15, or 20 years. The documentation for this warranty has to be secured and presented to the owner. Other items that require warranty documents include: planting materials, major equipment such as pumps and motors, glass, window systems, and carpeting.

In addition to the documents, the owner is given all spare parts, spare ceiling tile, spare floor tile, carpet remnants, wrenches, spare sprinkler heads, spare electrical fuses and the like as required by contract and the owner.

System Shakedown

The start up of the building mechanical systems and other operating systems is called a shakedown. Before the shakedown, the various trades should have presented to the design/build company all appropriate maintenance manuals and operating instruction booklets for all of the systems.

The ongoing system testing is done by the subcontractor as part of the installation. The shakedown is done as part of the acceptance of the system. The designer of each system is in charge of the shakedown procedure for that particular system. The design/build company coordinates the timing and assures the adequate representation of the subcontractors.

The mechanical heating/cooling system is the largest to be viewed, but other systems also have to be examined. The fire alarm system is tested for responsiveness and accuracy; each sensing location is activated and observed. The elevator is tested for operational efficiency and safety. The sprinkler alarms are tripped and the results noted. Pumps are started and the design rpm tested. Cycling of pumps is also noted. (Cycling is the process of one pump running for a while, then another back-up pump taking over, then the original pump running again – in a cycle.)

Electric service is tested along with the other systems. The connected power and the use of motors are all by electric energy. In addition, circuits are shut to see if the designations labeled on the distribution boxes are correct.

Instruction Period

Once the design engineers and the design/build contractor are satisfied that the building systems function properly as designed, an instruction period is arranged for the maintenance personnel to become familiar with the systems. At this time, the operation and maintenance manuals are presented to the owner. The instructions are given by the subcontractor under the observation of the design/build company. Valves are activated, controls are monitored and changed, and a general hands-on session is conducted. The subcontractor is available for further questions so that maintenance personnel can become familiar with the building. Many building systems are self sustaining and require little monitoring or replacement of parts. Learning about the system is therefore not a lengthy process.

Punch List

The project engineer supervises the punch list work and verifies completion with the subcontractors. The owner is then asked to sign off the corrected work.

Final Payment

Final payment will most often hinge upon the completion of certain tasks, listed below:

1. Presentation of as-built drawings
2. Presentation of warranties and guarantees
3. Presentation of operation and maintenance manuals
4. Completion of the punch list
5. Submission of a final application for payment
6. Submission of the Release of Lien Documents
7. Securing of the Certificate of Occupancy
8. Securing of all operational and use permits

Items 1 through 4 have already been discussed. Items 5 through 8 are explained in the following paragraphs.

5. Final application for payment

To receive final payment, the design/build company must submit a final payment request. This final request for payment is to include the agreed cost of the work plus the change orders. If change orders are outstanding, they need to be resolved prior to submission of the payment request.

6. Submission of the Release of Lien Documents

A Release of Lien is a legal document which is required of all subcontractors by the design/build company and, in turn, required of the design/build contractor by the owner. This document releases the design/build company and the owner from any obligations and liabilities related to claims, demands, and liens, by any subcontractor, vendor, or individual. A mechanic's lien placed on a project will cloud ownership and title will be refused by the title insurance company. Lack of clear title would prevent the securing of the permanent mortgage.

The release of liens also requires the signer to certify that all charges for labor, material, and services have been paid. The purpose of this document is to assure that the construction money dispensed in the course of the job has been properly redistributed to creditors. Once final payment has been made, the owner or design/build company should no longer be held responsible for outstanding bills.

7. Securing of the Certificate of Occupancy

A certificate of occupancy is a permit to occupy the building. It is issued by the building inspection services group that issued the original building permit. Receipt of this document is predicated upon the acceptance of the work by the various inspecting parties and signing off of same. Included in this group are the building inspector, electrical inspector, plumbing inspector, sprinkler inspector, fire department, and elevator inspector, to name a few. Conformance to parking requirements by zoning, health certification of the drinking water, or on site sewage disposal systems, egress conformance, and the like must all be examined finally as part of the process, and certified. The architect must also attest that the building has been constructed according to the design and in conformance to the building code.

8. Securing of all permits

An elevator use permit is a condition of acceptance. The sewage system use permit can also be a requirement. Any other permits required for the particular building or locale must be addressed and secured.

Maintaining Client Relations

When all of the close-out tasks have been accomplished, the owner and design/build contractor agree that the work is complete. The close-out period can be one of the most difficult in the building construction process. Getting subcontractors to complete unfinished work, attending to punch list items, and continuing to man the job with adequate personnel is a challenge. The additional pressure of an owner anxious to occupy the building may try the patience of all parties and could destroy in a short period of time all of the goodwill built up between the design/build company and the client during prior months. It is of great importance that the close-out activities be managed effectively, but also with patience and civility.

Part Five

INSURANCE AND BONDING

Chapter Twenty-Two

BONDING

Bonds are issued by surety insurance companies and act as an extension of credit. They are not loans. By providing a bond, the surety agrees to indemnify the obligee against default or failure of the principal.

The obligee – usually the owner – is the recipient of the bond. The principal is the company for which the bond is issued. This might be the design/build contractor or subcontractors. Many times the bonds are dual obligee which means that in addition to the owner, the lending institution is also protected.

Types of Construction Bonds

There are three types of bonds common to the construction industry. They are the Bid Bond, the Performance Bond, and the Labor and Material Payment Bond.

The bid bond is required on many public projects; it guarantees the bid amount. If the low bidder is in default, surety will pay the difference between the low bid and the accepted bid. This bond face amount varies from 25% of the bid to 100% of the bid.

The performance bond and labor and material payment bond are two separate bonds, but are most often required together. The American Institute of Architects (AIA) prints both forms as one packet. The performance bond guarantees that the contract will be performed. The labor and material bond protects against liens resulting from unpaid materials and labor.

Need for Bonds

Bonds are necessary in construction to insure the completion of the project. Design/build, by its nature, often requires bonds in order to do the work. This is because it is a negotiated project based on concept plans and scope. Trust is an important element in such a project. If a client/owner did not insist upon a bond, then the lending institution would. Design/build construction is not singled out as the only method that requires bonds. Turnkey and construction management also require bonds.

Qualifying for Bonds

To qualify for bonding, the design/build company must meet the standards of the surety. There are several factors that influence this approval process. Prominent among these factors are *character*, *capacity*, and *capital*. Character is revealed by an examination of a company's reputation, integrity, professional ability, and key personnel. Capacity is determined by a review of prior jobs, equipment, physical plant, management systems, and financial systems. Capital is determined by an examination of the last three years' audited financial statements, bank credit, and relationship with suppliers.

When the company is accepted for bonding, the limit of bonding capacity is set. This bonding capacity, also known as the work program, is the maximum value of uncompleted work that the contractor can have on hand at one time. This value is computed from the value of new contracts plus the remaining work on active contracts.

The amount of the work program is determined by multiplying a factor of from 10 to over 30 times the net worth, depending upon the contractor and his surety company evaluation.

When a contractor applies to a bonding company for a bond, the surety will examine:

- essential characteristics of the project, including hazards associated with the work, the identity and reputation of the owner, and the owner's ability to pay.
- total amount of uncompleted work on hand (bonded and unbonded). This review is made to determine if the contractor is overextended in working capital, equipment, and organization.
- adequacy of working capital and availability of credit.
- amount of money spread between the contractor low bidder and the next bid (amount left on the table). If the spread is more than five to six percent, the surety would be concerned.
- if this is the largest contract the contractor has taken, and to what degree he should exceed the limit of his experience.
- terms of the contract.
- amount of work to be subcontracted and the quality of the subcontractors.

Bonding of Subcontractors

The bonding of subcontractors is a way of easing the concerns of the surety. The lending institutions will usually go along with this arrangement and accept the subcontractor's bonds in lieu of the contractor's bonds. Many design/build companies require a bond from certain subcontractors as a company rule. The rationale for requiring a bond may be as simple as establishing a dollar limit. For example, all subcontractors whose contract amount is over $30,000 may be required to be bonded. The process can be somewhat complicated if, for instance, the subcontractors are required to submit a financial statement or if a ratio analysis must be performed. Those subcontractors whose ratios fall outside of the established guidelines would be obliged to furnish bonds. The simplest approach of all is to require that all of the subcontractors (without exception) furnish bonds.

The cost of the bonds will vary through a sliding scale of percentages from 3/4 to 1-1/2% or more of the amount of the contract. When estimating the cost of the work, the subcontractor bond cost is usually included at cost without markup for overhead and profit.

Other Bonds

Construction bonds are not the only bonds available. Others worth mention are: the *Maintenance Bond* (obtained by the contractor), which assures warranty compliance, and the *Lien Bond* (obtained by the owner), which is needed to protect the owner against the filing of a mechanic's or materialman's lien. A mechanic's or materialman's lien when properly perfected, gives the unpaid supplier or contractor a security interest in the property itself, constituting a claim against the owner's title to the property to the extent of the amount owed for labor or materials. The lien bond assures the capital to settle the claim and thereby re-establishes clear title.

A *Fidelity Bond* is another type, the purpose of which is to protect the owner against dishonesty by the contractor's employees. Bonded bank employees are also under a Fidelity Bond.

Finally, a *Subdivision Bond* guarantees construction of all improvements and utilities that are required as part of the approval of the subdivision.

Chapter Twenty-Three

INSURANCE

The design/build company, like any other construction company, must carry liability insurance. As part of the contract agreement, the design/build company has to furnish a certificate of insurance to the owner attesting to the fact that he is insured. The limits of the insurance are stipulated by agreement and the certificate is issued by the insurance company. All of the subcontractors working on the project furnish certificates of insurance to the design/build company in the same limits as those furnished to the owner by the design/build company.

General Liability Insurance

General Liability is a term to include a number of coverages such as:

- Comprehensive Form
- Premises Operations
- (XCU) Explosion, Collapse and Underground Hazards
- Completed Operations/Products
- Contractual Insurance
- Broad Form Property Damage
- Personal Injury

Comprehensive Form

The comprehensive form of General Liability is what covers all of the other types stated.

Premises Operation

Premises operation refers to the protection for damages done on the contractor's or subcontractor's premises. All of the coverage that a construction project would require is for the construction site, not at the business premises of the constructors. This is usually an automatic coverage which should be included but is not essential.

(XCU) Explosion, Collapse, and Underground

This coverage is important and should be mandatory for all trades. Explosion is generally thought of in terms of blasting operations, but it can be any explosion. Collapse hazard coverage usually applies to excavation contractors. Underground usually pertains to excavation contractors as well, and relates to the hazards of underground work, such as for utilities. While the coverage may appear to be specifically targeted to excavators, there are many instances in which others could be involved. For example, a graphics contractor hanging a sign in the building could drill into a gas pipe and cause an explosion. XCU coverage is included with trades that have little exposure to this risk; it is usually excluded for the trades that have a high exposure risk. For the former, it is a matter of routine and for the latter, a matter of requesting this coverage.

Completed Operations/Products

Even after a project is completed and turned over to the owner, the design/build company may continue to have liability for defects in design, materials, or workmanship. Completed operations coverage applies to such claims. A loss may occur years, or even decades, after completion of the project. In most states the applicable statutes of limitations for personal injury or property loss do not begin to run until the loss occurs. In such cases, if legal liability can be established against the design/build company based upon defects in design, materials, or workmanship, the completed operations policy will defend such claims and cover such losses. In many states design professionals are afforded additional protection by so-called "statutes of repose" which cut off liability for design defects after the passage of a specified length of time, usually six years, after completion of the project. The products liability part of this coverage may be applicable to claims arising from defects in components installed in the structure.

Contractual Insurance

The indemnity-hold harmless clause in the subcontracts assumes the owner's and contractor's liability for their portion of the work. The subcontractors buy contractual insurance to cover liability that they cannot afford to assume themselves. This form of insurance is a must requirement of subcontractors, many of whom do not have enough financial resources to back up their contractual obligations.

Broad Form Property Damage

Broad form property damage expands the definition of what a person is working on. There is an exclusion in every policy for damage to that portion of the project that is being worked on. Builder's Risk insurance covers that work. That is why if the building burns down and the contractor is at fault, property damage liability will not pay for it. Builder's Risk insurance will. This coverage expands the definition of custody and control. This means that there is no coverage for areas of work that you control. If you are working on a specific wall, there is no coverage for it. There is just coverage for the rest of the building.

Personal Injury

Personal injury is an expanded definition of what was formerly known as "bodily injury". Bodily injury is self-explanatory. Personal injury applies to direct damage and also includes elements such as mental anguish, libel, slander, and claims brought by the spouse or other relatives of the injured person.

Motor Vehicle Insurance

The next type of insurance is automobile liability, which covers the use of vehicles. All subcontractors are to include this coverage. In addition, all vendors making deliveries to the project should demonstrate automobile coverage before being allowed to enter the job site. If a concrete truck should come onto the job and run into the building or a worker, the design/build company should be absolutely sure that the concrete vendor can show financial responsibility for such an occurrence through insurance coverage.

Where employees operate their own motor vehicles in the course of their employment and a loss occurs, the employer can often be held liable under the doctrine of *respondeat superior*, under which a master is made responsible for the negligent acts or omissions of his servant. For this reason the comprehensive motor vehicle policy, which applies to claims arising from the ownership, maintenance, and use of owned, hired, and non-owned motor vehicles, is recommended.

Worker's Compensation Insurance

Worker's Compensation insurance is required by federal law; the benefit amounts are also determined by law. Worker's Compensation has two sections. One is the worker's compensation and the other is the employer's liability. This type of insurance is intended to be the sole remedy for the employee who is injured during the course of his work, regardless of who is at fault. Worker's Compensation was the first no-fault insurance. Even if the worker is doing something absolutely contrary to instructions or safe work habits when he is injured, worker's compensation pays – no questions asked.

The employer's liability is included in order to take care of the instance where an employee waives his rights under worker's compensation and chooses instead to file a law suit. The limits of this coverage are small, in the range of $100,000 to $250,000, because of the necessity to prove gross negligence. This coverage is also included under the umbrella coverage in most policies.

Umbrella or Excess Insurance

The umbrella or excess liability coverage provides additional protection. When a general policy contains umbrella coverage, the total coverage of the policy is the sum of the two. For example, general liability is for one million dollars and the umbrella coverage is for four million dollars. The total coverage is for five million dollars.

Cancellation

Certificates of insurance contain cancellation clauses. The standard time allowed is 10 days, but the recommended time is 60 days. If a subcontractor gets into a bind and his insurance is cancelled, what can the design/build company do? The subcontractor cannot be allowed to continue work without insurance, yet the work must be completed. A 60-day period gives the subcontractor a chance to secure other insurance or the design/build contractor the opportunity to secure another subcontractor; in the meantime, the subcontractor continues to work. The subcontractor who has his insurance cancelled and cannot get other insurance is in non-compliance with his contract, so his contract can be cancelled. His bond can also be acted upon to satisfy any losses.

Limits of Insurance

The business of establishing insurance limits presents certain problems. Few have the foresight to be able to predict future influences on cost. The standard that is generally used is based on the estimated replacement cost of the building. A construction project the cost of which is $70,000,000 would have liability coverages of up to $18,000,000. Why not the full amount? The cost of this much insurance would be prohibitive. The amount is established by convention and probabilities.

Builder's Risk Insurance

Builder's risk insurance is carried by the owner. The coverage is for the building and is intended to insure everything that is not otherwise covered by liability insurance. Along with builder's risk, the owner carries a general liability policy.

Builder's risk is limited to the replacement value of the building. The deductible is usually in the range of $25,000 to $35,000. Claims and damage resulting from defective design, workmanship, and material are not covered. Flood and earthquake protection is usually added coverage.

The general liability includes premises/operations, products/completed operation, blanket contractual liability, broad form property liability, personal injury, worldwide products, incidental medical malpractice, and non-owned watercraft. The latter two are included as a matter of course.

Special conditions of the owner's insurance state that the owner is to be named as additional insured under the subcontractor's general liability policies and the design/build contractor's general liability policy.

Builder's risk coverage includes materials, supplies, machinery, equipment, fixtures, and temporary structures to be used in the construction, fabrication, installation, or erection of the property.

The policy does not cover any item not destined to become a permanent part of the building, except forms and scaffolding. Drawings and personal property such as tools are not covered.

Errors and Omissions

The design/build company is also required to carry errors and omissions insurance. This is a type of professional liability insurance that protects against the hazard of errors in the design, or omissions in the design and elements of the building. As a rule, only design professionals need to carry this form of insurance. However, since the design/build company takes responsibility for the design and is involved with coordination and other aspects, they also must carry it. A $1,000,000 coverage is standard.

Part Six

THE OWNER'S PERSPECTIVE

Chapter Twenty-Four

OWNER'S SELF HELP

The design/build construction method accomplishes many things from an owner's point of view. It provides the most economical means of construction because of the following factors:

- fixed construction costs
- early established construction financing rate
- earlier use of the building (fast track)
- one line of responsibility for design and build

Owner's Responsibility

A successful design/build process is accomplished through a cooperative partnership with the owner, whose responsibilities include providing:

- clear information
- identification of goals
- a realistic budget
- direct participation in the design

Clear Information

The more information that is available in the data gathering phase, the better the design concept and the conceptual estimate will be. Following are some examples of the information that an owner can furnish:

- A contour map of the property or the authorization to create one.
- Results of soil borings and test pits.
- Identification of various space needs.
- A count of current employees and an estimation of future growth.
- A product flow chart.
- Insurance underwriter requirements and fire protection.

Identification of Goals

The owner needs to identify his image of the company and ideas for future expansion. These concepts can be reflected in the design.

Realistic Budget

There is a natural tendency to have the wants far exceed the resources. The car buyer who looks at the luxury cars while only having the ability to finance an economy car is not unlike the owner of a building project in the conceptual stage. The amount of money available for the construction is, however, predetermined and fixed. Whether it is a commercial venture the cost limits of which are based on rental income, or a small manufacturer whose more efficient operation will bring in more profits to apply to development costs, they all want more building than they can afford. The establishment of a realistic budget of needs/wants versus resources has to be done and can be very sobering. An owner needs to arrive at that realistic budget.

Direct Participation in the Design

The owner should be able to provide quality levels for different parts of the building. The owner is the best source of the needs of his equipment. Issues of energy conservation, innovative design elements, and future changes will require input from the owner. An owner who is busy running his business may not have the time to be involved in the construction planning to the extent that may be needed. Yet, it is in the owner's best interest to be knowledgeable about the process, as he is ultimately required to make decisions. The owner, whose money is being spent, can be in control by being an active participant. To fill this role, the owner needs to be familiar with the process.

Selecting a Design/Build Company

The method for selecting a design/build company is the reverse of the process a design/build company uses to find a client. A list of names can be gathered from trade magazine ads, friends, associates, bankers, or insurance companies, etc. Once a list is made, then a selection process can be initiated.

The method of initial selection should be based upon interviews and the presentation of information. The interview should focus on past experience, years doing design/build construction, experience with buildings of the type proposed, and introduction of the personnel who may be assigned to the project. The presentation information consists of a list of past client references, a financial statement, and a statement on bonding capacity.

When the selection process is completed, the list should be narrowed down to one, two, three, or four candidates, depending on the confidence gained in the interview and the review of their past performance.

Managing the Proposal

Furnishing information to the candidates can be done in a collective meeting format. If secrecy is required, then each candidate is met with separately. If the selection has been limited to one design/build company, then secrecy is not an issue. If, on the other hand, more than one company is involved, then each meeting session has to be repeated.

All of the information that the design/build company might require is furnished during this period prior to the formal presentation. The resulting presentation finds the owner needing to make a decision and not having a way to make it. The owner has received from each design/build company a different design having different aesthetic features and different cost. Furthermore, the format of the presentations to each of the candidates will be different, and this makes comparisons difficult. To avoid this situation, the owner may devise a presentation format that, through a measure of uniformity, allows for some comparison. That presentation format can be as follows:

- General conditions
- Site Work
- Foundations
- Slabs
- Roofing
- Structural system
- Exterior skin
- Interior walls and finishes
- Equipment
- Specialties
- Plumbing
- HVAC
- Sprinkler
- Electrical
- Contingency
- Design Costs
- Fee

The estimate and the outline specifications are developed in this format. After the presentations by the various design/build companies, the owner has to examine all of the information. This examination involves the scrutiny of all of the documents.

The conceptual plans, the elevations, and the rendering are compared, and the features of each noted to see that they achieve the desired image and the intended goals.

A line-by-line comparison of the estimates is made and differences are found in the items included. A cost comparison of similar items is also made. The cost of the design service as a percentage of the total cost can be calculated in the same manner. The comparison of the line items cost per square foot exposes differences and is the basis for further questions. The alternates and allowances provided are also examined. Have they been completely explained?

In reviewing the outline specifications, the excluded items must be singled out and listed for further clarification. Remember that the design/build company is responsible for code compliance. Thus, the examination does not have to include consideration of the code nuances. Suggestions for other items to examine are as follows:

- Does the design/build company understand the site?
- Is the site cost all an allowance, or is the allowance only for rock excavation?
- Are the local zoning ordinances on parking, green spaces, and setbacks satisfied?
- Are the utility connects understood?
- Is there a size comparison of the foundations?
- Are special loading requirements included?
- What type of roof is included?
- What is the quality of the exterior skin? – cost per unit area?
- What are the types of interior finish?
- What are the types of interior walls?
- Can the number of toilet fixtures be counted? – how many?
- What are the sprinkler requirements? – number of sprinkler heads shown? – coverage in all of the areas?
- Is the electrical load of incoming service stated?
- What are the HVAC design criteria? – are special needs taken care of?

After this examination, lists of questions are made. These questions can be used in a second round of selection.

What if, as previously stated, the owner is a busy man and cannot devote the time that is necessary to this process? The owner does not need to participate in every step of this investigation. He can simply respond to questions and accept or reject the proposals.

If this is what is involved, why use the design/build method at all? The fact is that any other method requires the same design input time. The only difference would be in the analysis of the presentation. If one design/build company was selected, then only one presentation and proposal need be scrutinized for conformance to concept and cost. After the design, the other methods provide a consultant for the final selection, the architect, to help.

Using a Professional

The owner may want to hire a professional to work with him in the selection process. A professional can help an owner in one of two ways:

- He can translate the owner's requirements into an outline form which then can be used to guide the design/build companies in their concept design. This method will provide a more uniform submission and thereby, a simpler evaluation. It does, however, have a drawback in that it gives only one interpretation (the professional's), of a client's needs.
- The other method is for the professional to focus on the basic information exchange and then allow the design/build companies to make their own interpretation of the needs. This approach allows for different solutions and hopefully different creative designs. The resulting presentations are now at the same point of non-comparison. It is the professional's task to make the analysis for recommendation to the owner.

The professional will proceed through the same steps that were advocated for the owner – creating a presentation format and performing an analysis of comparison in the same manner. The difference is that the professional has the knowledge and experience to carry out this process efficiently with a greater degree of probable success.

Selecting a professional depends somewhat on the availability in a given area. Ideally, the best choice is a professional who has experience in all aspects of design/build.

Why would an engineer not be a suitable choice for this role? A practicing engineer's knowlege is limited regarding the total process; his role in such a process usually concerns one aspect.

Why not an architect? An architect who has not participated in the full design/build selection process is also not the first choice – for the same reason as the engineer. His role as a normal part of the design/build method is limited to the architectural design.

It is the owner who must ultimately make all of the critical decisions that affect the success of the building. Most find it more fun to be part of the process, caught up in the excitement and satisfaction of the achievement, rather than remaining on the sidelines. Owners who choose to forego that involvement have the options of trusting in the design/build company, or hiring a professional .

Part Seven

APPENDIX

Part VII

APPENDIX

Table of Contents

Appendix A
CASE STUDY

This section contains documentation for a case study design/build project. The case study serves as an illustration of the phases of a design/build project, and the methods used to gather and present cost and design information. These phases and methods are discussed in detail in Chapters 3-8 and 10-13.

Included in this study are the presentation documents and back-up documents based on information from the client and other sources. A code review and a specification-aid form to list initial requirements are also part of the case study documentation. The presentation documents include the presentation-quality drawings. Preliminary concept sketches are part of the support information. Elements of the case study are listed in presentation order as follows:

Presentation Documents

- Cover Sheet
- Area Study
- G.M.P. (Guaranteed Maximum Price) Summary of Costs
- Outline Specification
- Items for Consideration
- Conceptual Estimate
- Drawings
- Progress Schedule

Back-up Documents

- Code Review
- Spec-Aid
- Rough Sketches

The case study project is a proposed manufacturing facility consisting of office, light manufacturing-assembly, and warehouse space. The facility will be constructed on a 5-1/2 acre site.

The facility is designed to allow for future expansion.

Presentation Documents

The order in which these documents is presented may vary, depending on the format of the presentation. Normally, the documents – cover sheet through conceptual estimate – are bound in booklets, and each section is individually tabbed. The drawings and rendering, if included, are mounted on presentation boards and displayed on easels. The progress schedule is presented in a similar manner. Reduced size drawings may also be included in the bound folder.

Conceptual Cost Estimate

The original conceptual estimate is the basis for the GMP (Guaranteed Maximum Price). Some design/build companies include this document, in its entirety, as part of the presentation. If it is included, the conceptual cost estimate usually appears as the first document, following the cover sheet and the area study.

Outline Specifications

The outline specification for the case study project is the next document in the proposal package. The format of the outline specs is shown below.

Division 1	Foundations
Division 2	Substructures
Division 3	Superstructures
Division 4	Exterior Closure
Division 5	Roofing
Division 6	Interior Walls, Interior Finish
Division 8	Mechanical
Division 9	Electrical
Division 10	Supervision, General Conditions, Winter Conditions, Architecture, Engineering
Division 11	Special Construction
Division 12	Site Work

Schedule

A preliminary construction schedule is the next element included as a part of the presentation documents. The schedule is intended to show – graphically – that the design/build company has a scheme to meet the client's required dates of completion. A bar chart and a critical path method (CPM) schedule have been done for this case study project.

November 13, 1987

Research and Development Corporation
Design Street
Newtown, Massachusetts 01022

Gentlemen:

In accordance with the enclosed Project Summary, Floor Plan, and Rendering, we have arrived at a firm price of THREE MILLION, THREE HUNDRED EIGHTY-SEVEN THOUSAND DOLLARS ($3,387,000.00) to design, engineer, and construct your new building.

We project that an October 1, 1988 substantial completion date can be safely attained based upon a March 15, 1988 start of construction, which would require an early 1988 commitment by you in order to start engineering and order long-lead-time items.

We have developed a list of alternate items which we offer for your consideration.

We look forward to working with you and appreciate having the opportunity to submit this proposal.

Sincerely,

Design Build Company

Vice President

tn/VP

Enclosures

Proposal for
Research and Development Corporation
Newtown, Massachusetts

PROPOSED MANUFACTURING, WAREHOUSE
AND OFFICE FACILITY

November 13, 1987

Design Build Company

Proposed Manufacturing, Warehouse and Office Facility

Square Foot Totals

Assembly Area	24,590 S.F.
Warehouse and Staging	39,060 S.F.
Office	9,600 S.F.
Total	73,250 S.F.
Perimeter	1070' -8"

Research and Development Corporation
Newtown, Massachusetts

Proposed Manufacturing, Warehouse, and Office Facility

Cost Breakdown

November 13, 1987

1.0	Foundations	$ 97,600
2.0	Substructures	299,900
3.0	Superstructure	290,500
4.0	Exterior Closure	409,100
5.0	Roofing	188,000
6.0	Interior Wall/Finish	231,500
8.0	Mechanical	416,700
9.0	Electrical	365,800
11.0	Special Construction	60,800
12.0	Site Work	242,800
	General Conditions	376,100
		2,978,800
Design Fee 8%		238,300
		3,217,100
Fee 5%		160,900
Total Guaranteed Maximum Price		$ 3,378,000

1.0 FOUNDATIONS

- A reinforced concrete footing, pier, and foundation wall system as required to support the structural and building loads.
- The foundation wall shall be at dock height at the Shipping and Receiving Area and at grade in all other areas of the Building.
- The foundation system at the South wall of the Manufacturing-Warehouse Areas shall be designed to accommodate the future expansion.
- Perimeter foundation walls below grade at the Office Areas shall be insulated with styrofoam.
- All exposed concrete walls at the dock height areas shall be rubbed to a smooth finish.
- Concrete pads at the egress doors at grade level shall be provided.
- One (1) concrete stair shall be installed as shown on the Site Plan.
- The forming of five (5) pits for the installation of the dock levelers.
- A concrete truck dolley pad shall be provided at the Shipping and Receiving Area.
- All concrete shall have a compressive strength of 3,000 PSI.
- It is assumed that the soil will support a uniformly distributed load of 3,000 PSF.

2.0 SUBSTRUCTURES

- Fine grading for all floor slabs on grade.
- A six-inch (6") concrete slab on grade shall be provided at all areas of the Building excluding the Office Area.
- A three-inch (3") concrete slab on 28 gauge metal deck shall be provided at the Office Area.
- All concrete floors at the Manufacturing and Warehousing Areas that shall remain exposed shall have a smooth, steel-troweled finish.
- All slabs on grade shall be dowelled at construction joints and shall have sawcut control joints, to control shrinkage cracking.
- Construction joints and sawcut joints at the Manufacturing and Warehousing areas shall be filled with an epoxy material.
- A premolded expansion filler shall be installed at the perimeter of all slabs adjacent to exterior foundation walls and at the full height concrete block walls.
- All concrete shall be 4,000 PSI and 4,000 PSI light weight.

3.0 SUPERSTRUCTURES

- The structural steel system shall be steel column, beam and steel bar joist with steel floor and roof decking for the entire building, using Type 2C construction as defined by BOCA.
- Clear height to the underside of the low point of the steel at the roof shall be twenty-three feet, two inches (23'-2").
- The bay sizes shall be 30' x 40' as shown on the Floor Plan.
- The structural steel system for the roof shall be designed to support a uniform live load of 40 P.S.F.
- Mezzanine deck shall be 28-ga painted.
- Roof deck shall be 1 1/2" 22-ga FM-1 painted.
- The structural system at the South wall of the Manufacturing Warehouse Areas shall be designed to accommodate the future expansion.
- All structural steel shall conform to the latest design standards of the A.I.S.C.
- All steel joists shall conform with the "Standard Specification for Open Web Steel Joists" of the Steel Joists Institute.
- All structural steel, bar joists and metal roof deck shall be shop painted gray.
- We have <u>not</u> included any bridge cranes, hoists, crane runways, monorails, etc., nor the strengthening of any areas to accommodate same.

Miscellaneous Iron

- All dockleveler frames, lintels, sill angles, overhead and sliding door frames, concrete filled pipe bumpers at the overhead doors, ladders, roof frames for mechanical equipment and other items of trim or reinforcement shall be furnished and installed.
- Ladders shall be installed in accordance with OSHA specifications.
- One steel pan egress stair to mezzanine.
- One architectural treated open stair to mezzanine.

4.0 EXTERIOR CLOSURE

- The exterior walls shall be constructed of 4" brick with 8" concrete block backup. (Brick Allowance $250.00 per 1000 Brick).

- The expansion walls shall be painted 12" concrete block.

- At the Office Area a five foot (5') spandrel of tinted insulated glass shall be provided.

- The window at the Shipping Area shall be fixed.

- Overhead and egress doors as shown on the Floor Plan.

- The five (5) 8' x 9' doors at the Shipping and Receiving Area shall be steel panel overhead doors manually operated.

- The two (2) doors at the Main Entrance and Vestibule and the (1) door at the egress shall be aluminum and glass.

- Egress doors shall be 18-gauge hollow metal set in 16-gauge pressed metal frames and shall be equipped with panic bars. All doors shall be painted.

- Exterior hardware shall be heavy duty as manufactured by Sargent and Company.

<u>5.0 ROOFING</u>

- A Johns-Manville, No. 801, or equal, four-ply, Class I Factory Mutual approved, asphalt and gravel built-up roofing system, with one and five sixteenths inch (1-5/16") fiberglass insulation, shall be installed for the entire building.

- For a distance of four feet (4') from the edge of the roof the fiberglass insulation shall be nailed to the roof deck, and the joints of the insulation shall be taped.

- Gravel stops, necessary flashing, cant strips, expansion joints, and wood blocking as required.

- The roof shall be pitched at a rate of one-eighth inch (1/8") per foot to the interior roof drainage system.

- One (1) roof hatch shall be provided to gain access to the roof.

6.0 INTERIOR WALLS

Interior walls shall be constructed of the following materials:

- The walls in the Manufacturing Area and Warehouse Area shall be constructed of eight-inch (8") concrete block installed to ten feet (10') high.

- The common Warehouse/Office wall shall be constructed of eight-inch (8") concrete block installed to the underside of deck.

- The Assembly/Warehouse demising wall shall be constructed of eight-inch (8") concrete block installed to the underside of deck.

- The common mezzanine wall with the assembly area shall be constructed of 3 5/8" metal stud with 2 layers of 5/8" F.C. each side of the underside of deck, and include sound attenuation insulation.

- All walls in the Office Area shall be constructed of three and five-eights inch (3-5/8") metal stud and one-half inch (1/2") sheetrock, each side, nine feet (9') high through the finished acoustical ceiling.

- The interior surface of the exterior and interior masonry walls at the Office Area shall be furred and receive sheetrock.

- All remaining metal stud and sheetrock walls shall be painted.

- Walls in Warehouse toilets shall receive Tru-Glaze. Walls in Assembly and Office toilets shall be full height ceramic tile.

- Metal sash with one-quarter inch (1/4") tempered glass shall be installed at the Shipping Office and Guard Station.

- Four (4) 8' x 8' overhead doors in the demising wall.

- All pass doors within the masonry walls shall be 18-gauge hollow metal set in 16-gauge metal frames.

- All doors within the metal stud and sheetrock walls shall be particle-filled, solid-core birch set in 16-gauge metal frames.

6.0 INTERIOR WALLS (Cont'd.)

- All interior hardware shall be heavy duty.
- All doors and frames shall be painted.
- Ceiling and walls of the Assembly Area to be painted.

INTERIOR FINISH

- A suspended 2' x 4' Class "A" five-eighths inch (5/8") thick, noncombustible acoustical ceiling shall be installed at the Office Area and Offices in Warehouse Area.
- The floors in the Office Area shall be carpeted, complete with four inch (4") high rubber base. An Allowance of $15.00 per square yard has been included for this carpet.
- The floors for the Warehouse Offices shall be 12" x 12" x 3/32" vinyl composite floor tile (VCT) complete with 4" high rubber cove base.
- The floors in all toilets shall be unglazed 1" x 1" x 1/4" ceramic tile complete with ceramic tile cove base.

8.0 MECHANICAL

PLUMBING

The Plumbing System shall be designed to meet governing codes and shall consist of the following:

- An interior roof drainage system consisting of roof drains and piping to ten feet outside the foundation wall.
- An interior waste and vent system from plumbing fixtures and floor drains to ten feet outside the foundation wall.
- An interior domestic water system beginning ten feet outside the foundation wall, including a water meter, with piping distribution to serve plumbing fixtures.
- An interior domestic hot water system, including hot water heaters, with piping to serve plumbing fixtures.
- A gas piping system will be provided from the building side of the meter with piping distribution to serve mechanical equipment.
- The plumbing fixtures shall consist of the following, and be American Standard, Kohler, or Crane.
- Eleven wall hung vitreous china water closets with flushometers.
- Five wall hung vitreous china urinals with flushometers.
- Four wall hung vitreous china lavatories with centerset fittings.
- Six vitreous china lavatories to be set in a countertop.
- One vitreous china lavatory wall hung in shipping toilet.
- Two enameled cast iron service sinks.
- Four electric water coolers.
- A stainless steel sink will be provided in kitchen.
- Three exterior hose bibbs will be provided for exterior watering.
- Color coding of piping and hook-up of owner's equipment is not included.

8.0 MECHANICAL (Cont'd.)

INTERIOR FIRE PROTECTION

- A wet ordinary hazard sprinkler system shall be installed in all areas of the building.
- Two (2) eight-inch (8") sprinkler entries, complete with wall indicator valves, shall be provided.
- The Manufacturing and Warehousing Areas shall be protected by means of brass upright heads at a ratio of one (1) head per each one hundred (100) square feet of floor area.
- The Office Area and Warehouse Office Areas where suspended acoustical ceilings shall be installed, shall be protected by means of chrome pendant heads at a ratio of one (1) head per each one hundred and thirty (130) square feet of floor area.
- Six (6) hose stations complete with one hundred feet (100') of one and one-half inch of woven fabric hose shall be installed at convenient locations in the Manufacturing and Warehousing Areas.
- All tees of the sprinkler piping shall be painted red in compliance with OSHA specifications.
- Sixteen (16) four foot wide by eight foot long (4' x 8') smoke and heat vents for Warehouse/Assembly Areas are included as required by Local Code. Design criteria low heat release content, 1 s.f. per 150 s.f. of floor space.

ITEMS NOT INCLUDED

- A.D.T., booster pumps, rack sprinklers, draft curtains, hose carts, fire extinguishers, firewalls, fireproofing, field painting, or any such items.
- The above fire protection specificity is our best estimate of the requirements of the fire underwriters. The actual final requirements will be determined by negotiation between the owner and his insurance company. Any additions or deletions from the above requirements, inclusive of final building materials acceptance, will result in a price adjustment which will be fully documented and substantiated.

8.0 MECHANICAL (Cont'd.)

HEATING, VENTILATING & AIR CONDITIONING

DESIGN CONDITIONS:

A. The heating, ventilating and air conditioning systems shall be designed within the limits of the following conditions:

OFFICE AREA

Summer: Outside - 91° FDB 78° FWB

Inside - 75° FDB 50% RH at the above outside conditions

Winter: Outside - 0° FDB

Inside - 70° FDB

Ventilation: Ventilation shall conform to the applicable codes, and as required to offset the heat gains and losses of spaces being served. Summer ventilation shall be provided at a rate of air change to limit the space temperature rise to 20°F above the outside summer ambient temperature indicated above.

WAREHOUSE/ASSEMBLY AREA

Winter - 0° - 70°FDB

B. The following codes, organizations, jurisdictional bodies, and insurance organizations shall be considered applicable:

1) F.M./F.I.A.
2) OSHA
3) NFPA
4) BOCA (Mechanical Code Appendix)
5) ASHRAE
6) SMACNA

8.0 MECHANICAL (Cont'd.)

HEATING, VENTILATING & AIR CONDITIONING (Cont'd.)

C. Ventilation: Ventilation shall be accomplished by means of the appropriate air handling device applicable to the specific area. Ventilation shall be accomplished according to the following schedule, unless specified otherwise hereinafter:

Space		Number of Air Changes
Offices - General	-	4 - 6 Max.
Assembly	-	2 (Summer) - 1 (Winter)
Warehouse	-	2 (Summer) - 1 (Winter)
Toilets	-	10 air changes or 40CFM per fixture (exhaust)
Janitor's Closets	-	10 air changes or 40CFM per fixture (exhaust)

D. Miscellaneous:

1. The Conference Room shall be equipped with a supplementary 5 ton cooling coil and rooftop condensing unit to provide optimum environmental control.

2. The kitchen will be fitted with exhaust ductwork and an up-blast type exhaust fan.

8.0 MECHANICAL (Cont'd.)

HEATING, VENTILATING & AIR CONDITIONING (Cont'd.)

II. DESCRIPTION OF SYSTEM:

A. HEATING: Heating for the Office Area shall be accomplished by means of gas-fired rooftop air unit(s) equipped with fan, burner, filters, and air intake plenum with damper. A supply duct distribution system shall be furnished to provide optimum air distribution. Terminal air diffusers shall provide air diffusion at the specific area or room, and the ceiling void shall be used as a return air plenum.

Gas-fired unit heaters shall serve the heating needs of all other spaces including the Shipping & Receiving Area.

B. COOLING: Cooling shall be accomplished by means of a package rooftop air conditioning unit(s) equipped with fans, compressors (electric), condenser, filters, and outside air intake plenum with damper. A supply duct distribution system shall be furnished to provide optimum air distribution. Terminal air diffusers shall provide air diffusion at the specific area or room, and the ceiling void shall be used as a return plenum.

A total nominal 30 tons of cooling for the Office Area and 5 ton supplemental cooling for Conference Room.

9.0 ELECTRICAL

GENERAL:

The scope of work consists of the installation of all materials to be furnished under this Section, and without limiting the generality thereof, includes all equipment, labor, and services required for furnishing, delivering, and installing the principal items of work hereinafter listed.

STANDARDS:

All materials and equipment shall be new and comply with the applicable standards of the following authorities:

Massachusetts Electrical Code
Underwriters Laboratories, Inc.
National Electrical Manufacturers' Association
Institute of Electrical & Electronic Engineers
American Society for Testing Materials
United States American Standards Institute
National Board of Fire Underwriters
Insulated Power Cable Engineers Association
Applicable Insurance Underwriter
National Electrical Safety Code
Occupational Safety and Health Act

TEMPORARY LIGHT AND POWER:

A 200A, 208/120V 3-phase, 4-wire temporary service will be provided by the electrical subcontractor from a source provided by the utility company for electric light and power requirements while the building is under construction and until the permanent feeders have been installed and are in operation. Energy charges will be borne by the Design/Build Contractor.

9.0 ELECTRICAL (Cont'd.)

WORK INCLUDED

The electrical systems installed and work performed under this Division of the specifications shall include but not necessarily be limited to the following:

A. 480/277V and 208/120V, 3-phase, 4-wire normal and emergency secondary distribution and power systems including main distribution panel, local lighting and power panels, disconnect switches, raceways, cables, wiring, junction and pullboxes, terminal cabinets and wireways.

B. All lighting systems (indoor and outdoor, normal, night emergency and exit) including all fixtures, lamps, plaster and/or tile frames, standards, switches, outlets, wiring, raceways, and all other components and fittings required for complete lighting systems.

C. Wiring, including connections for all heating, ventilating, air conditioning, plumbing systems, and vending equipment.

D. Grounding systems, building steel static relief, neutral and equipment grounding.

E. Temporary service, lighting, and power systems.

F. 15KW diesel-fired standby emergency generating system including generator, automatic controls, transfer equipment, and all specified accessories.

G. Disconnect switches for all HVAC and plumbing system motors.

H. All other systems hereinafter specified, leaving ready an electrical system in perfect operating condition.

HIGH TENSION SERVICE:

High tension service and pad-mounted transformer will be furnished by the utility company.

9.0 ELECTRICAL (Cont'd.)

480/277V SERVICE:

A secondary metered underground 800A service shall be furnished and installed from the pad mounted transformer to a main panel.

A secondary metered underground service shall be furnished and installed from the pad-mounted transformer to the main switchboard.

MAIN DISTRIBUTION PANEL

Furnish and install an 800 Amp 3-phase, 4-wire solid neutral main distribution panel board with 1200A main fuse disconnect switch.

The distribution section shall consist of current limiting switch/fuse protective devices.

All devices shall have: NEMA 1 general purpose enclosures, handles that are padlockable in the "ON" and "OFF" positions; positive quick-make, quick-break mechanisms; defeatable door interlocks that prevent the door from opening when the operating handle is in the "ON" position; nameplates, front cover mounted, that contain a permanent record of current limiting switch and fuse type, front cover doors shall be padlockable in the closed position.

LIGHTING/RECEPTACLE AND DISTRIBUTION PANELBOARDS:

Furnish and install deadfront panelboards, incorporating switching and protective devices of the number, rating, and type noted herein. Panelboards shall have NEMA 1 general purpose enclosures and shall be surface or flush mounted as required.

All panelboards shall be rated for the intended voltage and shall be in accordance with the Underwriters' Laboratories, Inc. "Standard for Panelboards" and Standard for Cabinets and Boxes", and shall be so labeled. Panelboards shall also comply with NEMA Standard for Panelboards, National Electrical Code, and Federal Specification 115a (Power Distribution Panels) where applicable.

9.0 ELECTRICAL (Cont'd.)

CIRCUIT BREAKERS

The circuit breakers in the panelboards shall be quick-make, quick-break, trip-indicating, and be of the bolt-in type. The panelboard assembly shall permit removal of any breaker or breakers without disturbing the mounting of other breakers not to be removed. These breakers shall also be manually operated and closed at will and shall indicate the "ON" and "OFF" position of the handle. The minimum interruption rating of all molded case circuit breakers located in lighting/receptacle panels shall be 10,000 amperes at the rated system voltage. (208/120V or 480/277V). The minimum interruption rating of circuit breakers in all distribution panels shall be as required by system fault analysis.

- Panelboards and breakers shall be as manufactured by Westinghouse, General Electric, or ITE.
- Each panel shall have a minimum of 10% spare breaker capacity.
- All panels shall be rated at least 225A and have a minimum of 42 poles unless otherwise noted.

DRY TYPE TRANSFORMERS

- Dry type transformers shall be furnished and installed with Class H insulation suitable for operation up to 150C, rise, indoor type.
- The transformers shall be equipped with primary taps two (2) at 2.5 percent above and below normal.
- Primary windings shall be rated for 480V for use on 3-phase, 3-wire circuits. Secondary windings shall be rated for 120/208V, 3-phase, 4-wire circuits.
- The transformer sound levels shall not exceed 45 decibels as measured per NEMA Standard ST-14-11.
- Manufacturer shall be General Electric, Westinghouse, Sorgel, or approved equal.

9.0 ELECTRICAL (Cont'd.)

WIRE AND CABLES

- All wire, cable, and wiring materials necessary for the complete installation, wiring, and connecting of all electrical equipment and devices shall be furnished and installed complete as a part thereof, as specifically called for herein.

- Conductors shall be 600V insulation and shall conform to the following Underwriters' Laboratories approved types and specifications:

- INTERIOR

  - Branch Circuits - Type THWN
  - Feeders - Type THWN
  - Signal Systems - Minimum #14 AWG-Type TW

- The outer jacket of all interior wiring shall be color coded to denote polarity as follows:

208/120V System		480/277V System	
Phase A	Black	Phase A	Brown
Phase B	Red	Phase B	Orange
Phase C	Blue	Phase C	Yellow
Neutral	White	Neutral	White
Equip. Grounds	Green	Equip. Grounds	Green

- No wire smaller than No. 12 AWG shall be used for power or lighting and no wire smaller than No. 14 AWG shall be used in any control or signal circuit, unless smaller wire is specifically called for herein or on the Drawings.

- All wire shall be Simplex, Plastic Wire & Cable Corporation, General Electric, or equal.

- No foreign compounds or lubricants shall be applied to the wire during the pulling-in process.

- Conductors shall be continuous from outlet to outlet and no splices shall be made, except within outlet or junction boxes.

9.0 ELECTRICAL (Cont'd.)

CONDUIT AND FITTINGS

- Electric metallic tubing may be used throughout where applicable by code. Minimum size shall be three-fourths inch (3/4").

- All conduit and EMT shall be hot-dipped galvanized or sheradized inside and out.

- All conduit runs shall be as follows:

  1. Runs shall be straight and true; offsets and bends shall be uniform and symmetrical.

  2. Couplings, connectors and fittings shall be types specifically designed and manufactured for the purpose.

  3. Minimum size of conduits shall be three-fourths inch (3/4"). Where size is not given, the latest issue of the National Electrical Code shall be followed.

- Horizontal or cross runs in building-type partitions or sidewall should be avoided. All conduit to outlets in building-type partitions shall run down from ceiling into partition.

- A conduit expansion fitting shall be installed in each conduit run wherever it crosses an expansion joint in the concrete structure. The expansion fitting shall be installed on one side of the joint with its sliding sleeve end flush with the expansion joint equal to at least three times the nominal width of the joint.

- A conduit expansion fitting shall be provided in each conduit run which is mechanically attached to separate structures to relieve strain caused by shifting of one structure in relation to the other.

9.0 ELECTRICAL (Cont'd.)

LIGHTING SYSTEM

In general, lighting circuits in assembly and warehouse will be controlled directly from the panels. Private offices, offices, toilets, mechanical rooms and small rooms, will have individual toggle switch control.

Area	Average Maintained Footcandles
Assembly	50 FC throughout and 90 to 100 at Bench Work Stations via fluorescent industrial fixtures.
Warehouse	15 to 20 horizontal FC via fluorescent industrial fixtures.
General Office Area	90 to 100 FC via 2 x 4 recessed fluorescent fixtures.
Kitchen, Storage Rooms and Toilet Rooms.	25 to 30 FC via 2 x 4 recessed fluorescent fixtures.
Security Lighting	Via metal halide fixtures.
Site Parking Lot Lighting	.5 FC via metal halide pole-mounted fixtures.

EMERGENCY LIGHTING SYSTEM

A combination emergency and night security lighting system is included in areas illuminated with fluorescent equipment. In areas illuminated with high intensity discharge type of equipment, emergency lighting will be provided, via quartz instant-on lamps, integral with the luminaires.

9.0 ELECTRICAL (Cont'd.)

TEMPERATURE CONTROL WIRING AND EQUIPMENT

- Temperature control wiring - starters and contactors:
  - All power wiring shall be furnished and installed for the operation of HVAC and plumbing systems by the electrical subcontractor. All temperature control and interlock wiring, line and low voltage, including wiring of line 120V space thermostats, shall be furnished and installed by the HVAC and plumbing subcontractors.
  - The HVAC and plumbing subcontractors shall furnish to the electrical subcontractor all magnetic and manual motor starters and contactors required for the control of equipment furnished by them under the HVAC and PLUMBING Sections of this specification.

RECEPTACLES

- Duplex convenience outlets will be installed on walls and columns throughout the plant. All outlets will be 3-wire grounding type, except as otherwise noted.

STANDBY POWER PLANT - 15KW - 480/277V

- A 15KW generator standby power plant will be furnished complete with phase failure, automatic transfer controls to energize the emergency lighting system, upon loss of utility service. The engine generator and transfer controls and silencer will be as manufactured by Onan or approved equal. Fuel tank furnished and installed under other section.

FIRE ALARM SYSTEM

- Furnish and install a complete manual fire alarm system consisting of combination horn/light signal devices and manual pull station. Systems shall conform to Mass. State Building Code requirements.

TELEPHONE SERVICE

- Furnish and install telephone service conduit from telephone room extending out ten feet (10') from the building and capped for incoming telephone cable.

9.0 ELECTRICAL (Cont'd.)

WORK NOT INCLUDED

- Utility Company service charges
- Busduct required for Owner's process equipment.
- Wiring to or connection to Owner's production equipment.
- Raceways and/or wiring of interior telephone system.
- Building security system, i.e. ADT or like systems.
- Television systems.
- Battery charging equipment.
- Wiring to or connection of data processing equipment.
- Public address system.
- Wiring or connection to dust collection.
- Telephone Company interior raceway systems or wiring.
- Dock lights.
- Power, wiring, or connections to computer or data processing equipment.
- Under-floor duct system.
- Power or wiring for air conditioning system assembly and warehouse.
- Power or wiring for future addition (28,500 s.f.).

11.0 SPECIAL CONSTRUCTION

- There shall be five (5) mechanical docklevelers, installed at the Shipping and Receiving Area.
- Docklevelers shall be provided with panic stops and hard rubber bumpers.
- There shall be five (5) Series F foam-type dockseals, complete with a rubber drop flap running the full width of the door opening as manufactured by Weather Shield, or equal, installed at the Shipping and Receiving Area.
- Canopy over truck dock to be structural steel and metal dock with bronze anodized facia trim. The underside of the loading dock canopy to be stucco ceiling system.
- Toilet room partitions shall be ceiling-hung.
- Plastic laminated toilet vanities.
- Wood kitchen cabinets and counters/backsplashes.
- Miscellaneous millwork for stair railings, window treatment, and closet shelf and pole.

12.0 SITE WORK

SITE PREPARATION

- Stripping and storing of existing topsoil.
- Necessary cut and fill the "balanced" site to achieve the appropriate grades.
- Rough grading and compacting of the soil within the construction limits.
- Construction limits encompass approximately five and one-half (5 1/2) acres.
- Excavation and backfilling for building foundations, storm drainage system, sanitary and water piping.
- The placing and compacting of six inches (6") of bank-run gravel under all floor slabs on grade, at all areas of bituminous paving, and at all areas of exterior concrete.
- Preparation of future expansion building pad is included.
- Ledge removal is not anticipated in the excavation of the building and site and has not been carried in the cost or as part of the work. If rock is encountered, its definition, for removal, shall be as follows: A single piece larger than 1 cubic yard and/or a continuous strata hard enough to require blasting to break up. Any ledge removable by equipment (not blasted) will not be considered as ledge excavation. All required ledge excavation is to be considered as extra cost to the project.

12.0 SITE WORK (Cont'd.)

SITE UTILITIES

- The installation of a six-inch (6") vitrified clay gravity sanitary sewer system from the building to the existing municipal line.
- The installation of a storm drainage system inclusive of corrugated metal and/or concrete piping, catchbasins and manholes.
- The installation of an eight-inch (8") cement-lined, cast-iron, two-sided sprinkler loop from the existing municipal line inclusive of one (1) valve pit, two (2) eight-inch (8") entries, with post indicator valves, and two (2) hydrants.
- A four-inch (4") cement-lined, cast-iron domestic water line shall be installed from the valve pit to the building.
- The installation of a concrete pad for the main electrical transformer to be furnished and installed by the local power company.
- Parking lot lighting shall be pole mounted fixtures.
- A four-inch (4") gas main.

SITE IMPROVEMENTS

- The placing of two and one-half inches (2 1/2") of bituminous paving, laid in two (2) courses, for the automobile roadway and parking areas as shown on the Site Plan.
- Parking lot striping for the ninety-seven (97) total spaces shall be provided.
- Asphalt berms shall be installed in the parking lot and entrance roadway.
- An Allowance of $30,000.00 has been included for furnishing and spreading topsoil, inclusive of seeding, and for tree and/or shrub planting.
- The installation of a reinforced concrete pad for the compactor.

ARCHITECTURE/ENGINEERING

Architecture/Engineering, shall provide complete engineering service to include in general, but not limited to, the following:

- Site Engineering
- Architectural and Structural Design and Specifications
- Electrical Design Plans and Specifications
- Mechanical Design Plans and Specifications
- Plumbing Plans and Specifications
- Quality Control will be provided to check all prepared drawings and shop drawings and also provide on-site inspection of work in progress.
- A typical list of construction drawings would be as follows:

L-1	Site Plan
L-2	Utility Site Plan
A-1	General Arrangement Floor Plan
A-2	Office Area Floor Plan
A-3	Elevations
A-4,5	Wall Sections and Details
A-6	Finish Schedule
A-7	Door/Window and Hardware Schedule
A-8	Architectural Roof Plan
S-1	Roof Framing Plan
S-2	Structural Steel Details
F-1	Foundation Plans
F-2	Foundation Sections
F-3	Anchor Bolt Plan and Pier Details
F-4	Slab Placement Plan
P-1,2	Plumbing Plans
HVAC-1,2	Heating, Ventilating and Air Conditioning Plans
HVAC-3,4	HVAC Details
E-1,2	Lighting Plans
E-3	Power Plan
E-4	Power Distribution Diagrams
E-5	Schedules and Details

A complete set of "As-Built" sepia drawings.

SUPERVISION

- The project shall be supervised by a field superintendent stationed in the field. He will coordinate all subcontractor work, direct the field engineering activities, and be responsible to the Project Manager.

- The Project Manager shall be based in our home office and shall coordinate the engineering and construction of the facility. He will serve as a liaison between the designers, owner and our field superintendent.

GENERAL CONDITIONS

- Temporary utilities shall be provided as required.

- Field engineering and testing shall be provided during construction.

- Clean-up and trucking shall be performed.

- Final clean-up shall include washing of windows and the removal of all debris. The building shall be left broom-clean.

- A field office complete with telephone and toilet facilities shall be maintained at the site.

WINTER CONDITIONS

- Provisions for construction of this facility during the winter months have NOT been included.

ITEMS FOR CONSIDERATION

1.	Reduce the thickness of the Warehouse concrete slab from 6" to 5" using 4,000 PSI concrete.	DEDUCT $ 7,200.00
2.	Eliminate the Allowance included within the Proposal for loam and seeding and tree and/or shrubbery planting.	DEDUCT $30,000.00
2a.	Plus associated fine grading.	DEDUCT $ 1,500.00
3.	Delete one (1) overhead door, dockleveler, and dock seal.	DEDUCT $ 8,000.00
4.	Substitute battery packs for emergency lighting in lieu of the emergency generator included.	DEDUCT $ 665.00
5.	Eliminate full height ceramic wall tile in Office toilets and substitute Tru-Glaze.	DEDUCT $ 4,500.00
6.	Reduce electric service capacity from 1200 amps to 800 amps.	DEDUCT $ 2,500.00

PRELIMINARY
ESTIMATE (Cost Summary)

PROJECT	TOTAL AREA	SHEET NO.
LOCATION	TOTAL VOLUME	ESTIMATE NO.
ARCHITECT	COST PER S.F.	DATE
OWNER	COST PER C.F.	NO. OF STORIES
QUANTITIES BY: PRICES BY:	EXTENSIONS BY:	CHECKED BY:

NO.	DESCRIPTION	SUB TOTAL COST	COST/S.F.	%
1.0	**Foundation**	97615		
2.0	**Substructure**	299863		
3.0	**Superstructure**	290470		
4.0	**Exterior Closure**	409089		
5.0	**Roofing**	188024		
6.0	**Interior Construction**	231453		
7.0	**Conveying**	—		
8.0	**Mechanical System**	416722		
9.0	**Electrical**	365799		
10.0	**General Conditions (Breakdown)**	376070		
11.0	**Special Construction**	60753		
12.0	**Site Work**	242790		
	Building Sub Total	$ 2,978,647		

→ $ ______

Sales Tax ______ % × Sub Total $ ______ /2 = $ ______

General Conditions (%) ______ % × Sub Total $ ______ = ______

General Conditions $ ______

Sub Total "A" $ ______

Overhead ______ % × Sub Total "A" $ ______ = $ ______

Sub Total "B" $ ______

Profit ______ % × Sub Total "B" $ ______ = $ ______

Sub Total "C" $ ______

SEE COST BREAKDOWN

Location Factor ______ % × Sub Total "C" $ ______ =

Adjusted Building Cost $ ______

Architects Fee ______ % × Adjusted Building Cost $ ______ = $ ______

Contingency ______ % × Adjusted Building Cost $ ______ = $ ______

Total Cost ______

Square Foot Cost $ ______ / ______ S.F. = ______ $/S.F.

Cubic Foot Cost $ ______ / ______ C.F. = ______ $/C.F.

Page 6 of 6

COST ANALYSIS

SHEET NO.

PROJECT

ESTIMATE NO.

ARCHITECT

DATE

TAKE OFF BY: QUANTITIES BY: PRICES BY: EXTENSIONS BY: CHECKED BY:

DESCRIPTION	SOURCE/DIMENSIONS			QUANTITY	UNIT	MATERIAL		LABOR		EQ./TOTAL	
						UNIT COST	TOTAL	UNIT COST	TOTAL	UNIT COST	TOTAL
1.0 Foundations											
Spread Footings											
30 x 40 Bay - Int.	011	120	7350	27	Ea.	125	3375	160	4320	285	7695
Corner			7100	2	Ea.	30	60	54	108	84	168
Ext.			7150	17	Ea.	63	1071	94	1598	157	2669
20 x 30 Bay - Int.			7350	18	Ea.	125	2250	160	2880	285	5130
Corner			7100	3	Ea.	30	90	54	162	84	252
Ext.			7150	10	Ea.	63	630	94	940	157	1570
40 x 25 Bay - Int.			7350	1	Ea.	125	125	160	160	285	285
Corner			7100	1	Ea.	30	30	54	54	84	84
Ext.			7150	2	Ea.	63	126	94	188	157	314
Strip Footing - CMU	011	140	2100	368	LF	3.62	1332	9.55	3514	13.17	4846
	011	140	2500	944	LF	7.65	7222	16.90	15953	24.55	23175
Found. Walls 4' High	011	210	1560	860	LF	13.35	11481	23	19780	36.35	31261
8' High	011	210	5060	200	LF	27	5400	45	9000	72	14400
Waterproofing	011	292	1000	1060	LF	.32	339	2.01	2130	2.33	2469
Perimeter Insulation	057	101	2300	4020	SF	.63	2533	.19	764	.82	3297
Total											97615

COST ANALYSIS

SHEET NO.

PROJECT | ESTIMATE NO.

ARCHITECT | DATE

TAKE OFF BY: | QUANTITIES BY: | PRICES BY: | EXTENSIONS BY: | CHECKED BY:

DESCRIPTION	SOURCE/DIMENSIONS			QUANTITY	UNIT	MATERIAL		LABOR		EQ./TOTAL	
						UNIT COST	TOTAL	UNIT COST	TOTAL	UNIT COST	TOTAL
2.0 Substructures											
Slab on grade											
Light assembly	021	200	4520	24590	SF	1.81	44508	1.81	44508	3.62	89016
Warehouse	021	200	4560	39060	SF	2.16	84370	3.11	121477	5.27	205847
Dock Leveler Pits				5	Ea.	—	—	500	2500	500	2500
Frames				5	Ea.	300	1500	200	1000	500	2500
Total											299863

COST ANALYSIS

PROJECT	SHEET NO.
ARCHITECT	ESTIMATE NO.
	DATE

TAKE OFF BY: QUANTITIES BY: PRICES BY: EXTENSIONS BY: CHECKED BY:

DESCRIPTION	SOURCE/DIMENSIONS			QUANTITY	UNIT	MATERIAL		LABOR		EQ./TOTAL	
3.0 Superstructure						UNIT COST	TOTAL	UNIT COST	TOTAL	UNIT COST	TOTAL
Elev. Slabs											
30 x 20 Bays	035	460	6700 6800	9600	SF	5.14	49344	2.59	24864	7.73	74208
Roof Slab - 30 x 20	037	410	4600 4700	11400	SF	2.13	24282	.75	8550	2.88	32832
25 x 40			5800 5900	1000	SF	2.38	2380	.83	830	3.21	3210
30 x 40			5800 6900	49200	SF	2.38	117096	.83	40836	3.21	157932
25 x 20			3400 3500	500	SF	2.05	1025	.77	385	2.82	1410
Stairs (Quality 1.5)	039	100	0740	1	Ea.	4875	4875	998	998	5873	5873
	039	100	0740	1	Ea.	3250	3250	665	665	3915	3915
	039	100	0480	1	Ea.	1175	1175	1125	1125	2300	2300
Miscellaneous Metals				73250	SF					0.12	8790
Overhead Door Channels				9	Sets						
Masonry anchors				800	Ea.						
Roof Angle Frames				15	Ea.						
Angles @ Floor Openings				14	LF						
Pipe Bumpers				4	Ea.						
Exterior Railings				24	LF						
Bar Ladders				20	LF						
Releaving Angles				200	LF						
Total											290470

COST ANALYSIS

SHEET NO.

PROJECT | ESTIMATE NO.

ARCHITECT | DATE

TAKE OFF BY: | QUANTITIES BY: | PRICES BY: | EXTENSIONS BY: | CHECKED BY:

DESCRIPTION	SOURCE/DIMENSIONS			QUANTITY	UNIT	MATERIAL		LABOR		EQ./TOTAL	
4.0 Exterior Closure						UNIT COST	TOTAL	UNIT COST	TOTAL	UNIT COST	TOTAL
Brick & 8" Block	041	272	1240	22,985	SF	4.90	112627	9.40	216059	14.30	328686
Lintels				69	LF	45	3105	-	—	45	3105
12" Block	041	211	7510	3752	SF	2.60	9755	4.09	15345	6.69	25100
Paint Block	065	100	0320	3752	SF	.15	563	.63	2364	.78	2927
Storefront - Door	046	100	6900	3	Ea.	1150	3450	500	1500	1650	4950
Storefront	047	582	1750	765	SF	7.40	5661	7.10	5432	14.50	11093
Stucco Ceiling	067	100	4300	720	SF	1.19	857	3.67	2642	4.86	3499
Doors	046	100	3450	3	Ea.	685	2055	160	480	845	2535
			4800	5	Ea.	845	4225	405	2025	1250	6250
Batt Insulation	061	580	0900	720	SF	.37	266	.18	130	.55	396
Windows	047	582	1750	910	SF	7.40	6734	7.10	6461	14.50	13195
Rough Carpentry				26,737	SF	.13	3476	.145	3877	.275	7353
Total											409089

COST ANALYSIS

SHEET NO.

PROJECT

ESTIMATE NO.

ARCHITECT

DATE

TAKE OFF BY: QUANTITIES BY: PRICES BY: EXTENSIONS BY: CHECKED BY:

DESCRIPTION	SOURCE DIMENSIONS			QUANTITY	UNIT	MATERIAL		LABOR		EQ./TOTAL	
						UNIT COST	TOTAL	UNIT COST	TOTAL	UNIT COST	TOTAL
5.0 Roofing											
4-Ply-Asphalt	051	103	1600	63906	SF	.44	28119	.86	54959	1.30	83078
Roof Edge, 8" Duran.	051	520	1500	1060	LF	9.75	10335	5.00	5300	14.75	15635
Base Flash - C.U. 200Z	051	510	1800	400	LF	8.40	3360	8.20	3280	16.60	6640
Expansion Joint				203	LF	6.00	1218	7.65	1553	13.65	2771
Smoke Hatch, Alum.	058	100	3000	16	Ea.	1450	23200	110	1760	1560	24960
Roof Hatch	058	100	0500	1	Ea.	455	455	110	110	565	565
Insul. 1 1/16" Fiberglass	057	101	0500	63096	SF	.47	29655	.24	15143	.71	44798
Roof Openings				12	Ea.	100	1200	100	1200	200	2400
Roof Curb Flashing				12	Ea.	100	1200	170	2040	270	3240
Pitch Pockets				24	Ea.	20	480	46	1104	66	1584
Perimeter nailing				4020	SF	.10	402	.34	1367	.44	1769
Roof Drains				8	Ea.	23	184	50	400	73	584
Total											188024

COST ANALYSIS

SHEET NO.

PROJECT — ESTIMATE NO.

ARCHITECT — DATE

TAKE OFF BY: QUANTITIES BY: PRICES BY: EXTENSIONS BY: CHECKED BY:

DESCRIPTION	SOURCE/DIMENSIONS			QUANTITY	UNIT	MATERIAL		LABOR		EQ./TOTAL	
						UNIT COST	TOTAL	UNIT COST	TOTAL	UNIT COST	TOTAL
6.0 Interior Wall Construction											
Interior Finish											
Partitions:											
3 5/8" @ 16" o.c.											
1/2" Gyp. Bd. Both sides	061	510	5400	10,000	SF	.89	8900	1.54	15400	2.43	24300
Insulation	061	580	0880	10,000	SF	.24	2400	.15	1500	.39	3900
Painting	065	100	0140	20,000	SF	.10	2000	.24	4800	.34	6800
Base Board				2,200	LF	.50	1100	.71	1562	1.21	2662
Furring w/1/2" Gyp. Bd.				2900	SF	.61	1769	1.49	4321	2.10	6090
Insulation	061	580	0920	2900	SF	.55	1595	.24	696	.79	2291
Painting	065	100	0140	2900	SF	.10	290	.24	696	.34	986
Base Board				410	LF	.50	205	.71	291	1.21	496
Doors - 3/0 7/0				35	Ea.	180	6300	130	4550	310	10850
Locksets	064	292	0400	35	Ea.	55	1925	24	840	79	2765
Closers		292	0560	10	Ea.	65	650	40	400	105	1050
Butts		292	0160	105	Ea.	10		–	1050	10	1050
Block Walls, 6"	061	210	1500	3040	SF	1.09	3314	2.78	8451	3.87	11765
Suspended Ceiling	067	810	3260 2740	9600	SF	.66	6336	.66	6336	1.32	12672
Flooring - Carpet	066	100	0160	9600	SF	1.58	15168	.75	7200	2.33	22368
V.A.T			1580 1600	150	SF	.80	120	.72	108	1.52	228
Base				2500	LF	.50	1250	.71	1775	1.21	3025
O.H. Doors 8 x 8	064	100	5000	4	Ea.	680	2720	345	1380	1025	4100
Ceramic Tile - Fl	066	100	1720	890	SF	1.96	1744	2.23	1985	4.19	3729
Quarry Tile - Fl			1800 1820	500	SF	3.00	1500	3.00	1500	6.00	3000
Ceramic Tile - Wall	065	100	1940	2304	SF	1.38	3180	2.15	4954	3.53	8134
Ceramic Tile - Base				256	LF	2.00	512	3.00	768	5.00	1280
Fire Doors - Sliding				4	Ea.					2000	8000
Total											141541

COST ANALYSIS

PROJECT | ARCHITECT | SHEET NO. | ESTIMATE NO. | DATE

TAKE OFF BY: | QUANTITIES BY: | PRICES BY: | EXTENSIONS BY: | CHECKED BY:

DESCRIPTION	SOURCE/DIMENSIONS			QUANTITY	UNIT	MATERIAL		LABOR		EQ./TOTAL	
6.0 Interior Wall Construction Interior Finish (Cont.)						UNIT COST	TOTAL	UNIT COST	TOTAL	UNIT COST	TOTAL
Block, 8"	061	210	2000	6010	SF	1.30	7813	2.97	17850	4.27	25663
Painting	065	100	0320	18740	SF	.15	2811	.63	11806	.78	14617
Floor Insulation, 9" Batt				9600	SF					.76	7296
Fire wall				1740	SF	1.60	2784	2.32	4037	3.92	6821
Tru-glaze	065	100	1380	252	SF	.31	78	.51	129	.82	207
Spray Assembly Ceiling				9600	SF	.10	960	.10	960	.20	1920
Fixed Interior Sash				4	Ea.	200	800	100	400	300	1200
Interior Carpentry				12900	SF	.15	1935	.125	1613	.275	3548
Wood Ceiling				80	SF	5	400	5	400	10	800
Mezzanine Ceiling				9600	SF	.89	8544	2.01	19296	2.90	27840
Total											89912

COST ANALYSIS

SHEET NO.

PROJECT

ESTIMATE NO.

ARCHITECT

DATE

TAKE OFF BY: QUANTITIES BY: PRICES BY: EXTENSIONS BY: CHECKED BY:

DESCRIPTION	SOURCE/DIMENSIONS			QUANTITY	UNIT	MATERIAL		LABOR		EQ./TOTAL	
8.0 Mechanical						UNIT COST	TOTAL	UNIT COST	TOTAL	UNIT COST	TOTAL
Plumbing											
Water Closets	081	470	2120	11	Ea.	335	3685	350	3850	685	7535
Urinals		450	2000	5	Ea.	390	1950	385	1925	775	3875
Lavatories		433	2120	11	Ea.	260	2860	360	3960	620	6820
Drinking Fount. (Recessed)		460	2000	2	Ea.	840	1680	285	570	1125	2250
Drinking Fount. (Flr Mtd.)		460	2040	2	Ea.	420	840	235	470	655	1310
Kitchen Sink		431	2000	1	Ea.	380	380	490	490	870	870
4" Cast Iron Pipe	081	040	0880	400	LF	5.50	2200	10.95	4380	16.45	6580
Water Pipe 1"			1620	400	LF	1.61	644	7.45	2980	9.06	3624
1/2" H & C			1560	400	LF	.78	312	5.05	2020	5.83	2332
Floor Drains (Factor 1.05)	081	310	4200	5	Ea.	255	1275	390	1950	645	3225
4" Cast Iron Pipe	081	040	0880	580	LF	5.50	3190	10.95	6351	16.45	9541
Roof Drains 6"	081	310	4200	8	Ea.	255	2040	390	3120	645	5160
6" Cast Iron Pipe	081	040	0920	580	LF	10.15	5887	12.85	7453	23	13340
Hot Water Heater- Gas	081	170	1780	3	Ea.	935	2805	875	2625	1810	5430
Sprinkler- Ordinary H	082	110	1100	63650	SF	.45	28643	1.04	66196	1.49	94839
Office	082	110	1100	9600	SF	.45	4320	1.04	9984	1.49	14304
Hose Cab.	082	390	0800	6	Ea.	115	690	85	510	200	1200
HVAC- Office	084	220	3880	9600	SF	5.45	52320	3.84	36864	9.29	89184
Lobby	084	220	1280	1	Ea.	14728	14728	7453	7453	22181	22181
Heating- Mfg. & Storage	083	162	1960	63650	SF	.85	54103	.92	58558	1.77	112661
Gas Piping- 1"Φ	081	040	4030	550	LF	1.53	842	5.05	2778	6.58	3620
Water Meter				1	Ea.	5000	5000		—	5000	5000
Misc. Insulation				700	LF	.24	168	2.39	1673	2.63	1841
Total											416722

COST ANALYSIS

SHEET NO.

PROJECT

ESTIMATE NO.

ARCHITECT

DATE

TAKE OFF BY: QUANTITIES BY: PRICES BY: EXTENSIONS BY: CHECKED BY:

DESCRIPTION	SOURCE/DIMENSIONS			QUANTITY	UNIT	MATERIAL		LABOR		EQ./TOTAL	
						UNIT COST	TOTAL	UNIT COST	TOTAL	UNIT COST	TOTAL
9.0 Electrical											
Temporary Service 200 A	091	210	0620	1	Ea.					1135	1135
Feeder	091	310	0280	100	LF					16.15	1615
Panel Board (u)	163	500	0400	1	Ea.					1150	1150
Branch Power	ALLOWANCE										15000
Lighting											
Service 800 A	091	210	0400	1	Ea.					5450	5450
Add for 277/480 V			0570	1	Ea.					1363	1363
Switch Board - 800 A		410	0320	1	Ea.					10550	10550
Add for 277/480 V			0410	1	Ea.					2110	2110
Feeder 800 A		310	0400	100	LF					72	7200
Lighting											
Assembly - 50 FC	092	233	0400	12000	SF					2.33	27960
100 FC	092	236	1400	12000	SF					4.37	52440
Warehouse - 20 FC	092	233	0240	38400	SF					1.20	46080
Office 100 FC	092	213	0400	9600	SF					5.81	55776
Switches 2.5/1000 SF	092	542	0320	9600	SF					.24	2304
Security	092	236	1800	8000	SF					4.02	32160
Branch Receptacles											
Assembly	092	522	0480	24000	SF					1.24	29760
Warehouse	092	522	0240	38400	SF					.77	29568
Office	092	522	0640	9600	SF					1.90	18240
Misc. Power	092	582	0320	72000	SF					.12	8640
A/C Office	092	610	0250	9600	SF					.23	2208
Generator 15 KW-Gas	094	310	0280	1	Ea.					665	665
Fire Alarm (50 Detectors)	094	100	0400	1	Ea.					14425	14425
Total											365799

COST ANALYSIS

SHEET NO.

PROJECT

ESTIMATE NO.

ARCHITECT

DATE

TAKE OFF BY: QUANTITIES BY: PRICES BY: EXTENSIONS BY: CHECKED BY:

DESCRIPTION	SOURCE/DIMENSIONS			QUANTITY	UNIT	MATERIAL		LABOR		EQ./TOTAL	
						UNIT COST	TOTAL	UNIT COST	TOTAL	UNIT COST	TOTAL
11.0 Special Construction											
Dock Levelers (U)	111	200	3500	5	Ea.	4675	23375	350	1750	5025	25125
Seals (U)			3640	5	Ea.	1425	7125	495	2475	1920	9600
Bumpers (U)			3300	10	Ea.	44	440	9	90	53	530
Canopy (U)	111	100	1640	1	Ea.	3525	3525	1625	1625	5150	5150
Toilet Partitions	061	870	0400	10	Ea.	230	2300	120	1200	350	3500
Screens	061	870	1428	4	Ea.	150	600	47	188	197	788
Accessories				27	Ea.	110	2970		—	110	2970
Vanities				36	LF					100	3600
Kitchen Cabinets				1	Ea.					1000	1000
Counters				10	LF					200	2000
Millwork Railings				50	LF					30	1500
Window Stools				200	LF					6	1200
Window Treatment				910	SF					4	3640
Closet / Pole				15	LF					10	150
Total											60753

COST ANALYSIS

SHEET NO.

PROJECT

ESTIMATE NO.

ARCHITECT

DATE

TAKE OFF BY: QUANTITIES BY: PRICES BY: EXTENSIONS BY: CHECKED BY:

DESCRIPTION	SOURCE/DIMENSIONS			QUANTITY	UNIT	MATERIAL		LABOR		EQ./TOTAL	
12.0 Site Work						UNIT COST	TOTAL	UNIT COST	TOTAL	UNIT COST	TOTAL
Clear and Grub (U)	021	100	0250	5 1/2	Acres	–	–			3775	20763
Strip Topsoil (U)	023	160	0900	8200	CY	–	–	1.66	13612	1.66	13612
Cut & Fill (U)	023	160	0900	16400	CY	–	–	1.66	27224	1.66	27224
Utilities- water											
Trench & Backfill	123	110	3580	80	LF	–	–	18.60	1488	18.60	1488
Pipe	123	540	3130	80	LF	6.05	484	4.10	328	10.15	812
Sewer -											
Trench & Backfill	123	110	3580	80	LF	–	–	18.60	1488	18.60	1488
Pipe	123	510	4580	80	LF	5.60	448	4.59	367	10.19	815
Drainage:											
Trench & Backfill	123	110	3580	300	LF	-	–	18.60	5580	18.60	5580
Pipe	123	510	4580	300	LF	5.60	1680	4.59	1377	10.19	3057
Bedding	123	310	2680	460	LF	1.28	589	2.03	934	3.31	1523
Manholes/Basins	123	710	3240	5	Ea.	680	3400	1125	5625	1805	9025
Curb Cuts				4	Ea.	–	–	500	2000	500	2000
Roadways: 24' w --	125	111	1800	220	LF	47	10340	23	5060	70	15400
Curb-Granite-Straight	127	610	3100	150	LF	12.10	1815	3.85	578	15.95	2393
Curved (U)	026	220	1300	75	LF	19.50	1463	7.50	563	27	2026
Berm	127	610	1500	1860	LF	1.01	1879	1.24	2306	2.25	4185
Sidewalks 5' w	127	140	1680	810	LF	4.66	3775	4.93	3994	9.59	7769
10' w	127	140	1680	20	LF	9.32	186	9.86	197	19.18	383
Parking Lots	125	510	1600	53900	SF	0.98	52822	0.75	40425	1.73	93247
Loam (U)	028	250	0400	2000	CY						
Hand Grade (U)	028	250	0600	165	CY						
Seed (U)	028	450	1100	11000	SY						
Plant Beds (U)	028	350	0010	8000	SF	ALLOWANCE					30000
Shrubs (U)	028	650	6300	80	Ea.						
Trees (U)	028	650	5900	20	Ea.						
Total											242790

PROJECT OVERHEAD SUMMARY

PROJECT | SHEET NO.

ESTIMATE NO.

LOCATION | ARCHITECT | DATE

QUANTITIES BY: | PRICES BY: | EXTENSIONS BY: | CHECKED BY:

DESCRIPTION	QUANTITY	UNIT	MATERIAL/EQUIPMENT UNIT	MATERIAL/EQUIPMENT TOTAL	LABOR UNIT	LABOR TOTAL	TOTAL COST UNIT	TOTAL COST TOTAL
Job Organization: Superintendent	56	WKS					1400	78400
Project ~~Manager~~ Engineer	54	WKS					900	48600
Timekeeper & Material Clerk	50	WKS					700	35000
Clerical								
Safety, Watchman & First Aid	56	WKS					300	16800
Travel Expense: Superintendent								5000
Project ~~Manager~~ Engineer								2000
Engineering: Layout 3 man crew	3	WKS					3000	9000
Inspection/Quantities								
Drawings								
CPM Schedule	1	LS						6000
Testing: Soil		LS						10000
Materials		LS						10000
Structural		LS						3000
Equipment: Cranes								
Concrete Pump, Conveyor, Etc.								
Elevators, Hoists								
Freight & Hauling								
Loading, Unloading, Erecting, Etc.								
Maintenance								
Pumping								
Scaffolding								
Small Power Equipment/Tools								
Field Offices: Job Office	13	Mo.					360	4680
Architect/Owner's Office								
Temporary Telephones	14	Mo.					200	2800
Utilities		LS						2000
Temporary Toilets	14	Mo.					60	840
Storage Areas & Sheds		LS						4000
Temporary Utilities: ~~Heat~~ Allow	13	Mo.					200	2600
~~Light & Power~~								
~~Water~~								
PAGE TOTALS								240720

Page 1 of 2

DESCRIPTION	QUANTITY	UNIT	MATERIAL/EQUIPMENT		LABOR		TOTAL COST	
			UNIT	TOTAL	UNIT	TOTAL	UNIT	TOTAL
Totals Brought Forward								240720
Winter Protection: Temp. Heat/Protection								7500
Snow Plowing								1500
Thawing Materials								
Temporary Roads	Sub							12000
Signs & Barricades: Site Sign								
Temporary Fences (chain link)	1000	LF						5250
Temporary Stairs, Ladders & Floors								
Photographs								600
Clean Up								
Dumpster								
Final Clean Up								9000
Punch List								
Permits: Building 1%								30000
Misc.								
Insurance: Builders Risk								15000
Owner's Protective Liability								18000
Umbrella								
Unemployment Ins. & Social Security								6500
Taxes								
City Sales Tax								
State Sales Tax								
Bonds								
Performance								30000
Material & Equipment								
Main Office Expense								
Special Items								
TOTALS:								376070

Page 2 of 2

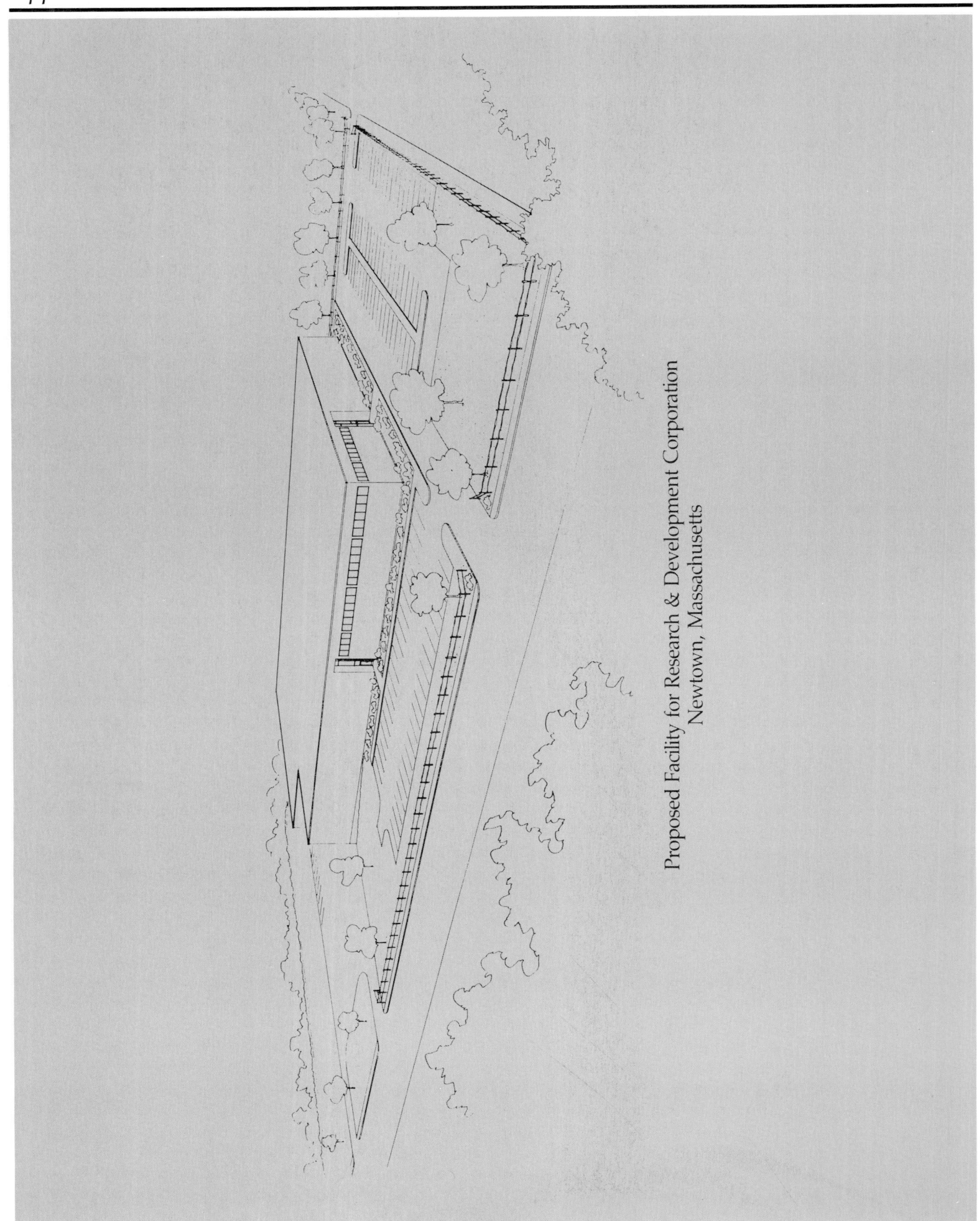
Proposed Facility for Research & Development Corporation
Newtown, Massachusetts

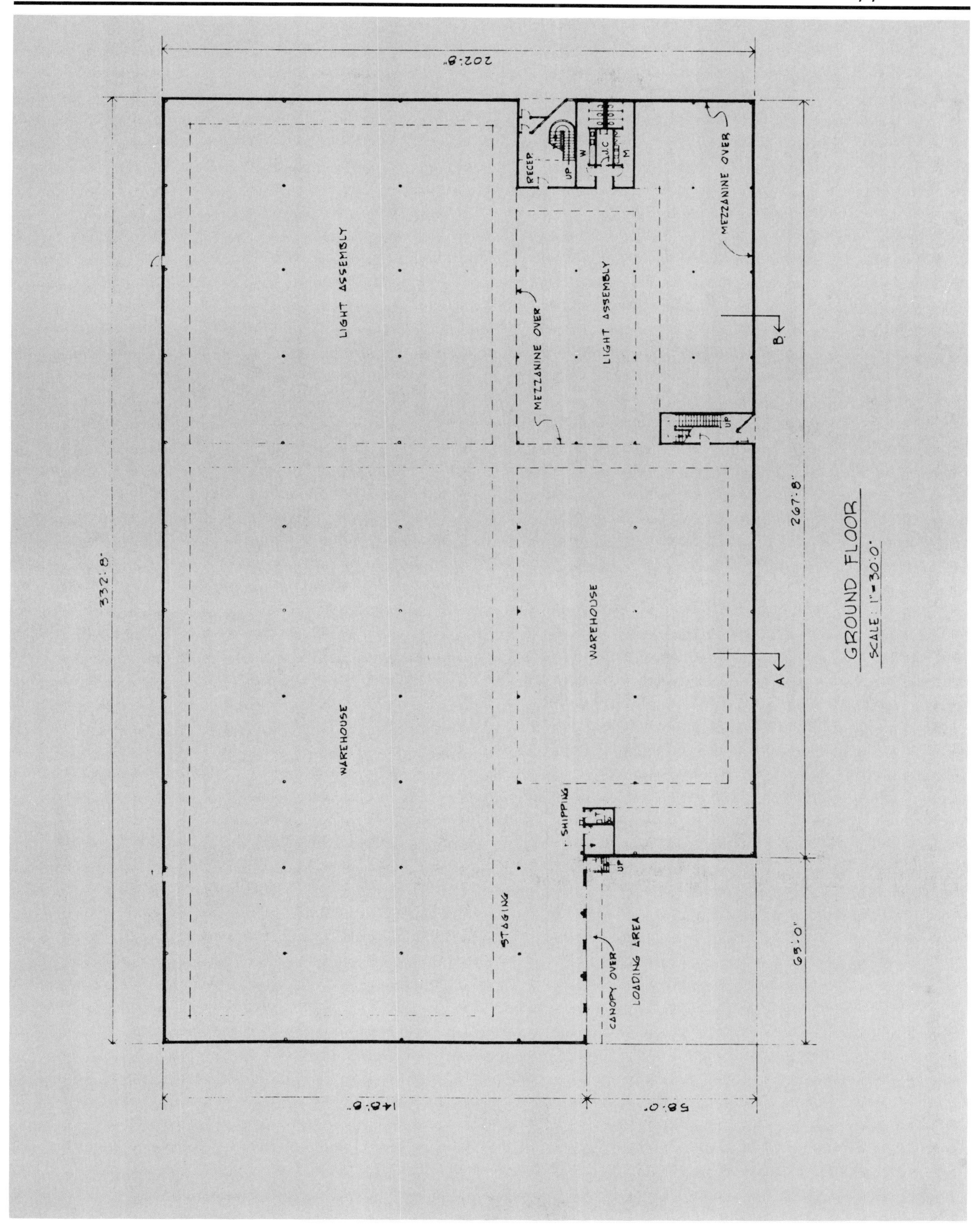
GROUND FLOOR
SCALE: 1" = 30'0"
LIGHT ASSEMBLY
LIGHT ASSEMBLY
MEZZANINE OVER
MEZZANINE OVER
WAREHOUSE
WAREHOUSE
STAGING
SHIPPING
CANOPY OVER
LOADING AREA
RECEP.
J.C
W
M
UP
A
B
202'8"
332'8"
267'8"
65'0"
148'8"
58'0"

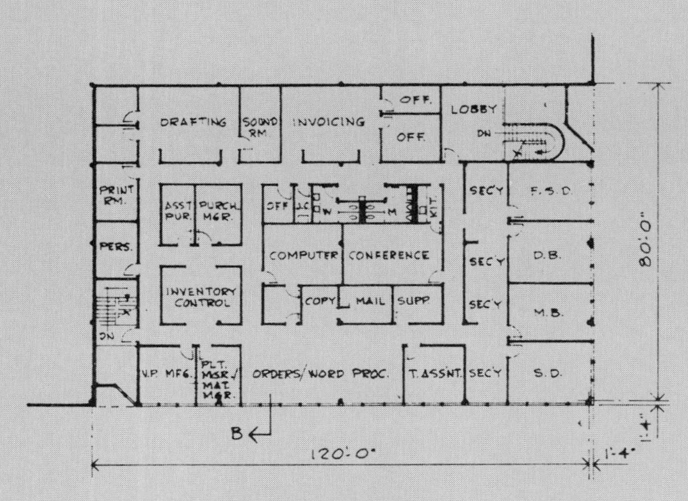
DRAFTING
SOUND RM
INVOICING
OFF.
OFF.
LOBBY
DN
PRINT RM.
ASST PUR.
PURCH. MGR.
OFF
M
KIT.
SEC'Y
F.S.D.
PERS.
COMPUTER
CONFERENCE
SEC'Y
D.B.
INVENTORY CONTROL
COPY
MAIL
SUPP
SEC'Y
M.B.
DN
V.P. MFG.
PLT. MGR./ MAT. MGR.
ORDERS/WORD PROC.
T. ASSNT.
SEC'Y
S.D.
B
120'-0"
1'-4"
80'-0"
MEZZANINE
SCALE: 1" = 30'-0"

R.S. MEANS COMPANY INC.

PLANTRAC (c) TIME NOW = 1 MAR 87 NETWORK = DB SELECTION 2 SEQUENCE 4 VERSION 22 PAGE NO. 1

SCHEDULE DESIGN BUILD MODEL 07/16/87 01:26:59

T Y	START EVENT	END EVENT	DESCRIPTION	DURATION (D)	RESP CODE	COST CODE	EARLIEST START	EARLIEST FINISH	LATEST START	LATEST FINISH	TOTAL FLOAT (D)	D I	RC	T	RESOURCES QTY.	FROM	FOR
S	212	603	ORDER & DELIVER STRUCTURAL STEEL	90.0	-	-	3/ 2/87	7/ 7/87	5/ 4/87	9/ 9/87	45.0						
S	212	726	ORDER & DELIVER MISC. IRON - STAIRS & H.M.FRAMES	80.0	-	-	3/ 2/87	6/22/87	9/14/87	1/ 8/88	137.0						
S	212	824	ORDER & DELIVER WINDOWS & STOREFRONT	80.0	-	-	3/ 2/87	6/22/87	10/27/87	2/22/88	167.0						
S	212	966	ORDER & DELIVER DOORS - MILLWORK & HARDWARE	100.0	-	-	3/ 2/87	7/21/87	10/20/87	3/14/88	162.0						
	81	321	ORDER & DELIVER REINFORCING	30.0	-	-	3/ 2/87	4/10/87	5/23/87	7/ 9/87	62.0						
	412	612	TEMPORARY SERVICES	5.0	-	-	3/ 2/87	3/ 6/87	5/ 8/87	5/14/87	49.0						
S	212	612	MOVE IN & LAYOUT	10.0	-	-	5/ 1/87	5/14/87	5/ 1/87	5/14/87	0.0						
	612	1012	CLEAR GRUB	8.0	-	-	5/15/87	5/27/87	5/15/87	5/27/87	0.0						
	1012	1212	STRIP TOPSOIL	12.0	-	-	5/28/87	6/12/87	5/28/87	6/12/87	0.0						
	1212	1412	START CUT & FILL	12.0	-	-	6/15/87	6/30/87	6/15/87	6/30/87	0.0						
	1412	301	EXCAVATE FOOTINGS	3.0	-	-	7/ 1/87	7/ 6/87	7/ 1/87	7/ 6/87	0.0						
	1412	1612	COMPLETE CUT & FILL	10.0	-		7/ 1/87	7/15/87	2/26/88	3/10/88	164.0						
	301	321	FORM CONTINUOUS FOOTING	3.0	-	-	7/ 7/87	7/ 9/87	7/ 7/87	7/ 9/87	0.0						
	321	361	FORM SPREAD FOOTINGS	3.0	-	-	7/10/87	7/14/87	7/10/87	7/14/87	0.0						
	321	381	REINFORCE CONTINUOUS FOOTING	1.0	-	-	7/10/87	7/10/87	7/14/87	7/14/87	2.0						
	381	421	CONCRETE CONTINUOUS FOOTING	1.0	-	-	7/13/87	7/13/87	7/15/87	7/15/87	2.0						
	361	401	REINFORCE SPREAD FOOTINGS	1.0	-	-	7/15/87	7/15/87	7/15/87	7/15/87	0.0						
	1612	1812	EXCAVATE UTILITIES	2.0	-	-	7/16/87	7/17/87	3/11/88	3/14/88	164.0						
	401	461	CONCRETE SPREAD FOOTINGS	1.0	-	-	7/16/87	7/16/87	7/22/87	7/22/87	4.0						
	421	441	START FORM & REINFORCE WALLS	5.0	-	-	7/16/87	7/22/87	7/16/87	7/22/87	0.0						
	1812	2012	UTILITY PIPING	5.0	-	-	7/20/87	7/24/87	3/15/88	3/21/88	164.0						
	1912	2212	EXCAVATE SITE DRAINAGE	2.0	-	-	7/20/87	7/21/87	3/16/88	3/17/88	165.0						
	2212	2612	PIPE & CATCH BASINS	5.0	-	-	7/22/87	7/28/87	3/18/88	3/24/88	165.0						
	441	501	START PLACE CONCRETE WALLS	1.0	-	-	7/23/87	7/23/87	8/ 7/87	8/ 7/87	11.0						
	461	521	COMPLETE FORM & REINF WALLS PITS	12.0	-	-	7/23/87	8/ 7/87	7/23/87	8/ 7/87	0.0						
	461	541	BACKFILL FOOTINGS	2.0	-	-	7/23/87	7/24/87	8/ 7/87	8/10/87	11.0						
	2012	2412	BACKFILL	3.0	-	-	7/27/87	7/29/87	3/22/88	3/24/88	164.0						
	2612	2812	BACKFILL	2.0	-	-	7/30/87	7/31/87	3/25/88	3/28/88	164.0						
	521	541	COMPLETE PLACE CONCRETE WALLS	1.0	-	-	8/10/87	8/10/87	8/10/87	8/10/87	0.0						
	541	5612	DAMPROOF & INSULATE	5.0	-	-	8/11/87	8/17/87	8/11/87	8/17/87	0.0						
	541	5812	UNDERSLAB PIPING	5.0	-	-	8/11/87	8/17/87	8/20/87	8/26/87	7.0						
	5612	5812	COMPLETE BACKFILL	7.0	-	-	8/18/87	8/26/87	8/18/87	8/26/87	0.0						
	5812	603	COMPACTED GRAVEL FILL	9.0	-	-	8/27/87	9/ 9/87	8/27/87	9/ 9/87	0.0						
	603	623	ERECT STRUCTURAL STEEL & DECK	35.0	-	-	9/10/87	10/29/87	9/10/87	10/29/87	0.0						
	623	702	MEZZ FLOOR SLAB	5.0	-	-	10/30/87	11/ 5/87	10/30/87	11/ 5/87	0.0						
	669	1009	ROUGH ELECTRICAL	50.0	-	-	10/30/87	1/13/88	1/11/88	3/21/88	47.0						

R.S. MEANS COMPANY INC.
=======================

PLANTRAC (c) TIME NOW = 1 MAR 87 NETWORK = DB SELECTION 2 SEQUENCE 4 VERSION 22 PAGE NO. 2

SCHEDULE DESIGN BUILD MODEL 07/16/87 01:28:10

T Y	START EVENT	END EVENT	DESCRIPTION	DURA-TION (D)	RESP CODE	COST CODE	EARLIEST START	EARLIEST FINISH	LATEST START	LATEST FINISH	TOTAL FLOAT (D)	C I	RESOURCES FC	T	QTY.	FROM	FOR
	645	804	INSULATION & ROOFING	40.0	-	-	10/30/87	12/29/87	12/15/87	2/10/88	38.0						
	688	1100	ROUGH PLUMBING	50.0	-	-	10/30/87	1/13/88	1/25/88	4/ 4/88	57.0						
	702	726	SLAB ON GRADE	15.0	-	-	11/ 6/87	11/30/87	12/17/87	1/ 8/88	27.0						
	784	804	BRICK & BLOCK	65.0	-	-	11/ 6/87	2/10/88	11/ 6/87	2/10/88	0.0						
	799	1106	H V A C	70.0	-	-	11/ 6/87	2/18/88	1/11/88	4/18/88	42.0						
	726	866	STEEL STUDS	7.0	-	-	12/ 1/87	12/ 9/87	2/ 9/88	2/18/88	48.0						
	743	866	ERECT STAIRS	6.0	-	-	12/ 1/87	12/ 8/87	2/10/88	2/18/88	49.0						
	760	1106	SPRINKLER SYSTEM	70.0	-	-	12/ 1/87	3/10/88	1/11/88	4/18/88	27.0						
	804	844	EXTERIOR BLOCK	5.0	-	-	2/11/88	2/18/88	2/11/88	2/18/88	0.0						
	966	946	FURRING INSULATION & GYPSUM BOARD	12.0	-	-	2/19/88	3/ 7/88	2/19/88	3/ 7/88	0.0						
	886	986	CEILING INSULATION	3.0	-	-	2/19/88	2/23/88	3/17/88	3/21/88	19.0						
	844	926	INTERIOR BLOCK	10.0	-	-	2/19/88	3/ 3/88	2/23/88	3/ 7/88	2.0						
	824	926	WINDOWS & STOREFRONT	10.0	-	-	2/19/88	3/ 3/88	2/23/88	3/ 7/88	2.0						
	904	926	OVERHEAD DOORS & DOCK LEVELORS	8.0	-	-	2/19/88	3/ 1/88	2/25/88	3/ 7/88	4.0						
E	2812	11410	PAVE WALKS TOPSOIL SEED & PLANT	35.0	-	-	3/ 7/88	4/22/88	3/29/88	5/16/88	16.0						
	946	1066	CERAMIC & QUARRY TILE	10.0	-	-	3/ 8/88	3/21/88	3/ 8/88	3/21/88	0.0						
E	1009	11410	LIGHT FIXTURES & COMPLETE ELECTRICAL	40.0	-	-	3/ 8/88	5/ 2/88	3/22/88	5/16/88	10.0						
	986	1026	CEILING SUSPENSION	8.0	-	-	3/ 8/88	3/17/88	3/22/88	3/31/88	10.0						
	966	1106	CARPENTRY DOORS & HARDWARE	25.0	-	-	3/ 8/88	4/11/88	3/15/88	4/18/88	5.0						
	967	1106	EXTERIOR SOFFIT	5.0	-	-	3/ 8/88	3/14/88	4/12/88	4/18/88	25.0						
	1026	1046	CEILING	12.0	-	-	3/18/88	4/ 4/88	4/ 1/88	4/18/88	10.0						
	1066	1106	PAINT & FINISH	20.0	-	-	3/22/88	4/18/88	3/22/88	4/18/88	0.0						
	10811	1106	KITCHEN	5.0	-	-	3/22/88	3/28/88	4/12/88	4/18/88	15.0						
	1108	1106	PLUMBING FIXTURES	10.0	-	-	3/22/88	4/ 4/88	4/ 5/88	4/18/88	10.0						
	1106	11210	RESILIENT FLOOR BASE & CARPET	10.0	-	-	4/19/88	5/ 2/88	4/19/88	5/ 2/88	0.0						
E	11210	11410	CLEANUP & MOVEOUT	10.0	-	-	5/ 3/88	5/16/88	5/ 3/88	5/16/88	0.0						

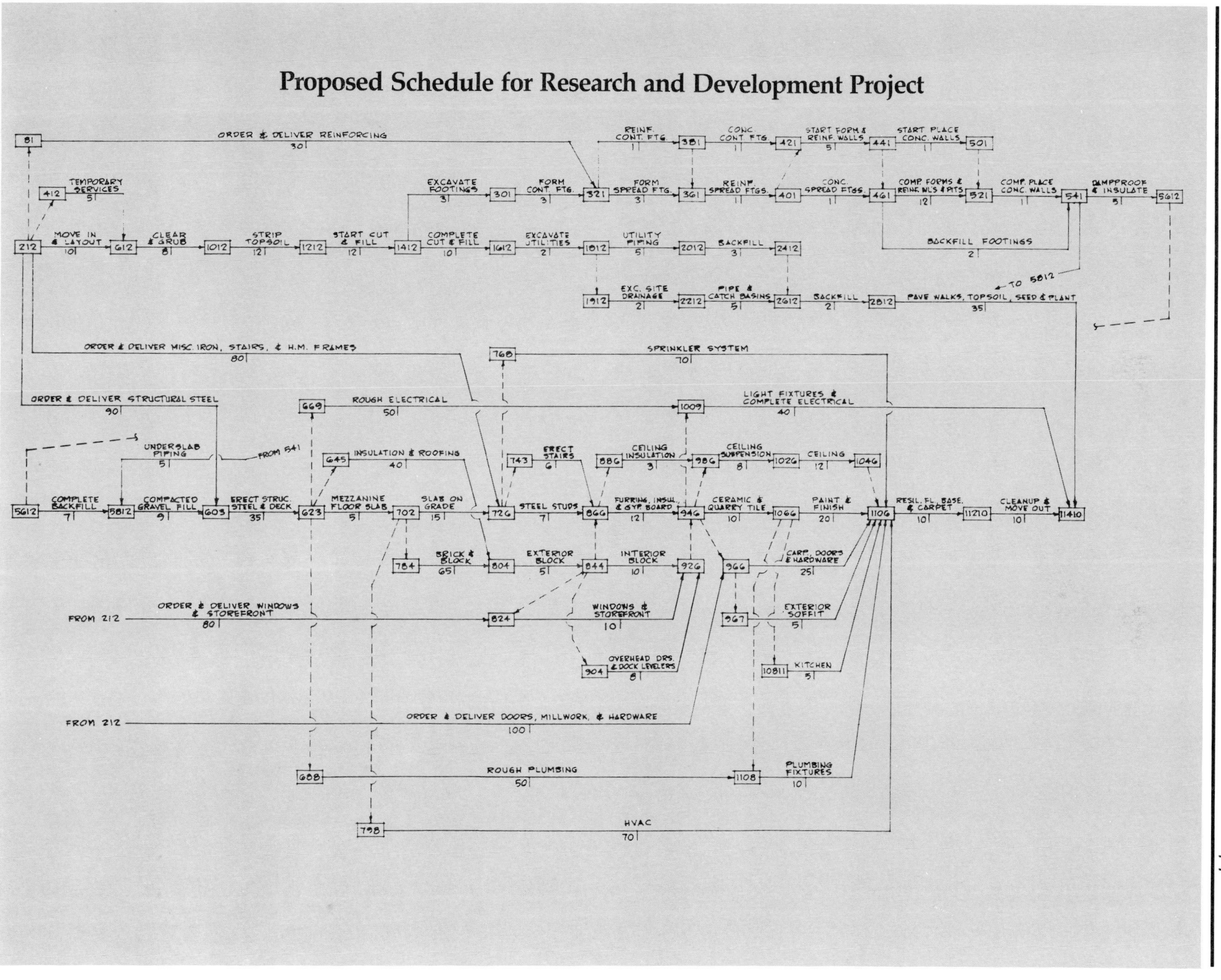

Proposed Schedule for Research and Development Project

Back-up Documents

The back-up documents for a design/build project can be in varying formats and may include a range of information. For the case study project, the following back-up documents are shown:

- Code Review
- Specification Aid
- Preliminary Sketches

Code Review

The code review for the case study also includes comments to further explain some of the articles. An actual code review would concern itself only with articles that pertain to the particular project at hand. Articles having no relevance to the design would be ignored.

The code source used for this illustration is BOCA (Building Officials and Code Administrators), 1986 edition. Many states utilize BOCA verbatim. Others refer to BOCA as a standard from which sections are modified to suit that state's particular requirements. Many regions use the same approach, as can be seen in the Southern Building Code.

Spec-Aid

The next back-up document shown is the Specification-Aid form (from *Means Forms* book). Reviewing and filling out this specification aid helps to ensure that all basic items are included. Because the categories are already listed, they need only be checked off. Thus, only minimal time is required to complete the Spec-Aid. The information recorded on the Spec-Aid is derived from several sources, including meetings with the owner and an investigation of his current facilities and the project site.

Preliminary Drawing

A preliminary drawing can be very rough and incomplete. Such drawings are based on broad assumptions and a limited knowledge of client requirements. For example, it may only be known that the client has certain square footage requirements for specific tasks. Based on these area requirements and the anticipated site, a concept is considered to test the given parameters of the building. A number of items must be considered at this time, including traffic access for cars and trucks, and such elements as the possible need for access to a railroad or helicopter landing.

Rough sketches of the project are shown as part of the case study support documentation. This example is used to illustrate the design evolution process of the building.

Case Study Code Review: BOCA Basic Building Code/1986

Article 1 – Administration and Enforcement

This article is concerned with the definition of structures and the various types of work that can be performed on them. It also deals with permitting the duties and powers of the building official.

Article 2 – Definitions

This article defines words and terms within the code.

Article 3 – Use Group Classification

Table 305.2 – Building Use Group

The use groups are as follows:

A Assembly
B Business
E Educational
F Factory and Industrial
H High Hazard
M Mercantile
R Residential
S Storage
U Utility & Miscellaneous

Select Use Group F–1, Moderate Hazard, Factory.
The case study building qualifies as Use Group F, and more specifically, as F-1, Moderate Hazard. If a conflict occurred in the use definition between B (Business due to the Office), or S (Storage due to the Warehouse), then the code stipulates:

Subsection 313.1, "The provisions of the code applying to each use shall apply to such parts of the building as come within the use group ..... conflicts shall defer to the greater public safety ....."

Article 4 – Types of Construction Classifications

The common construction for industrial buildings is structural steel and ballast joists. The fire resistance rating is considered non-combustible. In Table 401, type 2C non-combustible, unprotected is selected.

Section 403.1 states, ".... Type 2 construction are those in which walls, partitions, structural elements, floors, ceilings, roofs, and exits are constructed of approved non-combustible material ..."

Table 401 then proceeds to define fire resistance levels for a Type 2c building. Specifically, the following is the required fire resistance of the elements.

- Bearing exterior walls 1 hr.
- Fire walls/party walls 2 hrs.
- Fire enclosures of exits 2 hrs.
- Shafts 2 hrs.
- Exit corridors 1 hr.

The partition between the Office and the Assembly is considered a party wall. Similarly, the partition between the Assembly and the Warehouse is a party wall; both are to be of 2 hr. construction.

Article 5 – General Building Limitations

Table 501 General Area & Height Limitations:

For type 2C, Use Group F–1, this building is limited to two stories and a maximum of 30 feet in height. It also establishes a maximum floor area of 9600 S.F. per floor.

Section 502, Area Exceptions, subsection 502.2 Street frontage increase: "When a building has 25% of the perimeter fronting on a street ..... accessible from the street ..... the tabulation areas may be increased by 2% for each 1% excess."

The calculations are as follows:

$$\frac{\text{Frontage}}{\text{Perimeter}} \times 100 = \% - 25\% \times 2 = \%$$

$$\frac{535'\text{–}4''}{1070'\text{–}8''} \times 100 = 50.00\% - 25\% \times 2 = 50.00\%$$

Subsection 502.3 Automatic fire suppression system: "When a building ... is equipped throughout with an approved automatic fire suppression system, the tabulated areas may be increased 200% ....."

The allowable area by code may be found as follows:

Base 2 floors @ 9600 S.F. each	=	19,200 sf
Frontage multiplier 19,200 S.F. x 1.5000	=	28,800 sf
Sprinkler multiplier 28,800 S.F. x 3	=	86,400 sf

The height limitations remain at 30 feet and are in line with the allowable building height and size concept of 73,250 S.F.

Section 504.0 Unlimited Areas: this section allows one story buildings of Use Group F–1, moderate hazard, which does not exceed 85 feet in height and has an automatic fire suppression system.

"..... shall not be limited (in area) ..... (and) complying with the provisions of section 1702.0."

Section 1702.0 covers fire suppression systems.

Further conditions of this code allowance are concerned with travel distances to exitways. The case study building, as designed, conforms to these requirements.

The balance of Article 5 concerns itself with property and street line encroachment and special structures, such as Marquees.

Article 6 – Special Use and Occupancy Requirements

This article is concerned with the use of hazardous and explosive storage spaces, such as for oil and solvents. Included are other hazardous spaces such as storage areas, projection rooms, paint spray areas, grain silos, parking garages, tents, and similar types of structures.

Subsection 622.1 Mezzanines; "The aggregate area of a mezzanine ..... within a story shall not exceed one-third of the area of that story."

The area of the ground floor in the case study project is 63,650 S.F.

63,650 S.F. x 1/3 = 21,195

Proposed Mezzanine Area for case study: 9600 S.F., well within the code limits.

Article 7 – Interior Environment Requirement

Subsection 708.1 limits ceiling height to not less than 7′–6″.

Subsection 713.1.2 requires fire emergency ventilation systems in the corridors for Use Group F.

Article 8 – Means of Egress

Table 806 Maximum Occupancy Load:

Industrial	200 S.F. gross (area) per person
Office	100 S.F. gross (area) per person

Table 807 Length of Travel To Exit:

Use Group F	300 L.F. (with fire suppression)

From Table 806, the following is derived:

Industrial	63,650 S.F./200	=	318 people
Office	9,600 S.F./100	=	96 people

Table 809 Number of Exits per Occupancy Load:

For 500 persons or less, 2 exits per space are required (except as required by 807).

The balance of this article deals with ramps, waiting areas, stair construction, escalators, fire escapes, exit signs, and aisles.

Article 9 – Structural Load & Stress

Table 906 Minimum Live Loads:

Office Buildings	50 P.S.F.
Reading Rooms	150 P.S.F.
EDP Rooms	150 P.S.F.

Table 910 Roof Live Loads:

FLAT	20 P.S.F.

Table 911 Snow Loads 100 Year Storm:

Massachusetts	35 P.S.F.

The balance of Article 9 concerns load formulas, wind loading, earthquake load, and structural considerations and formulas. This article relates to design conditions which apply primarily to complicated multi-story buildings; though applicable to one-story industrial buildings, these conditions are not major factors due to the normal design safety factors.

Article 10 – Foundation System

Table 1001 Presumptive Surface Bearing Values

Line 8	loose, coarse sand and sand gravel mixture	3 tons/S.F.

The code in this article continues with standards for differing foundation systems. The more stringent requirements (based on code, structural design practice, or earthquake conditions) will govern. Most industrial buildings meet and exceed the requirements.

Article 11 – Materials and Tests

The standards for materials such as mortar, brick, cement masonry units, concrete units, glass, natural stone, plaster, gypsum board and plywood are stated in this article.

Article 12 – Steel, Masonry, Concrete, and Lumber Construction

This article describes the limitation imposed on the various structural systems and their design reference standards.

Article 13 – Building Enclosures, Walls, and Wall Thickness

Standards for enclosures, wind loading on windows and curtain wall systems, and wall veneers such as brick, stone, tile, and aluminum siding. This article also deals with vermin and rat proofing.

Article 14 – Fire-Resistance Construction Requirements

Table 1402 Fire Grading of Use Groups:

Use Group F	3 hrs.

Subsection 1413.3, Roofs 20 feet or higher, states: "When every part of the structural framework of roofs in building Type 1 and Type 2 is 20 feet or more above the floor ..... omission of all fire protection of the structural members (fire proofing) is permitted."

Section 1415.0 obliges that the separation between the Assembly space and the Warehouse space is a fire wall, with fire doors as a means of interconnection.

Section 1417.0 Fire Dampers:

If the air conditioning system was extended from the office to the assemblies area, then fire dampers would be required in the duct at the point where the duct intersects the fire wall.

Section 1420.0 Fire Stopping:

Between the main floor and the mezzanine there are openings that must be stopped up with a fire resistant material to prevent flame spread vertically from the lower floor to the mezzanine.

Fire stopping can be lumber in partitions, subsection 1420.2, ".....2 thicknesses of 1 inch lumber with broken lap-joints ....." Fire stopping material can also be fiberglass insulation stuffed into holes and slots.

Article 15 – Masonry Fireplaces

Article 16 – Mechanical Equipment and Systems

The article refers the reader to mechanical codes provided by ASHRAE (American Society of Heating, Refrigerating and Air Conditioning Engineers) and ASTM (American Society of Mechanical Engineers). The mechanical designers would be cognizant of these codes and base their designs accordingly.

Article 17 – Fire Protection Systems

Subsection 1702.1 states: "Fire suppression systems shall be installed and maintained ..... as indicated in Subsection 1702.2 through Subsection 1702.21."

Subsection 1702.9 requires fire suppression systems for Use Group F:

"1. when more than 12,000 in area
2. when more than three stories in height"

A fire suppression system was previously determined to be needed for the case study due to the area requirements in Article 5.

Article 18 – Precautions During Building Operations

This article is concerned with demolition, maintaining a building, selective remedial work in an occupied building, and life safety and health standards for the work involved.

Article 19 – Signs

All types of signs that are allowed are treated in this article. Included are wall signs, roof signs, pole-mounted signs, ground-mounted signs, and illuminated and marquee type signs.

Article 20 – Electric Wiring, Equipment, and System

This article refers to NFPA 70 (National Fire Protection Association), which is the National Electric Code.

Article 21 – Elevator, Dumbwaiter, and Conveyer

Reference is to ANSI A 17.1 (American National Standards Institute) primarily and deals with the installation, inspection, and operation of all forms of vertical conveyance.

Article 22 – Plumbing Systems

BOCA B/NPC-86 is the source reference for code compliance here. It deals with water supply, waste removal, plumbing fixtures, and venting.

Article 23 – Light-Transmitting Plastic Construction

ASTM is the reference for this article.

Subsection 2302.2, Use group F, allows glazing with approved plastic materials.

Subsection 2302.3.1 states, "The area of such glazing shall not exceed 25 percent of the wall face ....."

This article involves plastic roof coverings such as over pools, skylight assemblies, transparent partitions and light diffusing systems.

Article 24 – Energy Conservation

Formulas, charts, and other criteria are the bases for this article, which sets minimum energy conservation standards for the transmission of conditioned and outside air through perimeter exterior walls, windows, roofs, and openings. Article 24 also sets standards for mechanical and electrical systems.

SPEC-AID

DATE 11/13/87

DIVISION 1: GENERAL

PROJECT Proposed Facility LOCATION newtown, ma

Owner R&D Corp. **Architect** – **Project Mgr.** A. B.

Engineer: Structural ______ Plumbing ______

H.V.A.C. ______ Electrical ______

Contractor: General ______ Structural ______

Mechanical ______ Electrical ______

Building Type ______

Building Capacities: ______

Quality ☐ Economy ☐ Average ☑ Good ☐ Luxury Describe ______

Size

Ground Floor Area 5 1/2 Acres S.F.

Supported Levels (No.) 1 on grade x Area/Level 63650 S.F.

Supported Levels (No.) 2 mezzanine x Area/Level 9600 S.F.

Below Grade Area ______ S.F.

Other Area ______ S.F.

TOTAL GROSS AREA ______ S.F.

Floor to Floor Height: Maximum ______ Minimum ______ Average ______

Floor to ~~Ceiling~~ Clear Height: Maximum ______ Minimum ______ Average 23'-2"

Floor System Depth: Maximum ______ Minimum ______ Average ______

Building Codes ☐ City ______ ☐ County ______

☑ State ______ ☐ National ______

☐ Other ______ Seismic Zone ______

Zoning ☐ Residential ☐ Commercial ☑ Industrial ☐ None ☐ Other ______

Design Criteria Live Loads: Roof 55 psf. Total End Walls ______ psf. Window Openings ______ %

Supported Floor ______ psf. Side Walls ______ psf. Window Openings ______ %

Ground Floor ______ psf. WIND PRESSURE

Corridors ______ psf. ______ psf. from ______ to ______ ft.

Balconies ______ psf. ______ psf. from ______ to ______ ft.

Allow for Partitions ______ psf. ______ psf. from ______ to ______ ft.

Miscellaneous ______ psf. ______ psf. from ______ to ______ ft.

Comments ______

Typical Bay Spacing 30 x 40

Structural Frame ☐ Concrete ☑ Steel ☐ Wood ☐ Wall Bearing ☐ Other ______

Describe ______

Fireproofing ☑ None ☐ Columns ______ Hours ☐ Girders ______ Hours ☐ Beams ______ Hours ☐ Floor ______ Hours

Estimating Budget Estimate Due ______ 19 ______ Schematic Estimate Due ______ 19 ______

Preliminary Estimate Due ______ 19 ______ Final Estimate Due ______ 19 ______ at ______ % Working Drawings

Labor Market ☐ Highly Competitive ☑ Normal ☐ Non-Competitive ☐ Unreliable ☐ Union ☐ Non-union

Describe ______

Taxes Tax exempt ☑ No ☐ Yes State ______ % County ______ % City ______ % ~~Other~~ Town ______ % 25/1000

Bond ☑ Not Required ☐ Required ______

Bidding Date n/A Start Date ______ Construction Duration ______ Months

☐ Open Competitive ☐ Selected Competitive ☑ Negotiated ☐ Filed Bids ______

Contract ☑ Single ☐ Multiple Describe ______

Multiple Type asigned to General Contractor ☐ No ☐ Yes ______

Page 2 of 30

SPEC-AID DATE ____________

DIVISION 2: SITEWORK

PROJECT ____________ LOCATION ____________

Demolition Site: ☑ No ☐ Yes Allowance ____________ ☐ Separate Contract
Interior: ☑ No ☐ Yes ☐ Allowance ____________ ☐ Separate Contract
Removal From Site: ☐ No ☐ Yes Dump Location ____________ Distance ______
Topography ☐ Level ☑ Moderate Grades ☐ Steep Grades Describe ____________
Subsurface Exploration ☐ Borings ☐ Test Pits ☐ USDA Maps ☐ Other None Available
Performed by: ☐ Owner ☐ Engineer ☐ Contractor ____________
Site Area: Total 5.5 Acres to Clear ______ Acres To Thin ______ Acres Open ______ Acres
Clearing and Grubbing: ☐ No ☑ Light ☐ Medium ☐ Heavy ____________
Topsoil: ☐ No ☑ Strip ☐ Stockpile ☑ Dispose on Site ☐ Dispose off Site ______ Miles ☐ Furnish
Existing ______ Inches Deep Final Depth ______ Inches Describe ____________

Soil Type: ☑ Gravel ☑ Sand ☐ Clay ☐ Silt ☐ Rock ☐ Peat ☐ Other ____________
Rock Expected: ☑ No ☐ Ledge ☐ Boulders ☐ Hardpan ☐ Describe ____________
How Paid ____________
Ground Water Expected: ☑ No ☐ Yes Depth or Elevation ____________
Disposal by ☐ Pumping ☐ Wells ☐ Wellpoints ☐ Other ____________
Excavation: ☑ Grade and Fill on Site ☐ Dispose off Site ______ Miles ☐ Borrow Expected ______ Miles
Quantity Involved ____________
Describe ____________
Sheeting Required: ☑ No ☐ Yes Describe ____________
Protect Existing Structures: ☑ No ☐ Yes Describe ____________
Backfill: ☐ No ☑ Yes Area ______ Material ______ Inches Deep ______ % Compaction
Landscape Area ☐ No ☐ Yes Area ______ Material ______ Inches Deep ______ % Compaction
Building Area ☐ No ☐ Yes Area ______ Material ______ Inches Deep ______ % Compaction
Source of Materials ____________
Water Control: ☐ Ditching ☐ Sheet Piling ☐ Pumping ☐ Wells ☐ Wellpoints ☐ Pressure Grouting
☐ Chemical Grouting ☐ Other ____________
Describe ____________
Termite Control: ☑ No ☐ Yes Describe ____________
Special Considerations: ____________

Piles: ☑ No ☐ Yes ☐ Friction ☐ End Bearing ☐ Concrete ☐ Pipe, Empty ☐ Pipe, Concrete Filled ☐ Steel
☐ Step Tapered ☐ Tapered Thin Shell ☐ Wood ☐ Capacity ______ Tons
Size ______ Length ______ Number Required ____________
Caissons: ☑ No ☐ Yes ☐ Cased ☐ Uncased Capacity ____________
Size ______ Length ______ Number Required ____________
Pressure Injected Footings: ☑ No ☐ Yes ☐ Cased ☐ Uncased Capacity ____________
Size ______ Length ______ Number Required ____________
Special Considerations: ____________

Storm Drains: ☐ No ☑ Yes ☐ Asbestos Cement ☐ Bituminous Fiber ☑ Concrete ☑ Corrugated Metal ☐ ______
Size and Length ____________
Headwall: ☐ No ☑ Yes Type ______ Number ______
Catch Basins: ☐ No ☑ Yes ☐ Block ☐ Brick ☐ Concrete ☐ Precast Size ______ Number ______
Manholes: ☐ No ☑ Yes ☐ Block ☐ Brick ☐ Concrete ☐ Precast Size ______ Number ______
Building Sub Drains: ☐ No ☐ Yes ☐ Type ______ Length ______
French Drains: ☐ No ☐ Yes Size ______ Length ______
Trenches: Swales, etc.: ☐ No ☐ Yes Describe ____________
Rip Rap: ☐ No ☐ Yes Describe ____________
Special Considerations: ____________

Page 3 of 30

SPEC-AID

DIVISION 2: SITEWORK

Water Supply Existing Main: ☐ No ☑ Yes Location ______ Size 20"
Service Piping: By Utility ☐ By Others ☑ This Contract ______ Size ______
Wells: ☐ No ☐ Yes ☐ By Others ☐ This Contract ______ Capacity ______
Water Pumping Station: ☑ No ☐ Yes Type ______ Capacity ______
Sewers: ☐ No ☑ Yes ☐ By Others ☐ Asbestos Cement ☐ Concrete ☐ Plastic ☑ Vitrified Clay ☐ Exist 12"
Manholes: ☐ No ☑ Yes ☐ Block ☐ Brick ☐ Concrete ☐ Precast Size ______ Number ______
Sewage Pumping Station: ☐ No ☐ Yes Type ______ Capacity ______
Sewage Treatment: ☐ No ☐ Yes ☐ By Others ☐ This Contract ☐ Septic Tank ☐ Package Treatment Plant
Describe ______
~~**Special** Considerations:~~ Existing: gas piping 6"; Storm piping 36"

Driveways: ☐ No ☑ Yes ☐ By Others ☑ Bituminous ☐ Concrete ☐ Gravel ☐ ______ Thickness 2½"
Parking Area: ☐ No ☑ Yes ☐ By Others ☑ Bituminous ☐ Concrete ☐ Gravel ☐ ______ Thickness 2½"
Base Course: ☐ No ☑ Yes ☐ By Others ☑ Gravel ☐ Stone ☐ ______ Thickness 6"
Curbs: ☐ No ☑ Yes ☐ By Others ☑ Bituminous ☑ Concrete ☑ Granite ☐ ______ Size ______
Parking Bumpers: ☐ No ☐ Yes ☐ By Others ☐ Concrete ☐ Timber ______
Painting Lines: ☐ No ☑ Yes ☐ By Others ☐ Paint ☐ Thermo Plastic ☑ Traffic Lines ☑ Stalls ☐ ______
Guard Rail: ☐ No ☐ Yes ☐ By Others ☐ Cable ☐ Steel ☐ Timber ______
Sidewalks: ☐ No ☑ Yes ☐ By Others ☑ Bituminous ☐ Brick ☑ Concrete ☐ Stone
Width ______ Thickness ______
Steps: ☐ No ☑ Yes ☐ Brick ☑ Concrete ☐ Stone ☐ Timber ______
Signs: ☐ No ☐ Yes ☐ Stock ☐ Custom ______
Traffic Signals: ☐ No ☐ Yes ☐ By Others ______
Special Considerations: 150 parking spaces

Fencing: ☐ No ☑ Yes ☐ By Others ☐ Chain link ☐ Aluminum ☐ Steel ☐ Other wood
Height ______ Length ______ Gates ______
Fountains: ☐ No ☐ Yes ☐ By Others ______
Planters: ☐ No ☐ Yes ☐ By Others ☐ Asbestos Cement ☐ Concrete ☐ Fiberglass ☐ ______
Playground Equipment: ☐ No ☐ Yes ☐ By Others ☐ Benches ______ ☐ Bleachers ______
☐ Bike Rack ______ ☐ Goal Posts ______ ☐ Posts ______
☐ Running Track ______
☐ See Saw ______ ☐ Shelters ______ ☐ Slides ______
☐ Swings ______ ☐ Whirlers ______ ☐ ______
Playing Fields: ☐ No ☐ Yes ☐ By Others ______
Railroad Work: ☑ No ☐ Yes ☐ By Others Weight ______ lb. per L.Y. ☐ New ☐ Relay Length ______
Turnout: ☐ No ☐ Yes ______ ☐ Bumpers ______ ☐ Derails ______
Wheel Stops ______ ☐ Others ______
Retaining Walls: ☑ No ☐ Yes ☐ By Others ☐ Gravity Concrete ☐ Cantilever Concrete ☐ Steel Bin ☐ Cribbing
☐ Timber ☐ Other ______ Height ______ Length ______
Irrigation System: ☐ No ☑ Yes ☐ By Others ______
Tennis Courts: ☐ No ☐ Yes ☐ By Others Type ______ Number ______
Trash Closures: ☐ No ☐ Yes Size ______
Lawns & Planting: ☐ No ☑ Yes ☐ By Others ______ Allowance ✓
Topsoil: ☐ No ☑ Yes ☐ By Others Depth ______ Inches Source ______
Shrubs: ☐ No ☑ Yes ☐ By Others Describe ______ Allowance ______
Trees: ☐ No ☑ Yes ☐ By Others Describe ______ Allowance ______
Seeding: ☐ No ☑ Yes ☐ By Others Describe ______
Sodding: ☐ No ☑ Yes ☐ By Others Describe ______ Thickness ______
☐ Ground Cover ______ ☐ Edging ______ ☐ Mulching ______
Special Considerations: ______

Page 4 of 30

SPEC-AID

DATE ______

DIVISION 3: CONCRETE

PROJECT ______ LOCATION ______

Foundations Bearing on: ☐ Rock ☑ Earth ☐ Piles ☐ Caissons ☐ Other ______

Bearing Capacity ______

Footings Pile Caps: ☑ No ☐ Yes ______ psi Size ______

Forms ______ Reinforcing ______ Waterproofing ______

Spread Footings: ☐ No ☑ Yes ___ psi Size ______ Soil Bearing Capacity ______

Forms ______ Reinforcing ______ Waterproofing ______

Continuous Footings: ☐ No ☑ Yes ______ psi Size ______

Forms ______ Reinforcing ______ Waterproofing ______

Grade Beams: ☑ No ☐ Yes ______ psi Size ______

Forms ______ Reinforcing ______ Waterproofing ______

Piers: ☐ No ☑ Yes ______ psi Size ______

Forms ______ Reinforcing ______ Finish ______

Anchor Bolts: ☐ No ☑ Yes Size ______

Grout Column Base Plates: ☐ No ☑ Yes ______

Underslab Fill: ☐ No ☑ Yes Material 6" Gravel Depth ______

Vapor Barrier: ☐ No ☑ Yes Material ______ Thickness ______

Perimeter Insulation: ☐ No ☑ Yes Material 1" Rigid Board Dimensions ______

Slab on Grade: ☐ No ☑ Yes ______ psi 4000 Thickness 6"

Forms: ☐ Cold Keyed ☐ Expansion ☐ Other ______ Spacing ______

Reinforcing: ☐ No ☑ Mesh ☐ Bars ______ Type ______

Finish: ☐ Screed ☐ Darby ☐ Float ☐ Broom ☑ Trowel ☐ Granolithic ______

Special Finish: ☐ No ☐ Hardner ☐ Colors ☐ Abrasives ☐ ______

Columns: ☐ No ☐ Round ☐ Square ☐ Rectangular ☐ Precast ☑ Steel ☐ Encased Steel ☐ Lightweight

______ psi Size ______

Forms: ☐ Optional ☐ Framed Plywood ☐ Plywood ☐ Fiber Tube ☐ Steel ☐ Round Fiberglass ☐ ______

Reinforcing: ☐ No ☐ Square Tied ☐ Spirals Grade ______ Bar Sizes ______ Type Splice ______

Finish: ☐ Break Fins ☐ Rubbed ☐ Other ______

Elevated Slab System: ☐ No ☐ Flat Plate ☐ Flat Slab ☐ Domes ☐ Pans ☐ Beam & Slab ☐ Lift Slab ☑ Composite

☑ Floor Fill ☐ Roof Fill ☐ Standard Weight ☑ Lightweight Concrete Strength 4000 psi

Forms: ☐ Optional ☐ Plywood ☐ Other ______ Ceiling Height ______

Reinforcing: ☑ Mesh ☐ Bars Grade ______ Size: ______

Post-tension: ☐ No ☐ Simple Spans ☐ Continuous Spans ______ Depth ______

☐ Grouted ☐ Ungrouted Perimeter Conditions ______

Slab Finish: ☐ Screed ☐ Darby ☐ Float ☐ Broom ☑ Trowel ☐ Granolithic ☐ ______

Special Finish: ☐ No ☐ Hardener ☐ Colors ☐ Abrasives ☐ ______

Ceiling Finish: ☐ No ☐ Break Fins ☐ Rubbed ☐ Other Deck

Beams: ☐ No ☑ Steel ☐ Encased Steel ☐ Precast ☐ Regular Weight ☐ Lightweight ☐ Steel Composite ______

Description: ______

Forms: ☐ Optional ☐ Framed Plywood ☐ Plywood ☐ Steel ☐ Other ______ Ceiling Height ______

Reinforcing: ☐ Conventional ☐ Post-tension ☐ Simple Span ☐ Continuous Spans ______ Depth ______

☐ Grouted ☐ Ungrouted Perimeter Conditions ______

Walls: ☐ No ☐ Precast ☐ Tilt up ☐ Regular Weight ☐ Lightweight ______ psi Thickness ______

Forms: ☐ Optional ☐ Framed Plywood ☐ Plywood ☐ Steel ☐ Slipform ☐ ______

Reinforcing: ☐ No ☐ Bars Grade ______ Clear Height ______

Finish: ☐ No ☐ Break Fins ☐ Rubbed ☐ ______

Stairs: ☐ No ☐ Precast ☐ Ground Cast ☐ Form Cast ☑ Pan Fill Treads ☐ ______

Forms: ☐ Plywood ☐ Steel ☐ Prefab Steel, Left in Place ☐ ______

Reinforcing: ☐ Conventional Grade ______

Finish: ☐ All Surfaces ☐ Treads ☐ Risers ☐ Abrasives ☑ Nosings ______

Page 5 of 30

SPEC-AID

DIVISION 3: CONCRETE

Reinforcing Splices: ☐ No ☐ Yes ☐ Lap Type ☐ Compression Only ☐ 125% Yield ☐ Full Tension
☐ Horizontal ☐ Vertical ☐ Special ______

Gunite: ☐ No ☐ Yes ______

Cast in Place Special Considerations: ______

Copings: ☐ No ☐ Yes Size ______ Finish ______

Curbs: ☐ No ☐ Yes Size ______ Finish ______

Joists: ☐ No ☐ Yes Live load ______ psf. Span ______ Size ______
Describe ______

Lift Slab: ☐ No ☐ Yes No. Slabs ______ Thickness ______ Inches Columns Spacing ______
Story Height ______ ☐ Conventional Reinforcing ☐ Post-tension ______

Lintels: ☐ No ☑ Yes ☑ Doors ☑ Windows ☐ Other ______
☐ Conventional Reinforcing ☐ Prestressed ______

Prestressed Precast Floors: ☐ No ☐ Yes ☐ Plank ☐ Multiple Tee Depth ______ Span ______ Width ______
Roofs: ☐ No ☐ Yes ☐ Plank ☐ Double Tee ☐ Single Tee Depth ______ Span ______ Width ______
Supporting Beams: ☐ No ☐ Yes ☐ Cast in Place ☐ Precast Describe ______

Columns: ☐ No ☐ Yes ☐ Steel ☐ Cast in Place ☐ Precast Size ______
Walls: ☐ No ☐ Yes ☐ Multiple Tee ☐ Other Thickness ______ Height ______ Width ______

Stairs: ☐ No ☐ Yes ☐ Treads Only ☐ Tread & Riser Units ☐ Complete Stairs ☐ Other ______
Describe ______

Tilt Up Walls: ☐ No ☐ Yes Size ______ Finish ______

Wall Panels: ☐ No ☐ Yes ☐ Insulated ☐ Regular Weight ☐ Lightweight Panel Size ______
Finish: ☐ Gray ☐ White ☐ Exposed Aggregate ☐ Other ______
Reinforcing: ☐ Conventional ☐ Prestressed ☐ Plain ☐ Galvanized ______
Erection: ☐ No. Stories ______ Maximum Lift ______ Overhangs, etc. ______

Window Section: ☐ No ☐ Yes ☐ Size ______ Finish ______

Window Sills: ☐ No ☐ Yes Size ______ Finish ______

Precast Special Considerations: ______

Concrete Decks: ☐ No ☐ Yes ☐ Cast in Place ☐ Plank ☐ Topping ☐ Cement Fiber ☐ Channel Slab ☐ ______
Depth ______ Span ______ Sub Purlins ______ Roof Pitch ______

Concrete Fill: ☐ No ☐ Yes ☐ Regular Weight ☐ Lightweight Type ______ Depth ______

Formboard: ☐ No ☐ Yes ☐ Type ______ Depth ______ Spans ______
Sub Purlins: ☐ No ☐ Yes Span ______ Describe ______

Gypsum Decks Floor Plank: ☐ No ☐ Yes Depth ______ Span ______ Underlayment ______
Roofs: ☐ No ☐ Yes ☐ Cast in Place ☐ Plank Depth ______ Span ______ Pitch ______

Other Cementitious Decks: ☐ No ☐ Yes Describe ______

Cementitious Decks: Special Considerations ______

General Notes: Concrete Section ______

Page 6 of 30

SPEC-AID

DATE ____________

DIVISION 4: MASONRY

PROJECT ____________ LOCATION ____________

Exterior Walls: ☐ No ☑ Yes ☐ Load Bearing ☑ Non Load Bearing Story Height ____________
Describe ____________
Interior Walls: ☐ No ☑ Yes ☐ Load Bearing ☑ Non Load Bearing Ceiling Height ____________
Describe ____________
Mortar: ☐ Optional ☐ Type K ☐ Type 0 ☐ Type N ☐ Type S ☐ Type M ☐ Thinset ☐ ____________
☑ Colors Exterior ☐ Other ____________
Cement Brick: ☑ No ☐ Yes ☐ Solid ☐ Cavity ☐ Veneer ☐ ____________
Describe ____________
____________ Compressive Strength ____________ psi. ASTM No. ____________
Size ____________ Bond ____________ Joints ____________ Reinforcing ____________ Ties ____________
Common Brick: ☑ No ☐ Yes ☐ Solid ☐ Cavity ☐ Veneer ☐ ____________
Describe ____________
____________ Compressive Strength ____________ psi. ASTM No. ____________
Size ____________ Bond ____________ Joints ____________ Reinforcing ____________ Ties ____________
Face Brick: ☐ No ☑ Yes ☐ Solid ☐ Cavity ☐ Veneer ☐ ____________
Describe $250/m
____________ Compressive Strength ____________ psi. ASTM No. ____________
Size ☑ Standard ☐ Jumbo ☐ Norman ☐ Roman ☐ Engineer ☐ Double ☐ ____________
Allowance: ☐ No ☑ Yes $ 250/m per M Delivered ☐ Unglazed ☐ Single Glazed ☐ Double Glazed
Bond: ☑ Running ☐ Common ☐ English ☐ Flemish ☐ Stack ☐ ____________ Headers Every ____________ Course.
Joints: ☐ Concave ☑ Struck ☐ Flush ☐ Raked ☐ Weathered ☐ Stripped ☐ ____________
Reinforcing: ☐ No ☑ Yes Describe ____________
Wall Ties: ☐ No ☑ Yes Describe ____________
Anchor Bolts: ☐ No ☑ Yes Size ____________
Chimneys: ☑ No ☐ Yes ☐ Regular Brick ☐ Radial Brick Size ____________
Columns: ☐ No ☑ Yes Size ____________
Control Joints: ☐ No ☑ Yes Spacing ____________ Material ____________
Copings ☐ No ☐ Yes ☐ Concrete ☐ Stone ☐ ____________ Describe ____________
Fire Brick: ☑ No ☐ Yes ☐ Low Duty ☐ High Duty Describe ____________
Fireplaces: ☑ No ☐ Yes Describe ____________
Accessories ____________
Flooring: ☑ No ☐ Yes ☐ Laid Flat ☐ Laid on Edge ☐ Pattern ____________
☐ Regular ☐ Acid Resisting Describe ____________
Insulating Brick: ☑ No ☐ Yes Describe ____________
Insulation: ☐ No ☑ Yes ☐ Board ☐ Poured ☐ Sprayed Material ____________
Thickness ____________ Describe ____________
Lintels: ☐ No ☑ Yes ☐ Block ☑ Precast ☐ Steel Describe ____________
Masonry Restoration: ☑ No ☐ Yes ☐ Cut ☐ Recaulk ☐ Repoint ☐ Stucco Finish ☐ ____________
Sand Blast: ☐ No ☐ Yes Describe ____________
Steam Clean: ☐ No ☐ Yes Describe ____________
Piers: ☐ No ☑ Yes Size ____________
Pilasters: ☐ No ☐ Yes Size ____________
Refractory Work: ☐ No ☐ Yes Describe ____________
Simulated Brick: ☐ No ☐ Yes Material ____________ Describe ____________
Steps: ☐ No ☐ Yes Describe ____________
Vent Box: ☐ No ☐ Yes ☐ Aluminum ☐ Bronze Size ____________
Weep Holes: ☐ No ☐ Yes Spacing ____________ Describe ____________
Window Sills and Stools: ☐ No ☐ Yes ☐ Brick ☐ Concrete ☐ Stone ☐ ____________
Describe ____________

Page 7 of 30

SPEC-AID

DIVISION 4: MASONRY

Concrete Block: [ ] No [x] Yes [x] Exterior [x] Interior [x] Regular Weight [ ] Lightweight [ ] Solid [x] Hollow
[ ] Load Bearing [ ] Non Load Bearing Describe ______
______ Compressive Strength ______ psi. ASTM No. ______
Size: ______
Finish: [x] Regular [ ] Ground [ ] Ribbed [ ] Glazed [ ] ______
Bond: [x] Common [ ] Stack [ ] Other ______ Headers Every ______ Course.
Joints: [ ] Concave [x] Struck [ ] Flush [ ] Raked [ ] Weathered [ ] Stripped [ ] ______
Reinforcing: [ ] No [x] Yes ______ Strips Every ______ Course.
Wall Ties: [ ] No [x] Yes Describe ______
Bond Beams: [ ] No [x] Yes Size ______ Describe ______
Reinforcing ______ Grout ______
Lintels: [ ] No [x] Yes [x] Precast [ ] Steel [ ] Block Size ______ Describe ______
Reinforcing ______ Grout ______
Columns: [ ] No [x] Yes Size ______
Pilasters: [ ] No [ ] Yes Size ______

Glass Block: [x] No [ ] Yes [ ] Size ______ Type ______
Special Block ______
Describe ______

Glazed Concrete Block: [x] No [ ] Yes [ ] Solid [ ] Hollow Type ______
[ ] Non Reinforced [ ] Reinforced ______ Strips Every ______ Course.
Describe ______

Gypsum Block: [x] No [ ] Yes [ ] Solid [ ] Hollow Thickness ______ Describe ______
______

Grouting: [ ] No [x] Yes [x] Block Cores [ ] Bond Beams [ ] Cavity Walls [x] Door Frames [ ] Lintels [ ] ______
Describe ______

Insulation: [ ] No [x] Yes [x] Board [x] Poured [ ] Sprayed Material ______
Thickness ______ Describe ______

Solar Screen: [ ] No [ ] Yes Describe ______

Special Block: [ ] No [ ] Yes Describe ______
[ ] Parge Block [ ] Clean Cavity [ ] Spandrel Flashing ______

Special Considerations: ______

Ceramic Veneer: [ ] No [ ] Yes Describe ______

Structural Facing Tile: [ ] No [ ] Yes [ ] 6T Series [ ] 8W Series [ ] Other ______
Describe ______

Terra Cotta: [ ] No [ ] Yes [ ] Floors [ ] Partitions [ ] Fireproofing [ ] Load Bearing [ ] Non Load Bearing
Describe ______

Ashlar Stone: [ ] No [ ] Yes Type ______ Thickness ______
Describe ______

Rubble Stone: [ ] No [ ] Yes [ ] Coarsed [ ] Uncoarsed Type ______ Thickness ______
Describe ______

Cut Stone: [ ] No [ ] Yes [ ] Granite [ ] Limestone [ ] Marble [ ] Sand Stone [ ] Slate [ ] ______
[ ] Base ______ [ ] Columns ______ [ ] Coping ______
[ ] Curbs ______ [ ] Facing Panels ______
[ ] Flooring ______ [ ] Showers ______
[ ] Soffits ______ [ ] Stair Treads ______
[ ] Stairs ______ [ ] Thresholds ______
[ ] Window Sills ______ [ ] Window Stools ______
[ ] ______

Simulated Stone: [ ] No [ ] Yes Material ______ Describe ______

Special Stone ______

General Notes: Masonry ______
______
______

Page 8 of 30

SPEC-AID

DATE

DIVISION 5: METALS

PROJECT LOCATION

Design Criteria: In Division 1

Typical Bay Spacings: 30 x 40

Floor to Ceiling Heights:

Beam Depths:

Roof Slope: ☐ Flat ☐ Other 1/8" / Ft.

Eave Height: 26'-5"

Anchor Bolts: ☐ No ☐ By Others ☑ Yes Describe Number

Base Plates: ☐ No ☐ By Others ☑ Yes Describe Number

Metal Decking Floors: ☐ No ☐ By Others ☐ Cellular ☑ Non Cellular ☑ Painted ☐ Galv. Depth

☐ Acoustical ☐ Ventilating Gauge 28 Describe

Roof Deck: ☐ No ☐ By Others ☐ Cellular ☑ Non Cellular ☑ Painted ☐ Galv. Depth 1 1/2"

☐ Acoustical ☐ Ventilating Gauge 22 Describe

Structural System: ☐ Wall Bearing ☐ Free Standing ☐ Simple Spans ☑ Continuous Spans cantilever

☐ Conventional Design ☐ Plastic Design ☑ Field Welded ☑ Field Bolted ☐ Composite Design

☐ Other

Type Steel A 36 Grade

Estimated Weights: Beams Roof Frames Adjustable Spandrel Angles

Girders Girts Hanger Pods Bracing

Columns Connections Other

Paint: Shop ☐ No ☑ Yes prime Coats Material

Field Paint ☑ No ☐ Yes ☐ Brush ☐ Roller ☐ Spray Coats Material touch-up

Galvanizing: ☐ No ☐ Yes Thickness

Other:

Fireproofing: ☑ No ☐ Yes ☐ Beams ☐ Columns ☐ Decks ☐ Other Rating Hr.

☐ Concrete Encasement ☐ Spray ☐ Plaster ☐ Drywall ☐

Describe

Open Web Joists: ☐ No ☑ Yes ☑ H Series ☐ LH Series ☐

Estimated Weights:

Bridging: ☐ No ☑ Yes ☐ Bolted ☑ Welded ☐ Pod Bridging

Paint: Shop ☑ Standard ☐ Special Coats Field Paint ☑ No ☐ Yes Describe touch-up

Light Gauge Joists: ☐ No ☐ Yes Describe

Light Gauge Framing: ☑ No ☐ Yes Describe

Special Considerations:

Fasteners: Expansion Bolts High Strength Bolts ✓

Machine Screws Machinery Anchors

Nails Roof Bolts ✓

Sheet Metal Screws Studs

Timber Connectors Toggle Bolts

Welded Studs Other

Page 9 of 30

SPEC-AID

DIVISION 5: METALS

Area Walls: ☑ No ☐ Yes ______ Gratings ______ Caps ______
Bumper Rails: ☐ No ☑ Yes ______
Canopy Framing: ☐ No ☑ Yes ______
Checkered Plate: ☐ No ☐ Trench Covers ☑ Pit Covers ☐ Platforms ☐ ______
Columns: ☐ No ☐ Aluminum ☐ Steel ☐ Square ☐ Rectangular ☐ Round ☐ ______
Construction Castings: ☐ No ☐ Chimney Specialties ______ ☐ Column Bases ______
☐ Manhole Covers ______ ☐ Wheel Guards ______
☐ ______
Corner Guards: ☑ No ☐ Yes ______
Crane Rail: ☑ No ☐ Yes ______
Curb Angles: ☑ No ☐ Straight ☐ Curved ______
Decorative Covering: ☐ No ☐ Stock Sections ☐ Custom Sections ______
Doors ______ Walls ______
Door Frames: ☐ No ☑ Yes O.H. Doors ______ ☐ Protection: ☐ No ☐ Yes ______
Expansion Joints Ceilings: ☐ No ☐ Yes ______ Cover Plates: ☐ No ☐ Yes ______
Floors: ☐ No ☐ Yes ______ Cover Plates: ☐ No ☐ Yes ______
Walls: ☐ No ☐ Yes ______ Cover Plates: ☐ No ☐ Yes ______
Fire Escape: ☑ No ☐ Yes ______ Size ______
Stairs ______ Ladders ______ Cantilever ______
Floor Grating: ☑ No ☐ Aluminum ☐ Steel ☐ Fiberglass ☐ Platforms ☐ Stairs ☐ ______
Type ______ Weight ______
Special Finish ______
Ladders: ☐ No ☐ Aluminum ☑ Steel ☐ ______
☐ With Cage ☑ No Cage ______ ☐ Inclined Type ______
Lamp Posts: ☐ No ☐ Yes ______
Lintels: ☐ No ☑ Yes ☐ Plain ☐ Built-up ☐ Painted ☐ Galvanized ______
______
Louvers: ☐ No ☑ Yes ______
Manhole Covers: ☐ No ☑ Yes ______
Mat Frames: ☐ No ☐ Yes ______
Overhead Supports: ☐ No ☑ Toilet ☐ Partitions ☐ ______
Pipe Bumpers: ☐ No ☑ Yes ______
Pipe Supports: ☐ No ☐ Yes ______
Railings: ☐ No ☑ Yes ☐ Aluminum ☐ Steel ☐ Pipe ☐ ______
Balconies: ☐ No ☐ Yes ______
Stairs: ☐ No ☐ Yes ______
Wall: ☐ No ☐ Yes ______
Solar Screens: ☐ No ☐ Yes ______
Stairs: ☐ No ☑ Yes ☐ Aluminum ☑ Steel ☐ Stock ☐ Custom ☐ ______
Size ______ Landings ______
Spiral: ☐ No ☐ Aluminum ☐ Steel ☐ Stock ☐ Custom ☐ ______
Pre-erected: ☐ No ☐ Yes ______
Stair Treads: ☐ No ☐ Yes ______
Trench Covers: ☐ No ☐ Yes ______
Weather Vanes: ☐ No ☐ Yes ______
Window Guards: ☐ No ☐ Bars ☐ Woven Wire ☐ ______
Wire: ☐ No ☐ Yes ______
Wire Rope: ☐ No ☐ Yes ______
Special Considerations: ______
______
______

Page 10 of 30

SPEC-AID
DIVISION 6: CARPENTRY

DATE ____________

PROJECT ____________ LOCATION ____________

Framing: Type Wood ____________ Fiber Stress ____________ psi.

Beams: ☐ Single ☐ Built Up ____________ Grade ____________

Bracing: ☐ No ☐ Let In ☐ ____________

Bridging: ☐ Steel ☐ Wood ____________

Canopy Framing: ____________

Columns ____________ Fiber Stress ____________ psi.

Door Bucks: ☐ No ☐ Treated ☐ Untreated ____________

Floor Planks ____________ Grade ____________

Furring: ☐ Metal ☐ Wood ____________

Grounds: ☐ No ☐ Casework ☐ Plaster ☐ On Wood ☐ On Masonry ____________

Joists: ☐ No ☐ Floor ☐ Ceiling ____________ Grade ____________

Ledgers: ☐ No ☐ Bolted ☐ Nailed ____________

Lumber Treatment: ☐ No ☐ Creosote ☐ Salt Treated ☐ Fire Retardant ____________
☐ Kiln Dry ____________

Nailers: ☐ No ☑ Treated ☐ Untreated ____________

Plates: ____________ **Platform Framing:** ____________

Plywood Treatment: ☑ No ☐ Salt Treated ☐ Fire Retardant ____________

Posts & Girts: ____________

Rafters: ☐ No ☐ Ordinary ☐ Hip ____________

Roof Cants: ☐ No ☐ Yes ____________ **Roof Curbs:** ☐ No ☑ Yes ____________

Roof Decks: ☐ No ☐ Yes ________ Inches Thick

Roof Purlins: ☐ No ☐ Yes ____________

Roof Trusses: ☐ No ☐ Timber Connectors ☐ Nailed ☐ Glued Spaced ________ O.C. Span ________ feet

Sheathing, Roof: ☐ No ☐ Plywood ☐ Boards ☐ Wood Fiber ☐ Gypsum ____________
Wall: ☐ No ☐ Plywood ☐ Boards ☐ Wood Fiber ☐ Gypsum ____________

Siding Hardboard: ☐ Plain ☐ Primed ☐ Stained ____________
Particle Board: ____________ Wood Fiber ____________
Plywood: ☐ Cedar ☐ Fir ☐ Redwood ☐ Marine ☐ Natural ☐ Stained ☐ Plastic Faced ☐ ____________

Wood: ☐ Cedar ☐ Redwood ☐ White Pine ☐ Bevel ☐ Board & Batten ☐ Channel ☐ T & C ☐ Shiplap ________
☐ Natural ☐ Stained ____________

Sills: ____________ **Sleepers:** ____________

Soffits: ☐ No ☐ Open ☐ Vented ☐ Plywood ____________

Stressed Skin Plywood Box Beams: ____________ Depth ____________
Floor Panels: ☐ No ☐ Yes ____________ Depth ____________
Roof Panels: ☐ No ☐ Straight ☐ Curved ____________ Depth ____________
Folded Plate: ☐ No ☐ Yes ____________ Depth ____________

Studs: ☐ No ☐ Yes ____________ Grade ____________

Subfloor: ☐ No ☐ Plywood ☐ Boards ☐ Wood Fiber ☐ ____________

Suspended Ceiling Framing: ☐ No ☐ Yes ____________

Underlayment: ☐ No ☐ Particle Board ☐ Plywood ☐ Wood Fiber ☐ Hardboard ____________

Special Considerations: wood ceiling at watchman loading dock station

Laminated Framing: ☐ Beams ☐ Straight ☐ Curved ____________ Span ____________
☐ Bowstring Trusses ☐ Radial Arch ☐ Tudor Arch ☐ Columns ____________
Span ________ Height ____________
☐ Industrial Grade ☐ Premium Grade ☐ Exterior Glue ☐ Stain ☐ Varnish ☐ Treated ☐ ____________

Laminated Roof Deck: ☐ No ☐ Yes ____________ Thickness ________

Special Considerations: ____________

Page 11 of 30

SPEC-AID

DIVISION 6: CARPENTRY

Base: ☐ No ☐ One Piece ☐ Built up ☐ Pine ☐ Hardwood ____

Cabinets: ☐ No ☐ Corner ☑ Kitchen ☐ Toilet Room ☐ Other ____
☑ Stock ☐ Custom ____
☐ Unfinished ☑ Prefinished ____
Base Cabinets: ☐ Softwood ☐ Hardwood ☐ Drawer Units ____
Wall Cabinets: ☐ Softwood ☐ Hardwood ____
Tall Cabinets: ☐ Softwood ☐ Hardwood ____
Special: ____

Casings: ☐ No ☐ Doors ☐ Windows ☐ Beams ☐ Others ____
☐ Softwood ☐ Hardwood ____

Ceiling Beams: ☐ No ☐ Cedar ☐ Pine ☐ Fir ☐ Plastic ____

Chair Rail: ☐ No ☐ Pine ☐ Other ____

Closets: ☐ No ☑ Pole ☑ Shelf ☐ Prefabricated ____

Columns: ☐ No ☐ Square ☐ Round ☐ Solid ☐ Built up ☐ Hollow ☐ Tapered ____
Diameter ____ Height ____

Convector Covers: ☐ No ☐ Yes ____

Cornice: ☐ No ☐ 1 Piece ☐ 2 Piece ☐ 3 Piece ☐ Pine ☐ Cedar ☐ Other ____

Counter Tops: ☐ No ☑ Plastic ☐ Ceramic Tile ☐ Marble ☐ Suede Finish ☐ Other ____
☑ Stock ☐ Custom ____
☐ No Splash ☐ Square Splash ☑ Cover Splash ____
☐ Self Edge ☐ Stainless Edge ☐ Aluminum Edge ____
Special ____

Cupolas: ☐ No ☐ Stock ☐ Custom ☐ Wood ☐ Fiberglass ☐ Square ☐ Octagonal Size ____
☐ Aluminum Roof ☐ Copper Roof ☐ Other ____

Doors and Frames: See Division 8 ✓ ____

Door Moldings: ☐ No ☐ Yes ____

Door Trim: ☐ No ☐ Yes ____

Fireplace Mantels: ☐ No ☐ Beams ☐ Moldings ____
Size ____

Moldings: ☐ No ☐ Softwood ☐ Hardwood ☐ Metal ☐ Other ____

Paneling Hardboard: ☐ No ☐ Tempered ☐ Untempered ☐ Pegboard ☐ Plastic Faced ____
Plywood, Unfinished: ☐ No ☐ Veneer Core ☐ Lumber Core Grade ____ Thick ____
Plywood, Prefinished: ☐ No ☐ Stock ☐ Architectural Finish ____
Size ____
Wood Boards: ☐ No ☐ Softwood ☐ Hardwood ____

Railings: ☐ No ☐ Stock ☐ Custom ☐ Softwood ☐ Hardwood ____
☐ Stairs ☐ Balcony ☐ Porch ☐ Wall ☐ Other ____

Shelving: ☐ No ☐ Prefinished ☐ Unfinished ☐ Stock ☐ Custom ☐ Plywood ☐ Particle Board ☐ Boards ____
☐ Book Shelves ____ ☐ Linen Shelves ____
☐ Storage Shelves ____ ☐ Other ____

Stairs: ☐ No ☐ Prefabricated ☐ Built in Place ☐ Softwood ☐ Hardwood ____
☐ Box ☐ Open ☐ Circular ____

Thresholds: ☐ No ☐ Interior ☐ Exterior ____

Wainscot: ☐ No ☐ Boards ☐ Plywood ☐ Moldings ____

Windows and Frames: See Division 8 ____

Window Trim: ☐ No ☐ Yes ____

Special Considerations: ____

Page 12 of 30

SPEC-AID DATE ______

DIVISION 7: MOISTURE PROTECTION

PROJECT ______ LOCATION ______

Bentonite: ☐ No ☐ Panels ☐ Granular ______
Bituminous Coating: ☐ No ☐ Brushed ☐ Sprayed ☐ Troweled ☐ 1 Coat ☐ 2 Coat ☐ Protective Board ______
Building Paper: ☐ No ☐ Asphalt ☐ Polyethylene ☐ Rosin ☐ Kraft ☐ Foil Backed ☐ ______
☐ Roof Deck Vapor Barrier ______
Caulking: ☐ No ☑ Gun Grade ☐ Knife Grade ☐ Plain ☐ Colors ______
☑ Doors ☑ Windows ☐ ______
Cementitious: ☐ No ☐ 1 Coat ☐ 2 Coat Thickness ______ Inches Mix ______
Control Joints: ______ **Expansion Joints:** ______
Elastomeric Waterproofing: ☐ No ☐ EPDM ☐ Neoprene ☐ PVC ☐ Urethane ☐ ______
Liquid Waterproofing: ☐ No ☐ Silicone ☐ Stearate ☐ ______
Membrane Waterproofing: ☑ 1 Ply ☐ 2 Ply ☐ 3 Ply ☐ Felt ☐ Fabric ☐ Elastomeric ☐ Foundations
Metallic Coating: ☐ No ☐ Walls ______ in. Thick ☐ Floors ______ in. Thick ______
Preformed Vapor Barrier: ☐ No ☐ Yes ______
Sealants: ☐ No ☐ Butyl ☐ Polysulfide ☐ PVC ☐ Urethane ☐ ______
☐ Doors ☐ Windows ☐ ______
Special Waterproofing ______

Building Insulation: Rigid: ☐ No ☐ Fiberglass ☑ Polystyrene ☐ Urethane ☐ ______
Non Rigid: ☐ No ☑ Fiberglass ☐ Mineral Fiber ☐ Vermiculite ☐ Perlite ☐ ______
Form Board: ☐ No ☐ Acoustical ☐ Asbestos Cement ☐ Fiberglass ☐ Gypsum ☐ Mineral Fiber ☐ Wood Fiber
☐ Other ______ ☐ Sub Purlins ______ Span ______
Masonry Insulation: ☐ No ☐ Cavity Wall ☐ Block Cores ☑ Poured ☐ Foamed Type ______
Perimeter Insulation: ☐ No ☑ Yes Type ______ Thickness ______
Roof Deck Insulation: ☐ No ☐ Fiberboard ☑ Fiberglass ☐ Foamglass ☐ Polystyrene ☐ Urethane ☐ ______
Thickness 1 5/16" ______ ☐ **Cants** ______ Size ______
Sprayed: ☐ No ☐ Fibrous ☐ Cementitious ☐ Urethane ☐ ______
Special Insulation ______

Shingles: Aluminum: ☐ No ☐ Yes ______ Asbestos: ☐ No ☐ Yes ______
Asphalt: ☐ No ☐ Class C ☐ Class A ☐ ______ Weight ______ lb. per Sq. ______
Clay Tile: ☐ No ☐ Plain ☐ Glazed ☐ Spanish ☐ ______ Weight ______ lb. per Sq. ______
Concrete Tile: ☐ No ☐ Yes ______ Porcelain Enamel: ☐ No ☐ Yes ______
Slate: ☐ No ☐ Yes Type ______ Color ______ Exposure ______
Wood: ☐ No ☐ Roofing ☐ Siding ☐ Fire Retardant Type ______ Grade ______ Exposure ______
Shingle Underlayment: ☐ No ☐ Asbestos ☐ Asphalt ☐ ______ Weight ______
Special Shingles: ______

Aluminum: ☐ No ☐ Roofing ☐ Siding ☐ Painted ☐ Insulated ☐ Sandwich ☐ ______
Thickness ______ Type ______
Asbestos Cement: ☐ No ☐ Roofing ☐ Siding ☐ Flat ☐ Corrugated ☐ Natural ☐ Painted ☐ Sandwich
☐ Fire Rated Thickness ______ Type ______
Epoxy Panels: ☐ No ☐ Solid ☐ Plywood Back ☐ Hardboard Back ☐ Exposed Aggregate ☐ ______
Fiberglass Panels: ☐ No ☐ Roofing ☐ Siding ☐ Flat ☐ Corrugated ☐ ______ Thickness ______
Metal Facing Panels: ☐ No ☐ Field Assembled ☐ Factory Made Insulation ______
Outside Face ______ Inside Face ______
Protected Metal: ☐ No ☐ Roofing ☐ Siding Type ______ Gauge ______
Steel: ☐ No ☐ Roofing ☐ Siding ☐ Painted ☐ Galvanized ☐ Insulated ☐ Sandwich ☐ ______
Type ______ Gauge ______
Vinyl Siding: ☐ No ☐ Plain ☐ Insulated Type ______
Special Roofing & Siding: ______

Page 13 of 30

SPEC-AID

DIVISION 7: MOISTURE PROTECTION

Built Up Roofing: ☐ No ☐ Tar & Gravel ☑ Asphalt & Gravel ☐ Felt ☐ Mineral Surface ☐ Aggregate
☐ 1 Ply ☐ 2 Ply ☐ 3 Ply ☑ 4 Ply ☐ 5 Ply ☐ Bonded ____ years Roof Pitch ________ Type Deck metal
Underlayment: ☐ No ☐ Rosin Paper ☑ Vapor Barrier ☐ ________
Elastic Sheet Roofing: ☐ No ☐ Butyl ☐ Neoprene ☐ ________ Thickness ________
Describe ________
Fluid Applied Roofing: ☐ No ☐ Hypalon Neoprene ☐ Silicone ☐ Vinyl ☐ ________ Thickness ________
Describe ________
Roll Roofing: ☐ No ☐ Smooth ☐ Granular ________ Weight ________ lbs. per Sq.
Special Membrane Roofing: ________

Downspouts: ☐ No ☐ Aluminum ☐ Copper ☐ Lead Coated Copper ☐ Galvanized Steel ☐ Stainless Steel
☐ Steel Pipe ☐ Vinyl ☐ Zinc Alloy ☐ Stock ☐ Custom ☐ ________ Size ________
Describe ________
Expansion Joints: ☑ No ☐ Roof ☐ Walls ☐ No Curbs ☐ Curbs ☐ Rubber ☐ Metallic ☐ ________
Describe ________
Fascia: ☐ No ☐ Yes Describe ________ Thickness ________
Flashing: ☐ No ☑ Aluminum ☐ Asphalt ☑ Copper ☑ Fabric ☐ Lead ☑ Lead Coated Copper ☐ PVC ☐ Rubber
☐ Stainless Steel ☐ Terne ☐ Zinc Alloy ☐ Paper Backed ☐ Mastic Backed ☐ Fabric Backed ☐ ________
Describe ________ Thickness ________

Gravel Stop: ☐ No ☑ Aluminum ☐ Copper ☐ PVC ☐ Stainless Steel ☐ ________
☑ With Fascia ☐ No Fascia ☐ Natural ☑ Painted Thickness ________ Face Height 7"
Gutters: ☐ No ☐ Aluminum ☐ Copper ☐ Lead Coated Copper ☐ Galvanized Steel ☐ Stainless Steel ________
☐ Vinyl ☐ Wood ☐ Zinc Alloy ☐ ________ Thickness ________
☐ Box Type ☐ K Type ☐ Half Round ☐ Stock ☐ Custom ☐ ________ Size ________
Louvers: ☐ No ☑ Yes ________
Mansard: ☐ No ☐ Yes ________ Thickness ________
Metal Roofing: ☐ No ☐ Copper ☐ Copper Bearing Steel ☐ Lead ☐ Lead Coated Copper ☐ Stainless Steel
☐ Terne ☐ Zinc Alloy ☐ ________ Size ________ Thickness ________
☐ Standing Seam ☐ Flat Seam ☐ Batten Seam ☐ ________ Weight ________ lbs. per Sq.

Underlayment: ☐ No ☐ 15 lb. Felt ☐ 30 lb. Felt ☐ Rosin Paper ☐ ________
Reglet: ☐ No ☐ Aluminum ☐ Copper ☐ Galvanized Steel ☐ Stainless Steel ☐ Zinc Alloy ☐ ________
Thickness ________ Counter Flashing: ☐ No ☐ Yes ________ Thickness ________
Soffit: ☐ No ☐ Yes ________ Thickness ________
Special Sheet Metal Work: ________

Ceiling Hatches: ☐ No ☐ Steel ☐ Galvanized ☐ Painted ☐ Aluminum ☐ ________
Size ________
Roof Drains: ☐ No ☑ In Plumbing ☑ Yes ________
Roof Hatches: ☐ No ☐ Steel ☑ Galvanized ☐ Painted ☐ Aluminum ☐ ________
☐ Insulated ☐ Not Insulated ☑ With Curbs ☐ No Curbs ☐ ________ Size ________
Smoke Hatches: ☐ No ☑ Yes ________
Snow Guards: ☐ No ☐ Yes ________
Skylights: ☐ No ☐ Domes ☐ Vaulted ☐ Ridge Units ☐ Field Fabricated ☐ Glass ☐ Plastic ☐ Single
☐ Double ☐ Sandwich Panels ☐ With Curbs ☐ No Curbs ☐ ________ Size ________
Smoke Vents: ☐ No ☑ Yes ________
Skyroofs: ☐ No ☐ Yes ________
Ventilators: ☐ No ☐ In Ventilating ☐ Stationary ☐ Spinners ☐ Motorized ________
Special Roof Accessories ________

Page 14 of 30

SPEC-AID DATE ______

DIVISION 8: DOORS, WINDOWS & GLASS

PROJECT ______ LOCATION ______

Hollow Metal Frames: ☐ No ☑ Baked Enamel ☐ Galvanized ☐ Porcelain Enamel ______
Hollow Metal Doors: ☐ No ______ ☑ Core ______ ☑ Labeled ______
Aluminum Frames: ☐ No ☐ Clear ☑ Bronze ☐ Black ______
Aluminum Doors and Frames: ☐ No ☑ Yes ______ Frames ______
Wood Frames: ☐ No ☐ Exterior ☐ Interior ☐ Custom ☐ With Sill ☐ Vinyl Covered ☐ Pine ☐ Oak
Wood Doors: ☐ No solid Core ☐ Labeled no Frames no
Interior Door Frames: ☐ No ☐ Aluminum ☑ Hollow Metal ☐ Steel ☐ Wood ☐ Prehung ☐ Stock
☐ Custom ☐ ______
Custom Doors: ☑ No ☐ Swing ☐ Bi-Passing ☐ Bi-Folding ______ Frames ______
Accordion Folding Doors: ☑ No ☐ Yes ______ Frames ______
Acoustical Doors: ☑ No ☐ Yes ______ Decibels ______ Frames ______
Cold Storage: ☑ No ☐ Manual ☐ Power ☐ Sliding ☐ Hinged ______
Counter Doors: ☑ No ☐ Aluminum ☐ Steel ☐ Wood ______ Frames ______
Dark Room Doors: ☑ N0 ☐ Revolving ☐ 2 Way ☐ 3 Way ______
Floor Opening Doors: ☑ No ☐ Aluminum ☐ Steel ☐ Single ☐ Double ☐ Commercial ☐ Industrial ______
Glass Doors: ☑ No ☐ Sliding ☐ Swing ______ Frames ______
Hangar Doors: ☑ No ☐ Bi-Fold ☐ Other ☐ Electric ______
Jalousie Doors: ☑ No ☐ Plain Glass ☐ Tempered Glass ______
Kalamein: ☑ No ☐ Yes ☐ Labeled ______ Frames ______
Kennel Doors: ☑ No ☐ 2 Way Swing ______
Overhead Doors: ☐ No ☐ Regular Duty ☑ Heavy Duty ☐ Stock ☐ Custom ☐ One Piece ☑ Sectional
☑ Manual ☐ Electric ☐ Aluminum ☐ Fiberglass ☑ Steel ☐ Wood ☐ Hardboard ☐ Commercial
☐ Residential Size ______
Rolling Doors Exterior: ☐ No ☐ Manual ☐ Electric ☐ Labeled ______
Rolling Doors Interior: ☐ No ☐ Manual ☐ Electric ☐ Labeled ______ Frames ______
Rolling Grilles: ☐ No ☐ Manual ☐ Electric ☐ Aluminum ☐ Steel ______
Service Door Frames: ☐ No ☐ Aluminum ☐ Hollow Metal ☐ Steel ☐ Wood ☐ Stock ☐ Custom ______
Service Doors: ☐ No ☐ Stock ☐ Custom ______ ☐ Transoms ______ ☐ Sidelights ______
☐ Aluminum ______ ☐ Hollow Metal ______ ☐ Core ______ ☐ Kalamein ______
☐ Steel ______ ☐ Wood ______ ☐ Core ______ ☐ Labeled ______ ☐ Special Finish ______
Shock Absorbing Doors: ☐ No ☐ Flexible ☐ Rigid ______ Frames ______
Sliding Doors: ☐ No ☐ Glazed ☐ Unglazed Aluminum ☐ Steel ☐ Wood ______
Swing Doors: ☐ No ☐ Single ☐ Double ______
Telescoping Door: ☐ No ☐ Manual ☐ Electric ______
Tinclad Doors: ☐ No ☐ Manual ☐ Electric ______
Vault Front Doors: ☐ No ☐ Stainless Steel ☐ Time Lock ☐ 1 Hr. Test ☐ 2 Hr. Test ☐ 4 Hr. Test ______
Special Exterior Doors: ☐ No ☐ Yes ______
Special Interior Doors: ☐ No ☐ Yes ______
Balanced Doors: ☐ No ☐ Economy ☐ Premium ☐ Aluminum ☐ Stainless Steel ______
Revolving Doors: ☐ No ☐ Stock ☐ Custom ☐ Manual ☐ Electric ☐ Diameter ______
Entrance Units: ☐ No ☐ Aluminum ☐ Bronze ☐ Glass ☐ Hollow Metal ☐ Stainless Steel ☐ Wood ☐ Steel
☐ Stock ☐ Custom ☐ Balanced ☐ Sidelights ☐ Transoms Special Finish ______
Entrance Frames: ☐ No ☑ Aluminum ☐ Hollow Metal ☐ Steel ☐ Wood ☐ Stainless Steel ☐ Stock
☐ Custom ______
Store Fronts: ☐ No ☐ Sliding ☑ Fixed ☐ Institutional Grade ☐ Monumental Grade ☐ Commercial Grade ______
Windows: ______ % of Exterior Walls ______
Projected: ☑ No ☐ Glazed ☐ Unglazed ☐ Aluminum ☐ Steel ☐ Wood Fixed 5' spandrel tinted
Single Hung: ☐ No ☐ Glazed ☐ Unglazed ☐ Aluminum ☐ Steel ☐ Wood ______
Sliding: ☐ No ☐ Glazed ☐ Unglazed ☐ Aluminum ☐ Steel ☐ Wood ______
Security Windows: ☐ No ☐ Yes ______

Page 15 of 30

SPEC-AID

DIVISION 8: DOORS, WINDOWS & GLASS

Casement: ☐ No ☐ Fixed ________% Vented ☐ Aluminum ☐ Steel ☐ Wood
Picture Window: ☐ No ☐ Glazed ☐ Unglazed ☐ Aluminum ☐ Steel ☐ Wood
Double Hung: ☐ No ☐ Glazed ☐ Unglazed ☐ Aluminum ☐ Steel ☐ Wood
Special Windows: ☐ No ☑ Yes shipping
Screens: ☐ No ☐ Aluminum ☐ Steel ☐ Wood

Finish Hardware Allowance: ☐ No ☑ Yes
Exterior Doors
Interior Doors
Automatic Openers: ☐ No ☐ 1 Way ☐ 2 Way ☐ Double Door ☐ Activating Carpet
Automatic Operators: ☐ No ☐ Sliding ☐ Swing ☐ Controls
Bumper Plates: ☐ No ☐ U Channel ☐ Teardrop
Door Closers: ☐ No ☑ Regular ☐ Fusible Link ☑ Concealed ☐ Heavy Use
Door Stops: ☐ No ☑ Yes
Floor Checks: ☐ No ☐ Single Acting ☐ Double Acting
Hinges: ☐ No ☑ Butt ☐ Pivot ☐ Spring ☐ Frequency
Kick Plates: ☐ No ☑ Yes
Lock Set: ☐ No ☑ Cylindrical ☑ Mortise ☐ Heavy Duty ☑ Commercial ☐ Residential
Panic Device: ☐ No ☑ Yes ☑ Exit Only ☐ Exit & Entrance
Push-Pull Device: ☐ No ☑ Yes ☐ Bronze ☑ Aluminum ☐ Other
Cabinet Hardware: ☐ No ☐ Yes
Window Hardware: ☐ No ☐ Yes
Special Hardware: ☐ No ☐ Yes
Threshold: ☐ No ☑ Yes
Weather Stripping Doors: ☐ No ☐ Zinc ☐ Bronze ☐ Stainless Steel ☐ Spring Type ☑ Extruded Sections
Windows: ☐ No ☐ Zinc ☐ Bronze

Acoustical Glass: ☐ No ☐ Yes Thickness
Faceted Glass: ☐ No ☐ Yes Thickness
Glazing: ☐ No ☐ Putty ☐ Flush ☐ Bead ☐ Gasket ☐ Butt ☐ Riglet ☐
Insulated Glass: ☐ No ☑ Standard ☐ Non-Standard Thickness
Laminated Glass: ☐ No ☐ Yes Thickness
Mirrors: ☐ No ☑ Plate ☐ Sheet ☐ Transparent ☐ Incl. Frames ☐ No Frames ☐
Door Type ________ Wall Type
Obscure Glass: ☐ No ☐ Yes Thickness
Plate Glass: ☐ No ☐ Clear ☐ Tinted ☐ Tempered Thickness
Plexiglass: ☐ No ☐ Masked ☐ Unmasked Thickness
Polycarbonate: ☐ No ☐ Masked ☐ Unmasked Thickness
Reflective: ☐ No ☐ Clear ☐ Tinted Thickness
Sand Blasted: ☐ No ☐ Yes Thickness
Sheet or Float Glass: ☐ No ☐ Clear ☐ Gray Thickness
Spandrel Glass: ☐ No ☐ Plain ☐ Insulated ☐ Sandwich Thickness
Stained Glass: ☐ No ☐ Yes
Vinyl Glazing: ☐ No ☐ Yes Thickness
Window Glass: ☐ No ☐ DSA ☐ DBS ☐ Tempered Thickness
Wire Glass: ☐ No ☐ Yes Thickness
Special Glazing: ☐ No ☐ Yes
Curtain Walls: ☐ No ☐ Yes
Window Walls: ☐ No ☐ Yes

Page 16 of 30

SPEC-AID

DATE ______

DIVISION 9: FINISHES

PROJECT ______ LOCATION ______

Furring: Ceiling: ☐ No ☐ Wired Direct ☐ Suspended ______
Partitions: ☐ No ☐ Load Bearing ☐ Non Load Bearing ______ Thickness ______
Walls: ☐ No ☐ Yes ______
Gypsum Lath: ☐ No ☐ Walls ☐ Ceilings ☐ Regular ☐ Foil Faced ☐ Fire Resistant ☐ Moisture Resistant ______
______ Thickness ______
Metal Lath: ☐ No ☐ Diamond ☐ Rib ☐ ______ Weight ______
☐ Painted ☐ Galvanized ☐ Paper Backed ______
☐ Walls ☐ Ceilings ☐ Suspended ☐ Partitions ☐ Load Bearing ☐ Non Load Bearing ______

Drywall Finishes: ☑ Taped & Finished ☐ Thin Coat Plaster ☐ Prime Coat ☐ Electric Heat Compound ☐ ______
Mountings: ☐ Nailed ☑ Screwed ☐ Laminated ☐ Clips ☐ ______
Beams: ☐ No ______ Layers ______ Thickness ______
Ceilings: ☐ No ☑ Standard ☐ Fire Resistant ☐ Water Resistance ______ Thickness ______
Columns: ☐ No ✓ Layers ______ Thickness ______
Partitions: ☐ No ☐ Wood Studs ☑ Steel Studs ______ Layers ______ Thickness ______
Prefinished: ☐ No ☐ Standard ☐ Fire Resistance ______ Thickness ______
Soffits: ☐ No ✓ Layers ______ Thickness ______
Sound Deading Board: ☐ No Type ______ Thickness ______
Walls: ☐ No ______ Layers ______ Thickness ______
Plaster Finishes: ☐ 1 Coat ☐ 2 Coat ☐ 3 Coat ☐ Gypsum ☐ Perlite ☐ Vermiculite ☐ Wood ☐ ______
Beams: ☐ No ______ Ceilings: ☐ ______
Columns: ☐ No ______ Soffits: ☐ No ______
Partitions: ☐ No ☐ Wood Studs ☐ Steel Studs ☐ Solid ☐ Hollow ______
Walls: ☐ No ______
Special Plaster: ______
Sprayed Acoustical: ☐ No ☐ Yes ______ Thickness ______
Fireproofing: ☐ No ☐ Yes ______ Thickness ______
Stucco: ☐ No ☐ On Mesh ☐ Masonry ✓ blue board.
Cast Stone: ☐ No ☐ Glazed ☐ Unglazed ☐ Waxed ______ Thickness ______
Ceramic Tile Base: ☐ No ☐ Cove ☐ Sanitary ☐ ______ Set ______ Height ______
Floors: ☐ No ☐ ✓ Set ☐ Natural Clay ☐ Porcelain ☐ Conductive ______ Color Group ______
Walls: ☐ No ☐ ✓ Set ☐ Interior ☐ Exterior ☐ Glazed ☐ Crystalline Glazed ☐ ______
☐ Unmounted ☐ Backmounted ______
Panels: ☐ No ☐ Yes ______
Glass Mosaics: ☐ No ☐ Yes ______ Color Group ______
Metal Tile: ☐ No ☐ Aluminum ☐ Copper ☐ Stainless Steel ______
Plastic Tile: ☐ No ☐ Yes ______ Thickness ______
Quarry Tile Base: ☐ No ☐ Cove ☐ Sanitary ______ Height ______
Floor ☐ No ______ Set Size ______ Color ______
Stairs: ☐ No ☐ Treads ☐ Risers ______
Wainscot: ☐ No ______ Set Size ______
Cast In Place Terrazzo Base: ☐ No ☐ Yes ______ Curb: ☐ No ☐ Yes ______
Floor: ☐ No ☐ Bonded ☐ Unbonded ☐ Gray Cement ☐ White Cement ☐ Conventional ☐ Venetian ______
☐ Conductive ☐ Monolithic ☐ Epoxy ☐ ______
Divider Strips: ☐ No ☐ Brass ☐ Zinc ______ Spacing ______
Stairs: ☐ No ☐ Yes ______ Wainscot: ☐ No ☐ Yes ______
Precast Terrazzo Base: ☐ No ☐ Yes Curb: ☐ No ☐ Yes ______
Floor Tiles: ☐ No Size ______ Thickness ______
Stairs: ☐ No ☐ Treads ☐ Risers ☐ Stringers ☐ Landings ______
Wainscot: ☐ No ☐ Yes ______ Thickness ______

Page 17 of 30

SPEC-AID

DIVISION 9: FINISHES

Barriers: ☐ No ☐ Aluminum ☐ Foil ☐ Mesh ☐ Lead ☐ Leaded Vinyl ______

Acoustical Barriers: ☐ No ☐ Yes ______

Ceilings: ☐ No ☐ Boards ☐ Tile ☐ Cemented ☐ Stapled ☑ On Suspension ☐ ______

☐ Fiberglass ☑ Mineral Fiber ☐ Wood Fiber ☐ Metal Pan ______

☐ Fire Rated ☐ Ventilating ☐ ______ Ceiling Height ______

☐ Luminous Panels ______ ☐ Access Panels ______

Suspension System: ☐ No ☑ T Bar ☐ Z Bar ☐ Carrier Channels ☐ ______

Strip Lighting: ☐ No ☐ Yes ______ Foot Candles ______

Special Acoustical ______

______

Brick Flooring: ☐ No ☐ Yes ______

Carpet ☐ No ☑ Yes ☐ With Padding ☐ With Backing ☐ ______ Allowance ______

Type: ☐ Acrylic ☐ Nylon ☐ Polypropylene ☐ Wool ☐ Tile ☐ ______ Face Weight ______

Padding: ☑ No ☐ Yes ______ Backing: ☑ No ☐ Yes ______

Composition Flooring: ☐ No ☐ Acrylic ☐ Epoxy ☐ Mastic ☐ Neoprene ☐ Polyester ☐ ______

☐ Regular Duty ☐ Heavy Duty ______ Thickness ______

Concrete Floor Topping: ☐ No ☑ In Concrete ☐ Yes ______

Resilient Floors: Base: ☐ No ☐ Rubber ☐ Vinyl ______ Height ______

Asphalt Tile: ☐ No ☐ Yes ______ Color Group ______

Conductive Tile: ☐ No ☐ Yes ______ Thickness ______

Cork Tile: ☐ No ☐ Yes ______ Thickness ______

Linoleum: ☐ No ☐ Yes ______ Thickness ______

Polyethylene: ☐ No ☐ Yes ______

Polyurethane: ☐ No ☐ Yes ______ Thickness ______

Rubber Tile: ☐ No ☐ Yes ______ Thickness ______

Vinyl: ☐ Sheet ☐ Tile ______ Thickness ______

Vinyl Asbestos Tile: ☐ No ☐ Yes ______ Color Group ______

Stair Covering: ☐ No ☐ Risers ☐ Treads ☐ Landings ☐ Nosings ☐ Rubber ☐ Vinyl ______

Steel Plates: ☐ No ☐ Cement Bed ☐ Epoxy Bed ______

Wood Floor: ☐ No ☐ Block ☐ Strip ☐ Parquetry ☐ Unfinished ☐ Prefinished ☐ Stock ☐ Custom ______

Fir: ☐ No ☐ Flat Grain ☐ Vertical Grain ______ Size ______

Gym: ☐ No ☐ Yes Type ______

Maple: ☐ No ☐ Yes Grade ______ Size ______

Oak: ☐ No ☐ Red ☐ White Grade ______ Size ______

Other: ☐ ______ Grade ______ Size ______

Finish Required: ☐ No ☐ Yes ______

Wood Block Floor: ☐ No ☐ Creosoted ☐ ☐ Natural ______ Thickness ______

______

Special Coatings: ☐ No ☐ Floor ☐ Wall ______

Painting: ☐ No ☑ Regular ☐ Fireproof ☐ Fire Retardant ☐ Brush ☑ Roller ☐ Spray ______

Casework: ☐ No ______ Coats ______ Ceilings: ☐ No ______ Coats ______

Doors: ☐ No ______ Coats ______ Trim: ☐ No ______ Coats ______

Walls, Exterior: ☐ No ______ Coats ______ Interior Walls: ☐ No ______ Coats ______

Windows: ☐ No ______ Coats ______ Piping: ☐ No ______ Coats ______

Other: ______

Structural Steel: ☐ No ☐ Yes ______ Miscellaneous Metals: ☐ No ☐ Yes ______

Wall Covering: ☐ No ☐ Cork Tile ______ ☐ Metal Foil ______

☐ Flexible Wood Veneers ______ ☐ Vinyl ______ Weight ______

Wall Paper ______ Vinyl ______ Murals ______

Other ______

Guards: Corner: ☐ No ☐ Rubber ☐ Steel ☐ Vinyl ______

Wall: ☐ No ☐ Rubber ☐ Steel ☐ Vinyl ______

Page 18 of 30

DATE ______

DIVISION 10: SPECIALTIES

PROJECT ______ LOCATION ______

Bathroom Accessories: ☐ No ☐ Curtain Rod ______ ☐ Dispensers ______ ☐ Grab Bar ______
☐ Hand Dryer ______ ☐ Medicine Cabinet ______ ☐ Mirror ______ ☐ Robe Hook ______
☐ Soap Dispenser ______ ☐ Shelf ______ ☐ Tissue Dispenser ______ ☐ Towel Bar ______
☐ Tumbler Holder ______ ☐ Wall Urn ______ ☐ Waste Receptical ______ ☑ ______
Bulletin Board: ☐ No ☐ Cork ☐ Vinyl Cork ☐ Unbacked ☐ Backed ☐ Stock ☐ Custom
☐ Tan ☐ Framed ☐ No Frames ☐ Changeable Letter ☐ ______ Thickness ______

Canopies: ☐ No ☐ Free Standing ☐ Wall Hung ☐ Stock ☐ Custom ______
Chalkboard: ☐ No ☐ Cement Asbestos ☐ Hardboard ☐ Metal ______ Ga. ☐ Slate ______ in. Thick ☐ Tempered Glass ______
☐ Treated Plastic ☐ ______ ☐ Unbacked ☐ Backed with ______
☐ No Frames ☐ Frames ☐ Chalk Tray ☐ Map Rail ☐ ______
☐ Built in Place ☐ Prefabricated ______
☐ Portable ☐ Reversible ☐ Swing Wing ☐ Sliding Panel ______

Chutes Linen: ☐ No ☐ Aluminum ☐ Aluminized Steel ☐ Stainless Steel ☐ ______ Ga. Diameter ______
☐ Bottom Collector ☐ Sprinklers ______
Mail: ☐ No ☐ Aluminum ☐ Bronze ☐ Stainless ☐ ______ Size ______ ☐ Bottom Collector
Package: ☐ No ☐ Aluminum ☐ Bronze ☐ Stainless ______
Rubbish: ☐ No ☐ Aluminum ☐ Aluminized Steel ☐ Stainless Steel ______ Ga. Diameter ______
☐ Bottom Collector ☐ Sprinklers ______
Compartments & Cubicles: ☐ No ☐ Hospital ______ ☐ Office ______
☐ Shower ______ ☐ Toilet ______ ☐ ______
Control Boards: ☐ No ☐ Yes ______
Decorative Grilles and Screens: ☐ No ☐ Yes ______
Directory Boards: ☐ No ☐ Exterior ☐ Interior ☐ Aluminum ☐ Bronze ☐ Stainless ☐ Lighted ______
Describe ______
Disappearing Stairs: ☐ No ☐ Stock ☐ Custom ☐ Manual ☐ Electric ______ Ceiling Height ______
Display Cases: ☐ No ☐ Economy ☐ Deluxe ______
Fire Extinguishers: ☐ No ☐ CO_2 ☐ Dry Chemical ☐ Foam ☐ Pressure Water ☐ Soda Acid ☐ ______
☐ Aluminum ☐ Copper ☐ Painted Steel ☐ Stainless Steel ☐ ______ Size ______
Cabinets: ☐ No ☐ Aluminum ☐ Painted Steel ☐ Stainless Steel ☐ ______
Hose Equipment: ☐ No ☐ Blanket ☐ Cabinets ☐ Hose ______ Size ______
Protection System: ☐ No ☐ Yes ______
Fireplace, Prefabricated: ☐ No ☐ Economy ☐ Deluxe ☐ Wall Hung ☐ Free Standing ______

Flagpoles: ☐ No ☐ Aluminum ☐ Bronze ☐ Fiberglass ☐ Stainless ☐ Steel ☐ Wood ☐ Tapered ☐ Sectional ______
☐ Ground Set ☐ Wall Set ☐ Counterbalanced ☐ Outriggers ______ Height ______
Bases: ☐ No ☐ Economy ☐ Deluxe ______
Foundation: ☐ No ☐ Yes ______
Folding Gates: ☐ No ☐ Scissors Type ☐ Vertical Members ☐ Stock ☐ Custom ______ Opening ______
Lockers: ☐ No ☐ No Locks ☐ Keyed ☐ Combination ______ Tier Size ______ Height ______
Athletic: ☐ No ☐ Basket ☐ Ventilating ☐ Overhead ______ Size ______
Benches: ☐ No ☐ Yes ______
Special Lockers: ______
Mail Specialties Boxes: ☐ No ☐ Front Loading ☐ Rear Loading ☐ Aluminum ☐ Stainless ☐ ______
Size ______
Letter Slot: ☐ No ☐ Yes ______ Counter Window: ☐ No ☐ Yes ______
Directory: ☐ No ☐ Yes ______ Key Keeper ______
Other: ______

Page 19 of 30

SPEC-AID
DIVISION 10: SPECIALTIES

Accordion Folding Partitions: ☐ No ☐ Acoustical ☐ Non Acoustical ______ Weight ______ psf.
Ceiling Height ______ Describe ______
Folding Leaf Partitions: ☐ No ☐ Acoustical ☐ Non Acoustical ______ Weight ______ psf.
Ceiling Height ______ Describe ______
Hospital Partitions: ☐ No ☐ Metal ☐ Curtain Track ______

Movable Office Partitions: ☐ No ☐ Acoustical ☐ Non Acoustical ☐ Asbestos Cement ☐ Hardboard
☐ Laminated Gypsum ☐ Plywood ☐ ______
☐ With Glass ☐ No Glass Describe ______ Partition Height ______
Special Finish: ______
Doors: ☐ No ☐ Yes Type ______ Finish ______ Size ______
Operable Partitions: ☐ No ☐ Yes Type ______
Portable Partitions: ☐ No ☐ Acoustical ☐ Non Acoustical ______ Weight ______ psf.
Partition Height ______ Describe ______
Shower Partitions: ☐ No ☐ Fiberglass ☐ Glass ☐ Marble ☐ Metal ☐ ______ Finish ______
☐ Stock ☐ Custom ☐ Economy ☐ Deluxe Size ______
Doors: ☐ No ☐ Glass ☐ Tempered Glass ☐ Plastic ☐ Curtain Only ______ Size ______
Receptors: ☐ No ☐ Concrete ☐ Metal ☐ Plastic ☐ Terrazzo ______ Size ______
Tub Enclosure: ☐ No ☐ Stock ☐ Custom ☐ Economy ☐ Deluxe ______ Size ______

Toilet Partitions: ☐ No ☐ Fiberglass ☐ Marble ☑ Metal ☐ Slate ☐ Wood ☐ ______
☐ Floor Mounted ☐ Wall Hung ☑ Ceiling Hung ______
Special Finish ______
Doors: ☐ No ☐ Yes ______
Screens: ☐ No ☐ Full Height ☐ Urinal ☐ Floor Mounted ☐ Wall Hung ☐ Ceiling Hung ______

Woven Wire Partitions: ☐ No ☐ Walls ☐ Ceilings ☐ Panel Width ______ Height ______
Doors: ☐ No ☐ Sliding ☐ Swing ______ Windows: ☐ No ☐ Yes ______
☐ Painted ☐ Galvanized ______
Other Partitions: ______

Parts Bins: ☐ No ☐ Yes ______
Scales: ☐ No ☐ Built in ☐ Portable ☐ Beam Type ☐ Dial Type ______ Capacity ______
Platform Size ______ Material ______ Foundations ______
Accessory Items ______
Shelving, Storage: ☐ No ☐ Metal ☐ Wood ______
Signs: Individual Letters: ☐ No ☐ Aluminum ☐ Bronze ☐ Plastic ☐ Stainless ☐ Steel ☐ ______
☐ Cast ☐ Fabricated Describe ______
Plaques: ☐ No ☐ Aluminum ☐ Bronze ______
Signs: ☐ No ☐ Metal ☐ Plastic ☐ Lighted ______

Sun Control Devices: ☐ No ☐ Yes ______
Telephone Enclosures: ☐ No ☐ Indoor ☐ Outdoor ______
Turnstiles: ☐ No ☐ Yes ______
Vending Machines: ☐ No ☐ Yes ______
Wardrobe Specialties: ☐ No ☐ Yes ______
Other Specialties: ______

Page 20 of 30

SPEC-AID DATE ____________

DIVISION 11: ARCHITECTURAL EQUIPMENT

PROJECT ____________ LOCATION ____________

Appliances, Residential: ☐ No ☐ Yes Allowance ____________ ☐ Separate Contract
☐ Cook Tops ____ ☐ Compactors ____ ☐ Dehumidifier ____ ☐ Dishwasher ____
☐ Dryer ____ ☐ Garbage Disposer ____ ☐ Heaters, Electric ____
☐ Hood ____ ☐ Humidifier ____ ☐ Ice Maker ____ ☐ Oven ____
☐ Refrigerator ____ ☐ Sump Pump ____ ☐ Washing Machine ____ ☐ Water Heater ____
☐ Water Softener ____ ☐ ____

Automotive Equipment: ☐ No ☐ Yes Allowance ____________ ☐ Separate Contract
☐ Hoists ____ ☐ Lube ____ ☐ Pumps ____ ☐ ____

Bank Equipment: ☐ No ☐ Yes Allowance ____________
☐ Counters ____ ☐ Safes ____ ☐ Vaults ____ ☐ Windows ____
☐ ____

Check Room Equipment: ☐ No ☐ Yes Allowance ____________ ☐ Separate Contract
Describe ____________

Church Equipment: ☐ No ☐ Yes Allowance ____________ ☐ Separate Contract
☐ Altar ____ ☐ Baptistries ____ ☐ Bells & Carillons ____ ☐ Confessionals ____
☐ Organ ____ ☐ Pews ____ ☐ Pulpit ____ ☐ Spires ____
☐ Wall Cross ____ ☐ ____

Commercial Equipment: ☐ No ☐ Yes Allowance ____________ ☐ Separate Contract
Describe ____________

Darkroom Equipment: ☐ No ☐ Yes Allowance ____________ ☐ Separate Contract
Describe ____________

Data Processing Equipment: ☐ No ☐ Yes Allowance ____________ ☐ Separate Contract
Describe ____________

Dental Equipment: ☐ No ☐ Yes Allowance ____________ ☐ Separate Contract
☐ Chair ____ ☐ Drill ____ ☐ Lights ____ ☐ X-Ray ____
☐ ____

Dock Equipment: ☐ No ☑ Yes Allowance ____________ ☐ Separate Contract
☑ Bumpers ____ ☐ Boards ____ ☑ Door Seal ____ ☑ Levelers ____
☐ Lights ____ ☐ Shelters ____ ☐ ____

Food Service Equipment: ☐ No ☐ Yes Allowance ____________ ☐ Separate Contract
☐ Bar Units ____ ☐ Cooking Equip. ____ ☐ Dishwashing Equip. ____ ☐ Food Prep. ____
☐ Food Serving ____ ☐ Refrigerated Cases ____ ☐ Tables ____ ☐ ____

Gymnasium Equipment: ☐ No ☐ Yes Allowance ____________ ☐ Separate Contract
☐ Basketball Backstops ____ ☐ Benches ____ ☐ Bleachers ____
☐ Divider Curtain ____ ☐ Gymnastic Equip. ____ ☐ Mats ____ ☐ Scoreboards ____
☐ ____

Industrial Equipment: ☐ No ☐ Yes Allowance ____________ ☐ Separate Contract
Describe ____________

Laboratory Equipment: ☐ No ☐ Yes Allowance ____________ ☐ Separate Contract
☐ Casework ____ ☐ Counter Tops ____ ☐ Hoods ____ ☐ Sinks ____
☐ Tables ____ ☐ ____

Laundry Equipment: ☐ No ☐ Yes Allowance ____________ ☐ Separate Contract
☐ Dryers ____ ☐ Washers ____ ☐ ____

Library Equipment: ☐ No ☐ Yes Allowance ____________ ☐ Separate Contract
☐ Book Shelves ____ ☐ Book Stacks ____ ☐ Card Files ____ ☐ Carrels ____
☐ Charging Desks ____ ☐ Racks ____ ☐ ____

Medical Equipment: ☐ No ☐ Yes Allowance ____________ ☐ Separate Contract
☐ Casework ____ ☐ Exam Room ____ ☐ Incubators ____ ☐ Patient Care ____
☐ Radiology ____ ☐ Sterilizers ____ ☐ Surgery Equip. ____ ☐ Therapy Equip. ____
☐ ____

Page 21 of 30

SPEC-AID

DIVISION 11: ARCHITECTURAL EQUIPMENT

Mortuary Equipment: ☐ No ☐ Yes Allowance ______ ☐ Separate Contract
Describe ______

Musical Equipment: ☐ No ☐ Yes Allowance ______ ☐ Separate Contract
Describe ______

Observatory Equipment: ☐ No ☐ Yes Allowance ______ ☐ Separate Contract
Describe ______

Parking Equipment: ☐ No ☐ Yes Allowance ______ ☐ Separate Contract
☐ Automatic Gates ______ ☐ Booths ______ ☐ Control Station ______
☐ Ticket Dispenser ______ ☐ Traffic Detectors ______
☐ ______

Playground Equipment: In Division 2 ______

Prison Equipment: ☐ No ☐ Yes Allowance ______ ☐ Separate Contract
☐ Ceiling Lining ______ ☐ Wall Lining ______ ☐ Bar Walls ______ ☐ Doors ______
☐ Bunks ______ ☐ Lavatory ______ ☐ Water Closet ______ ☐ ______

Residential Equipment: ☐ No ☐ Yes Allowance ______ ☐ Separate Contract
☐ Kitchen Cabinets (Also Div. 6) ______ ☐ Lavatory Cabinets ______ ☐ Kitchen Equipment ______
☐ Laundry Equip. ______ ☐ Unit Kitchens ______ ☐ Vacuum Cleaning ______ ☐ ______
☐ ______

Safes: ☐ No ☐ Yes Allowance ______ ☐ Separate Contract
☐ Office ______ ☐ Money ______ ☐ ______ ☐ Rating ______
Describe ______

Saunas: ☐ No ☐ Yes Allowance ______ ☐ Separate Contract
☐ Built in Place ☐ Prefabricated Size ______ Describe ______
☐ Heater ______ ☐ Seats ______ ☐ Timer ______ ☐ ______

School Equipment: ☐ No ☐ Yes Allowance ______ ☐ Separate Contract
☐ Art & Crafts ______ ☐ Audio-Visual ______ ☐ Language Labs ______ ☐ Vocational ______
☐ Wall Benches ______ ☐ Wall Tables ______ ☐ ______
☐ ______

Shop Equipment: ☐ No ☐ Yes Allowance ______ ☐ Separate Contract
Describe ______

Stage Equipment: ☐ No ☐ Yes Allowance ______ ☐ Separate Contract
Describe ______
______

Steam Baths: ☐ No ☐ Yes Allowance ______ ☐ Separate Contract
Describe ______

Swimming Pool Equipment: ☐ No ☐ Yes Allowance ______ ☐ Separate Contract
☐ Diving Board ______ ☐ Diving Stand ______ ☐ Life Guard Chair ______ ☐ Ladders ______
☐ Heater ______ ☐ Lights ______ ☐ Pool Cover ______ ☐ Slides ______
☐ ______

Unit Kitchens: ☐ No ☐ Yes Allowance ______ ☐ Separate Contract
Describe ______

Vacuum Cleaning, Central: ☐ No ☐ Yes Allowance ______ ☐ Separate Contract
☐ ______ Valves Describe ______

Waste Disposal Compactors: ☐ No ☐ Yes ______
Incinerators: ☐ No ☐ Electric ☐ Gas Type Waste ______ Capacity ______

Special Equipment: ______
______
______
______
______
______

SPEC-AID DATE ______

DIVISION 12: FURNISHINGS

PROJECT ______ LOCATION ______

Artwork: ☐ No ☐ Yes Allowance ______ ☐ Separate Contract
☐ Murals ______ ☐ Paintings ______ ☐ Photomurals ______ ☐ Sculptures ______
☐ Stained Glass ______ ☐ ______

Interior Landscaping: ☐ No ☐ Yes Allowance ______ **Blinds, Exterior:** ☐ No ☐ Yes Allowance ______ ☐ Separate Contract
☐ Solid ☐ Louvered ☐ Aluminum ☐ Nylon ☐ Vinyl ☐ Wood ☐ ______
Describe ______

Blinds, Interior: ☐ No ☐ Yes Allowance ______ ☐ Separate Contract
Folding: ☐ No ☐ Stock ☐ Custom ☐ Wood ☐ ______
Describe ______
Venetian: ☐ No ☐ Stock ☐ Custom ☐ Aluminum ☐ Plastic ☐ Steel ☐ Wood ☐ ______
Describe ______
Vertical: ☐ No ☐ Aluminum ☐ Cloth ☐ Vinyl ☐ ______
Describe ______
Other: ______

Cabinets: ☐ No ☐ Yes Allowance ______ ☐ Separate Contract
☐ Classroom ______
☐ Dormitory ______
☐ Hospital ______
☐ ______

Carpets: In Division 9 ______

Dormitory Units: ☐ No ☐ Yes Allowance ______ ☐ Separate Contract
☐ Beds ______ ☐ Desks ______ ☐ Wardrobes ______ ☐ ______
☐ ______

Drapery & Curtains: ☐ No ☐ Yes Allowance ______ ☐ Separate Contract
Describe ______
______

Floor Mats: ☐ No ☐ Yes Allowance ______ ☐ Separate Contract
☐ Recessed ☐ Non Recessed ______
☐ Link ☐ Solid ______

Furniture: ☐ No ☐ Yes Allowance ______ ☐ Separate Contract
☐ Beds ______ ☐ Chairs ______ ☐ Chests ______ ☐ Desks ______
Sofas ______ ☐ Tables ______ ☐ ______
☐ ______

Seating Auditorium: ☐ No ☐ Yes Allowance ______ ☐ Separate Contract
Describe ______
Classroom: ☐ No ☐ Yes Allowance ______ ☐ Separate Contract
Describe ______
Stadiuim: ☐ No ☐ Yes Allowance ______ ☐ Separate Contract
Describe ______

Shades: ☐ No ☐ Yes Allowance ______ ☐ Separate Contract
☐ Stock ☐ Custom ☐ Lightproof ☐ Fireproof ______
☐ Cotton ☐ Fiberglass ☐ Vinyl ☐ Woven Aluminum ☐ ______
Describe ______

Wardrobes: ☐ No ☐ Yes Allowance ______ ☐ Separate Contract
☐ Classroom ______ ☐ Dormitory ______ ☐ Hospital ______ ☐ ______
Describe ______

Other Furnishings: ______
______
______

Page 23 of 30

SPEC-AID DATE ______

DIVISION 13: SPECIAL CONSTRUCTION

PROJECT ______ LOCATION ______

Acoustical Echo Chamber: ☐ No ☐ Yes Allowance ______ ☐ Separate Contract
Describe ______
Enclosures: ☐ No ☐ Yes Allowance ______ ☐ Separate Contract
Describe ______
Panels: ☐ No ☐ Yes Allowance ______ ☐ Separate Contract
Describe ______

Air Curtains: ☐ No ☐ Yes Allowance ______ ☐ Separate Contract
☐ Heated Air ☐ Unheated Air ☐ Recirculating ☐ Non Recirculating ☐ ______
Describe ______

Air Inflated Buildings: ☐ No ☐ Yes Describe ______ ☐ Separate Contract

Anechoic Chambers: ☐ No ☐ Yes Allowance ______ ☐ Separate Contract
Describe ______

Audiometric Rooms: ☐ No ☐ Yes Allowance ______ ☐ Separate Contract
Describe ______

Bowling Alleys: ☐ No ☐ Yes Allowance ______ ☐ Separate Contract
Describe ______

Broadcasting Studio: ☐ No ☐ Yes Allowance ______ ☐ Separate Contract
Describe ______

Chimneys: ☐ No ☐ Yes Allowance ______ ☐ Separate Contract
Concrete: ☐ No ☐ Unlined ☐ Lined ☐ ______ Diameter______ Height______
Metal: ☐ No ☐ Insulated ☐ Not Insulated ☐ U.L. Listed ☐ Not U.L. Listed ☐ ______
Describe ______ Diameter______ Height______
Radial Brick: ☐ No ☐ Unlined ☐ Lined ☐ ______ Diameter______ Height______
Foundation: ☐ No ☐ Yes ______

Clean Rooms: ☐ No ☐ Yes Allowance ______ ☐ Separate Contract
Describe ______

Comfort Stations: ☐ No ☐ Yes Describe ______

Dark Rooms: ☐ No ☐ Yes Allowance ______ ☐ Separate Contract
Describe ______

Domes, Observation: ☐ No ☐ Yes Allowance ______ ☐ Separate Contract
Describe ______

Garage: ☐ No ☐ Yes Describe ______ Cars ______

Garden House: ☐ No ☐ Yes Allowance ______ ☐ Separate Contract
Describe ______

Grandstand: ☐ No ☐ Yes Describe ______ Seats ______

Greenhouse: ☐ No ☐ Yes Allowance ______ ☐ Separate Contract
Describe ______

Hangars: ☐ No ☐ Yes Describe ______ Planes ______

Hyperbaric Rooms: ☐ No ☐ Yes Allowance ______ ☐ Separate Contract
Describe ______

Incinerators (See also Division 10): ☐ No ☐ Yes Allowance ______ ☐ Separate Contract
Describe ______ Capacity ______

Insulated Rooms: ☐ No ☐ Yes Allowance ______ ☐ Separate Contract
Doors: ☐ No ☐ Cooler ☐ Freezer ☐ Manual ☐ Electric ______
☐ Galvanized ☐ Stainless Describe ______
Coolers: ☐ No ☐ Yes Describe ______
Freezers: ☐ No ☐ Yes Describe ______
Partitions: ☐ No ☐ Yes ☐ Stock ☐ Custom Describe ______
Other: ______

Page 24 of 30

SPEC-AID

DIVISION 13: SPECIAL CONSTRUCTION

Integrated Ceilings: ☐ No ☐ Yes Module ____ Ceiling Height ____
Lighting: ☐ No ☐ Yes Describe ____ Foot Candles ____
Heating: ☐ No ☐ Yes Describe ____
Ventilating: ☐ No ☐ Yes Describe ____
Air Conditioning: ☐ No ☐ Yes Describe ____
Music Practice Rooms: ☐ No ☐ Yes Allowance ____ ☐ Separate Contract
Describe ____
Pedestal Floors: ☐ No ☐ Yes Allowance ____ ☐ Separate Contract
☐ Aluminum ☐ Plywood ☐ Steel ☐ ____ Panel Size ____ Height ____
☐ High Density Plastic ☐ Vinyl Tile ☐ V.A. Tile ☐ ____
Describe ____
Portable Booths: ☐ No ☐ Yes Allowance ____ ☐ Separate Contract
Describe ____
Prefabricated Structures: ☐ No ☐ Yes Allowance ____ ☐ Separate Contract
Describe ____
Radiation Protection, Fluoroscopy Room: ☐ No ☐ Yes ____
Nuclear Reactor: ☐ No ☐ Yes ____
Radiological Room: ☐ No ☐ Yes ____
X-Ray Room: ☐ No ☐ Yes ____
Other: ____
Radio Frequency Shielding: ☐ No ☐ Yes Allowance ____ ☐ Separate Contract
Describe ____
Radio Tower: ☐ No ☐ Yes Allowance ____ ☐ Separate Contract
☐ Guyed ☐ Self Supporting Wind Load ____ psf. ____ Height ____
Foundations ____
Saunas and Steam Rooms: ☐ No ☐ Yes Allowance ____ ☐ Separate Contract
Describe ____
Silos: ☐ No ☐ Yes Allowance ____ ☐ Separate Contract
☐ Concrete ☐ Steel ☐ Wood ☐ ____ Diameter ____ Height ____
Foundations ____
Squash & Hand Ball Courts: ☐ No ☐ Yes Allowance ____ ☐ Separate Contract
Describe ____
Storage Vaults: ☐ No ☐ Yes Allowance ____ ☐ Separate Contract
Describe ____
Swimming Pool Enclosure: ☐ No ☐ Yes Allowance ____ ☐ Separate Contract
Describe ____
Swimming Pool Equipment: In Division 11 ____
Swimming Pools: ☐ No ☐ Yes Allowance ____ ☐ Separate Contract
☐ Aluminum ☐ Concrete ☐ Gunite ☐ Plywood ☐ Steel ☐ ____
☐ Lined ☐ Unlined ____
Deck: ☐ No ☐ Concrete ☐ Stone ____ Size ____
Bath Houses: ☐ No ☐ Yes ____ Fixtures ____
Tanks: ☐ No ☐ Yes Allowance ____ ☐ Separate Contract
☐ Concrete ☐ Fiberglass ☐ Steel ☐ Wood ☐ ____ Capacity ____
☐ Fixed Roof ☐ Floating Roof ☐ ____ Height ____
Foundations: ____
Therapeutic Pools: ☐ No ☐ Yes Describe ____
Vault Front: ☐ No ☐ Yes Allowance ____ ☐ Separate Contract
Describe ____ Hour Test ____
Zoo Structures: ☐ No ☐ Yes Describe ____
Other Special Construction: ____

Page 25 of 30

SPEC-AID DATE ____

DIVISION 14: CONVEYING SYSTEMS

PROJECT ____ LOCATION ____

Ash Hoist: ☐ No ☐ Yes Allowance ____ ☐ Separate Contract
Describe ____

Conveyers: ☐ No ☐ Yes Allowance ____ ☐ Separate Contract
Describe ____

Correspondence Lift: ☐ No ☐ Yes Allowance ____ ☐ Separate Contract
Describe ____

Dumbwaiters: ☐ No ☐ Yes Allowance ____ ☐ Separate Contract
Capacity ____ Size ____ Number ____ Floors ____
Stops ____ Speed ____ Finish ____
Describe ____

Elevators, Freight: ☑ No ☐ Yes Allowance ____ ☐ Separate Contract
☐ Hydraulic ☐ Electric ☐ Geared ☐ Gearless ____
Capacity ____ Size ____ Number ____ Floors ____
Stops ____ Speed ____ Finish ____
Machinery Location ____ Door Type ____
Signals ____ Special Requirements ____

Elevators, Passenger: ☑ No ☐ Yes Allowance ____ ☐ Separate Contract
☐ Hydraulic ☐ Electric ☐ Geared ☐ Gearless ____
Capacity ____ Size ____ Number ____ Floors ____
Stops ____ Speed ____ Finish ____
Machinery Location ____ Door Type ____
Signals ____ Special Requirements ____

Escalators: ☑ No ☐ Yes Allowance ____ ☐ Separate Contract
Capacity ____ Size ____ Number ____ Floors ____
Story Height ____ Speed ____ Finish ____
Machinery Location ____ Incline Angle ____
Special Requirements ____

Hoists & Cranes: ☐ No ☐ Yes Allowance ____ ☐ Separate Contract
Describe ____

Lists: ☐ No ☐ Yes Allowance ____ ☐ Separate Contract
Describe ____

Material Handling Systems: ☐ No ☐ Yes Allowance ____ ☐ Separate Contract
☐ Automated ☐ Non Automated ☐ ____
Describe ____

Moving Stairs & Sidewalks ☐ No ☐ Yes Allowance ____ ☐ Separate Contract
Capacity ____ Size ____ Number ____ Floors ____
Story Height ____ Speed ____ Finish ____
Machinery Location ____ Incline Angle ____
Special Requirements ____

Pneumatic Tube System: ☐ No ☐ Yes Allowance ____ ☐ Separate Contract
☐ Automatic ☐ Manual ☐ ____ Size ____ Stations ____
Length ____ Special Requirements ____

Vertical Conveyer: ☐ No ☐ Yes Allowance ____ ☐ Separate Contract
☐ Automatic ☐ Non Automatic ☐ ____
Describe ____

Other Conveying: ____

Page 26 of 30

SPEC-AID DATE ____________

DIVISION 15: MECHANICAL

PROJECT ____________ LOCATION ____________

Building Drainage: Design Rainfall ____ ☑ Roof Drains ____ ☐ Court Drains ____
☑ Floor Drains ____ ☐ Yard Drains ____ ☐ Lawn Drains ____ ☐ Balcony Drains ____
☐ Area Drains ____ ☐ Sump Drains ____ Shower Drains ____ ☐ ____
☐ Drain Piping: Size ____ Describe ____
☐ Drain Gates ____ ☐ Clean Outs ____ ☐ Grease Traps ____

Sanitary System: ☐ No ☑ Yes ☐ Site Main ____ ☐ Manholes ____
☐ Sump Pumps ____ ☐ Bilge Pumps ____ ☐ Ejectors ____
☐ Soils, Stacks ____ ☐ Wastes, Vents ____ ☐ ____

Domestic Cold Water: ☐ No ☑ Water Meters ✓ ____ ☐ Law Sprinkler Connection ____
☐ Water Softening ____ ☐ Water Filtering ____
☐ Boiler Feed Water ____ ☐ Conditioning Apparatus ____
☑ Standpipe System ____ ☑ Hose Bibbs ____
☐ Pressure Tank ____ ☐ Booster Pumps ____
☐ Reducing Valves ____ ☐ ____

Domestic Hot Water: ☐ No ☐ Electric ☑ Gas ☐ Oil ☐ Solar ____
☐ Boiler ____ ☐ Conditioner ____ ☐ Fixture Connections ____
☑ Storage Tanks ____ Capacity ____
☐ Pumps ____

Piping: ☐ No ☐ Yes Material ____
☐ Air Chambers ____ ☐ Escutcheons ____ ☐ Expansion Joints ____
☐ Shock Absorbers ____ ☐ Hangers ____
☐ Valves ____ ☐ Paint ____

Special Piping: ☐ No ☐ Compressed Air ____ ☐ Vacuum ____
☐ Oxygen ____ ☐ Nitrous Oxygen ____
☐ Carbon Dioxide ____ ☐ Process Piping ____

Insulation Cold: ☐ No ☑ Yes Material ____ Jacket ____
Hot: ☐ No ☐ Yes Material ____ Jacket ____

Fixtures Bathtub: ☐ No ☐ C.I. ☐ Steel ☐ Fiberglass ☐ ____ Color ____
☐ Curtain ☐ Rod ☐ Enclosure ☐ Wall Shower ____
Drinking Fountain: ☐ No ☑ Yes ☐ Wall Hung ☐ Pedestal ____
Hose Bibb: ☐ No ☑ Yes Describe ____
Lavatory: ☐ No ☑ China ☐ C.I. ☑ Steel ☐ ____ Color ____
☑ Wall Hung ☐ Legs ☐ Acid Resisting ____
Shower: ☑ No ☐ Individual ☐ Group ☐ Heads ☐ ____ Size ____
Compartment: ☐ No ☐ Metal ☐ Stone ☐ Fiberglass ☐ ____ ☐ Door ☐ Curtain
Receptor: ☐ No ☐ Plastic ☐ Metal ☐ Terrazzo ☐ ____
Sinks: ☐ No ☑ Kitchen ____ ☑ Janitor ____
☐ Laundry ____ ☐ Pantry ____
☐ ____
Urinals: ☐ No ☐ Floor Mounted ☑ Wall Hung ____
Screens: ☐ No ☐ Floor Mounted ☑ Wall Hung ____
Wash Centers: ☐ No ☐ Yes Describe ____
Wash Fountains: ☐ No ☐ Floor Mounted ☐ Wall Hung ____ Size ____
Describe ____
Water Closets: ☐ No ☐ Floor Mounted ☑ Wall Hung Color ____
Describe ____
Water Coolers: ☐ No ☐ Floor Mounted ☐ Wall Hung ____ Capacity ____ gph.
☐ Water Supply ☐ Bottle ☐ Hot ☐ Compartment ____
Other Fixtures: ____

Page 27 of 30

SPEC-AID

DIVISION 15: MECHANICAL

Fire Protection: ☐ Carbon Dioxide System ______ ☐ Standpipe ______
☐ Sprinkler System ☑ Wet ☐ Dry ______ Spacing ______
☑ Fire Department Connection ______ ☐ Building Alarm ______
☐ Hose Cabinets ______ ☑ Hose Racks ______
☑ Roof Manifold ______ ☐ Compressed Air Supply ______
☑ Hydrants ______ ☐ ______

Special Plumbing ______

______

Gas Supply System: ☐ No ☑ Natural Gas ☐ Manufactured Gas ______
Pipe: Schedule ______ Fittings ______
Shutoffs: ______ Master Control Valve: ______
Insulation: ______ Paint: ______

Oil Supply System: ☐ No ☐ Tanks ☐ Above Ground ☐ Below Ground ______
☐ Steel ☐ Plastic ☐ ______ Capacity ______

Heating Plant: ☐ No ☐ Electric ☑ Gas ☐ Oil ☐ Solar ______
☐ Boilers ______ ☐ Pumps ______
☐ PRV Stations ______ ☐ Piping ______
☐ Heat Pumps ______

Cooling Plant: ☐ No ☑ Yes Electric ______ Tons ______
Chillers: ☐ Steam ☐ Water ☐ Air ______
Condenser—Compressor ☐ Air ☐ Water ______
Pumps ______ Cooling Towers ______

System Type: ______
☑ Single Zone ______ ☐ Multi-Zone ______
☐ All Air ______ ☐ Terminal Reheat ______
☐ Double Duct ______ ☐ Radiant Panels ______
☐ Fan Coil ______ ☐ Unit Ventilators ______
☐ Perimeter Radiation ______ ☐ ______

Air Handling Units: Area Served ______ Number ______
Total CFM ______ % Outside Air ______
Cooling, Tons ______ Heating, MBH ______
Filtration ______ Supply Fans ______
Economizer ______

Fans: ☐ No ☐ Return ☑ Exhaust ☐ ______
Describe ______

Distribution: Ductwork ______ Material ______
Terminals: ☐ Diffusers ______ ☐ Registers ______
☐ Grilles ______ ☐ Hoods ______
Volume Dampers: ______
Terminal Boxes: ☐ High Velocity ______ ☐ With Coil ______
☐ Double Duct ______ ☐ ______

Coils: ______
☐ Preheat ______ ☐ Reheat ______
☐ Cooling ______ ☐ ______

Piping: See Previous Page ______

Insulation: Cold: ☐ No ☑ Yes Material ______ Jacket ______
☐ Hot: ☑ No ☐ Yes Material ______ Jacket ______
Automatic Temperature Controls: ______
Air & Hydronic Balancing: ______

Special HVAC: ______

______

Page 28 of 30

SPEC-AID DATE ____________

DIVISION 16: ELECTRICAL

PROJECT ____________ LOCATION ____________

Incoming Service: ☐ Overhead ☑ Underground

	Primary	Secondary
Voltage		
Unit Sub-station & Size		
Number of Manholes		
Feeder Size		
Length		
Conduit		
Duct		
Concrete: ☐ No ☐ Yes		
Other		

Building Service: Size ____________ Amps Switchboard ____________

Panels: ☐ Distribution ____________ Lighting ____________ Power ____________

Describe ____________

Motor Control Center: Furnished by ____________

Describe ____________

Bus Duct: ☑ No ☐ Yes Size ____________ Amps Application ____________

Describe ____________

Cable Tray: ☐ No ☑ Yes Describe ____________

Emergency System: ☐ No ☑ Yes Allowance ____________ ☐ Separate Contract

Generator: ☐ No ☐ Diesel ☐ Gas ☐ Gasoline ____________ Size ____________ KW

Transfer Switch: ☐ No ☐ Yes Number ____________ Size ____________ Amps

Area Protection Relay Panels: ☐ No ☐ Yes ____________

Other ____________

Conduit: ☐ No ☑ Yes ☑ Aluminum ____________

☐ Electric Metallic Tubing ____________

☐ Galvanized Steel ____________

☐ Plastic ____________

Wire: ☐ No ☑ Yes ☐ Type Installation ____________

☐ Armored Cable ____________

☐ Building Wire ____________

☐ Metallic Sheath Cable ____________

☐ ____________

Underfloor Duct: ☑ No ☐ Yes Describe ____________

Header Duct: ☑ No ☐ Yes Describe ____________

Trench Duct: ☑ No ☐ Yes Describe ____________

Underground Duct: ☑ No ☐ Yes Describe ____________

Explosion Proof Areas: ☑ No ☐ Yes Describe ____________

Motors: ☐ No ☑ Yes Total H.P. ________ No. of Fractional H.P. ____________ Voltage ____________

☐ 1/2 to 5 H.P. ____________ ☐ 7-1/2 to 25 H.P. ____________ ☐ Over 25 H.P. ____________

Describe ____________

Starters: Type ____________

Supplied by: ____________

Page 29 of 30

SPEC-AID

DIVISION 16: ELECTRICAL

Telephone System: ☐ No ☑ Yes Service Size ______ Length ______
Manhole: ☐ No ☐ Yes Number ______ Termination ______
Concrete Encased: ☐ No ☐ Yes ☑ Rigid Galv. ☐ Duct ☐ ______

Fire Alarm System: ☐ No ☑ Yes Service Size ______ Length ______ Wire Type ______
Concrete Encased: ☐ No ☐ Yes ☑ Rigid Galv. ☐ Duct ☐ ______
☐ Stations ______ ☐ Horns ______ ☐ Lights ______ ☐ Combination ______
Detectors: ☐ Rate of Rise ______ ☐ Fixed ______ ☐ Smoke ______
Describe ______ Insulation ______ Wire Size ______
☐ Zones ______ ☐ Conduit ______ ☐ E.M.T. ______ ☐ Empty ______
Describe ______

Watchmans Tour: ☑ No ☐ Yes ☐ Stations ______ ☐ Door Switches ______
☐ Alarm Bells ______ ☐ Key Re-sets ______ ☐ ______
☐ Conduit ______ ☐ E.M.T. ______ ☐ Wire ______ ☐ Empty ______
Describe ______

Clock System: ☑ No ☐ Yes ☐ Electronic ☐ Wired ☐ ______
☐ Single Dial ______ ☐ Double Dial ______ ☐ Program Bell ______
☐ Conduit ______ ☐ E.M.T. ______ ☐ Empty ______
Describe ______

Sound System: ☑ No ☐ Yes Type ______ Speakers ______
☐ Conduit ______ ☐ Cable ______ ☐ E.M.T. ______ ☐ Empty ______
Describe ______

Television System: ☑ No ☐ Yes Describe ______
☐ Antenna ______ ☐ Closed Circuit ______ ☐ Teaching ______ ☐ Security ______
☐ Learning Laboratory ______ ☐ ______
☐ Conduit ______ ☐ E.M.T. ______ ☐ Wire ______ ☐ Empty ______

Lightning Protection: ☑ No ☐ Yes Describe ______

Low Voltage Switching: ☑ No ☐ Yes Describe ______

Scoreboards: ☑ No ☐ Yes Describe ______ Number ______

Comfort Systems: ☑ No ☐ Electric Heat ☐ Snow Melting ☐ ______
Describe ______

Other Systems: ______

______

______

Lighting Fixtures: ☐ No ☑ Yes ☐ Allowance ______ ☐ Separate Contract
☐ Economy ☑ Commercial ☐ Deluxe ☐ Explosion Proof ☐ ______
☑ Incandescent ______
______ Foot Candles ______
☑ Fluorescent ______
______ Foot Candles ______
☐ Mercury Vapor ______
______ Foot Candles ______
☐ ______ Foot Candles ______
☐ Step Lighting ______ ☐ Planter Lighting ______ ☐ Fountain Lighting ______
☐ Site Lighting ______ ☐ Poles ______ ☐ Area Lighting ______ ☐ Flood Lighting ______
Dimming System: ☐ No ☐ Yes ☐ Incandescent ☐ Fluorescent ______
Ceilings: ☐ T Bar ☐ Concealed Spline ☐ ______
Emergency Battery Units: ☐ No ☐ Lead Acid ☑ Nickel Cadmium ☐ 6 Volt ______ 12 Volt ______
Describe ______

Special Considerations: ______

______

______

Page 30 of 30

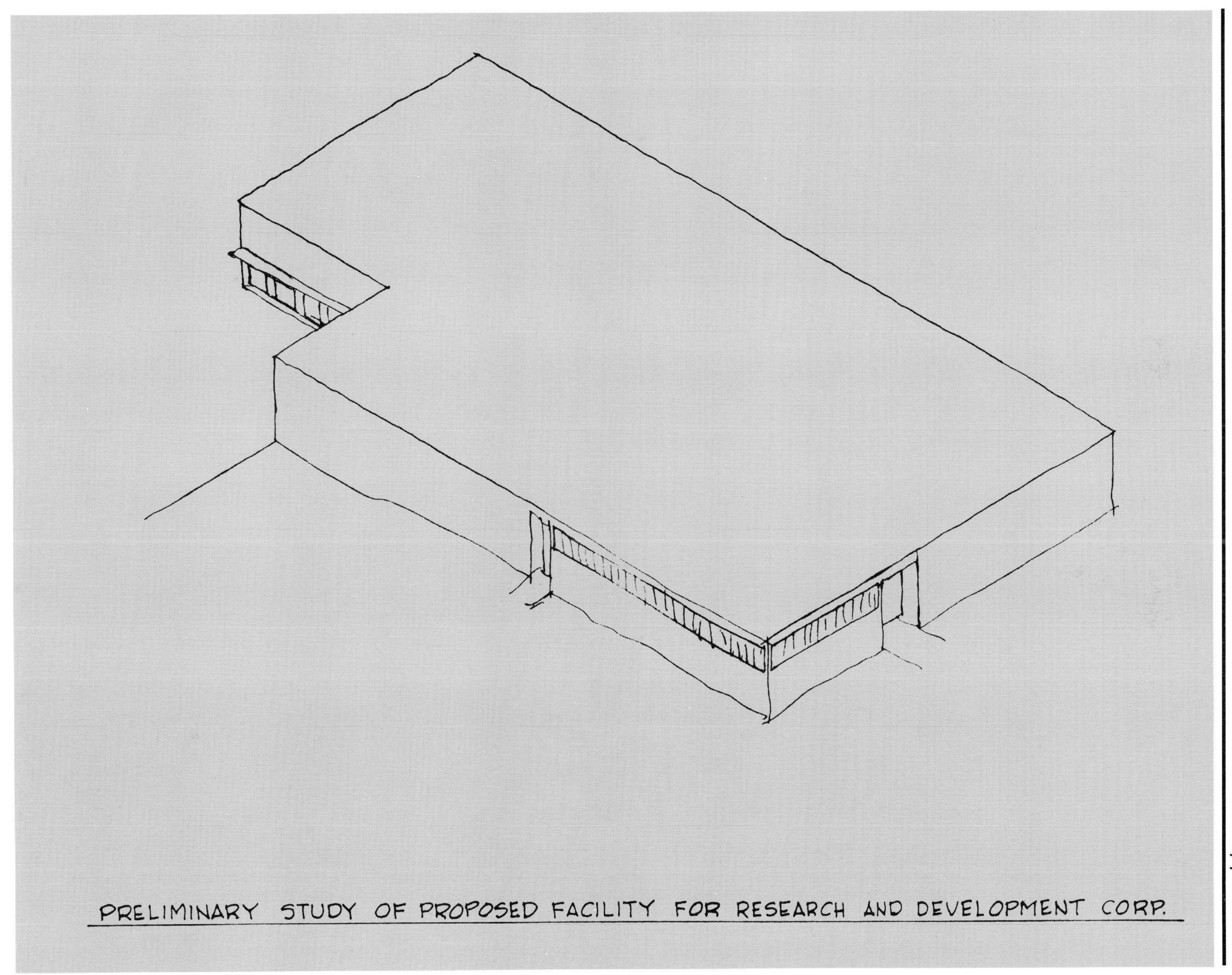
PRELIMINARY STUDY OF PROPOSED FACILITY FOR RESEARCH AND DEVELOPMENT CORP.

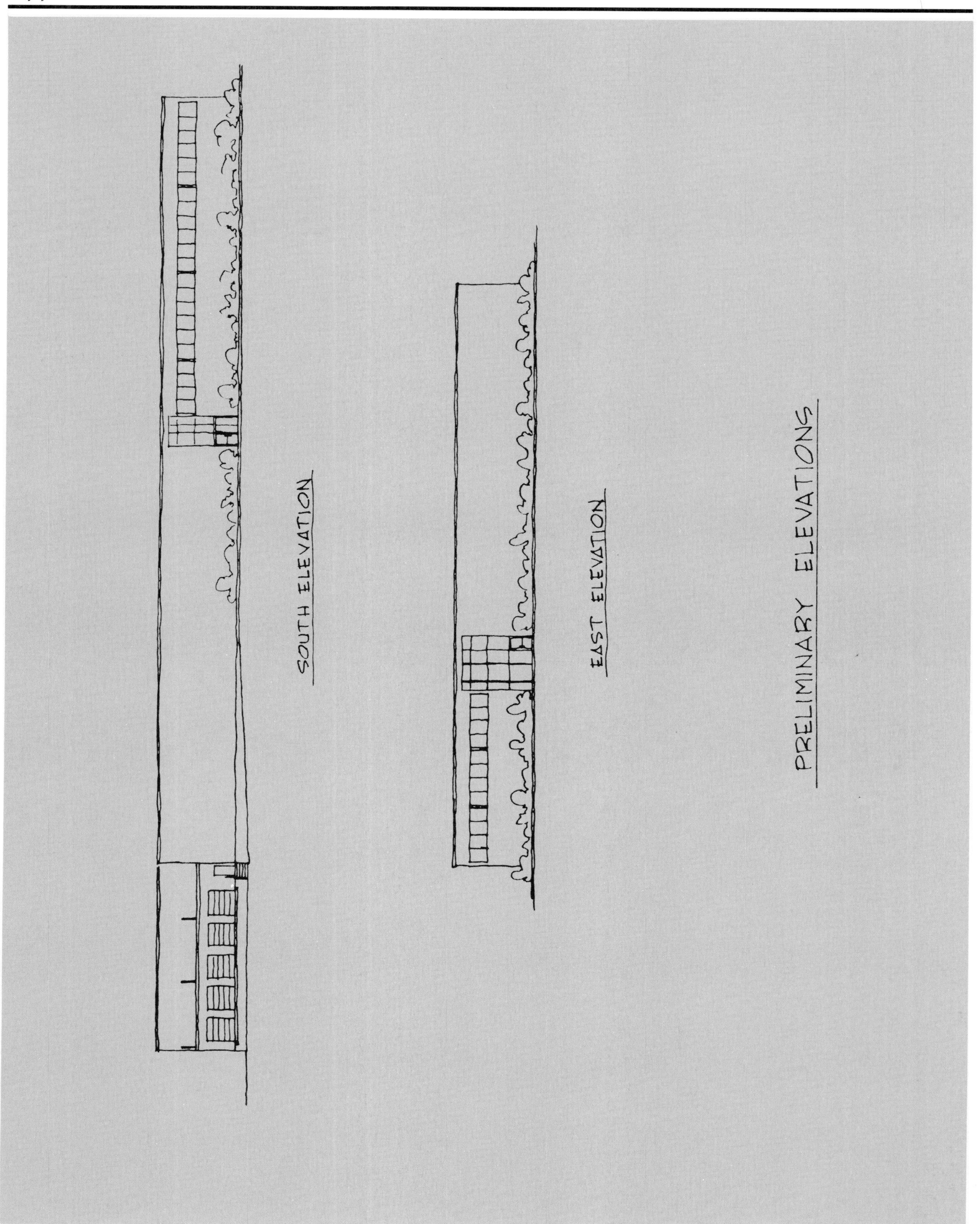
SOUTH ELEVATION
EAST ELEVATION
PRELIMINARY ELEVATIONS

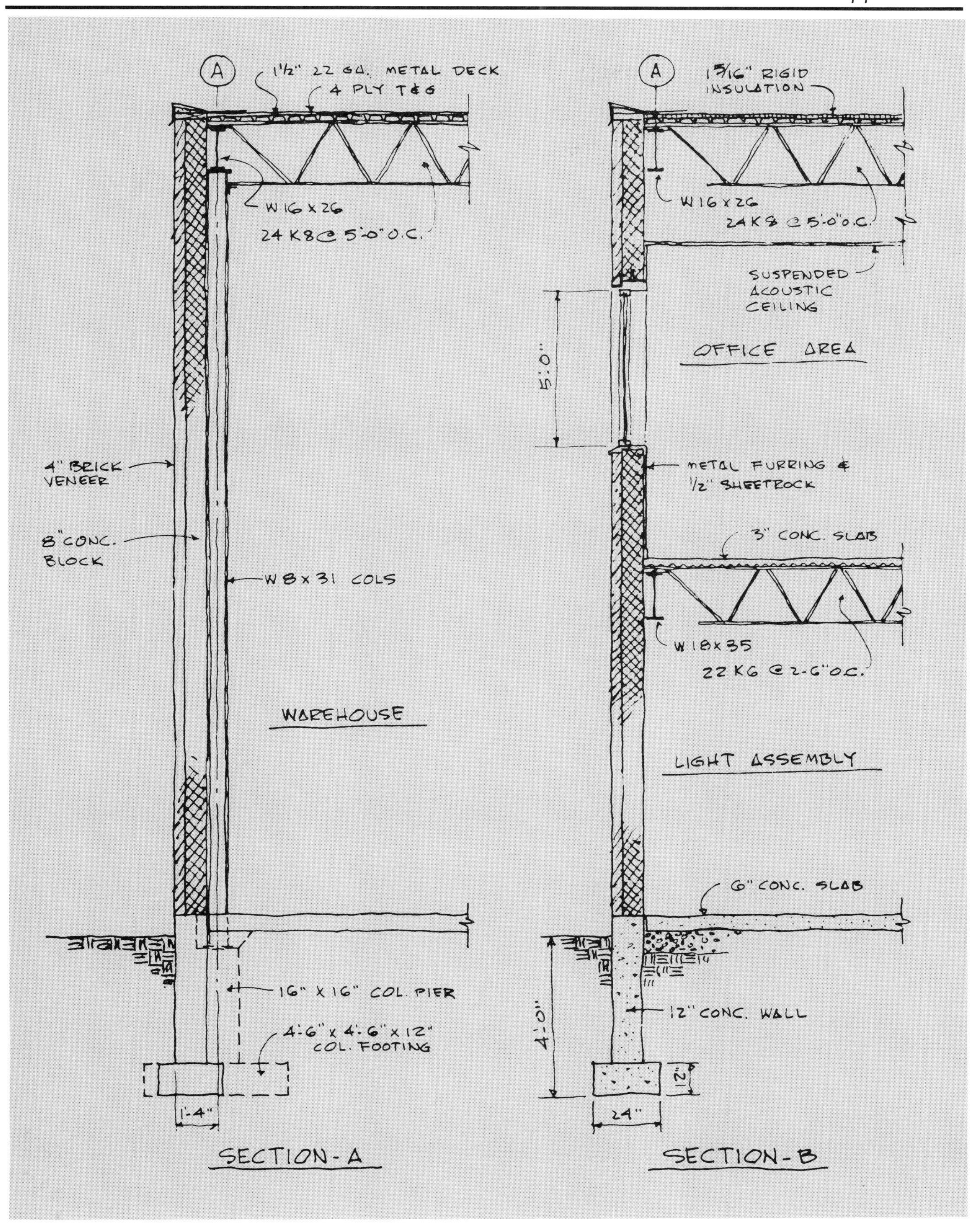

A
1½" 22 GA. METAL DECK
4 PLY T&G
W16x26
24K8 @ 5'-0" O.C.
4" BRICK VENEER
8" CONC. BLOCK
W8x31 COLS
WAREHOUSE
16" x 16" COL. PIER
4'-6" x 4'-6" x 12" COL. FOOTING
1'-4"
SECTION-A
1 5/16" RIGID INSULATION
W16x26
24K8 @ 5'-0" O.C.
SUSPENDED ACOUSTIC CEILING
OFFICE AREA
5'-0"
METAL FURRING & ½" SHEETROCK
3" CONC. SLAB
W18x35
22K6 @ 2'-6" O.C.
LIGHT ASSEMBLY
6" CONC. SLAB
12" CONC. WALL
4'-0"
12"
24"
SECTION-B

Appendix B

Request for Proposal

The following is an example Request for Proposal for HVAC. It includes an addenda (No. 1) and illustrates the kind of information and degree of detail that might be furnished for this purpose.

H. V. A. C.

REQUEST FOR PROPOSALS

1. Request for Bids

   a. Submit on or before December 1, 1987 at the office of Design/Build Construction Co., 80 Build Street, Boston, MA 02119, a written proposal for HVAC work.

   b. All bids are to be submitted upon your company's letterhead and signed by a person authorized to make the proposal.

   c. Bidders are to indicate the structure under which he conducts business and in what state.

2. Documents

   a. Specifications as listed on Exhibit "A."

   b. Drawings as listed on Exhibit "B."

3. Scope of Work

   a. See Exhibit "C," attached, for clarification to specifications.

   b. Metal Louvers

      The bidders are to include as part of their work all metal louvers attached to fans and/or ductwork and louvers hidden from view. See alternate for cost to furnish and install the architectural louvers shown.

   c. The contractor shall furnish all access doors necessary for his work. Doors to be located in drywall or masonry walls shall be installed by the respective trades. Doors necessary in sheetmetal or equipment are to be installed by this contractor.

   d. All motor control centers are to be furnished by the electrical subcontractor. The mechanical contractor shall furnish motor starters (installed by electrical) for all motors not shown as part of the motor control center.

H. V. A. C.

REQUEST FOR PROPOSALS (Cont'd.)

3. Scope of Work (Cont'd.)

e. All automatic temperature controls and all reference to same shall be carried as part of the mechanical work. The electrical contractor shall furnish power as required.

f. All references to contractor or mechanical contractor in the specifications shall be interpreted to refer to the mechanical subcontractor unless modified in Exhibit "C," Clarification to Specifications.

g. The bidders are to carry necessary temporary filters and rehab work to permanent building heating systems. It is anticipated that the building heating system shall be used for the '88 - '89 heating season. Furnish unit cost for fireman, see unit costs.

h. The cost of labor and material and performance bonds is to be included as an alternate.

i. It is the intent that the bid be for a completed system. Any area, item or subsystem lacking which prohibit this goal to be stated and a cost for same included as part of the alternates.

j. Scope of work for sprinkler, plumbing and electrical trades are included for coordination. It is the bidders responsibility to alert Design/Build Construction Company to any gaps or duplication.

4. Conditions of Site

a. The project is accessible from three sides. Erection cranes can travel on Main Avenue and on Access Parkway inside of the protective concrete barrier system, set by Design/Build Construction Co., and along the west side of the project.

b. There is limited storage at the project site. The contractor shall arrange for his own storage in order to maintain an orderly flow of the material to the project.

H. V. A. C.

REQUEST FOR PROPOSALS (Cont'd.)

4. Conditions of Site (Cont'd.)

   c. The contractor shall make a site visit to acquaint himself with field conditions. Mr. Smith is the Project Superintendent. Job telephone number is (617) 555-1111.

   d. The structural steel erection will go column tier by column tier and will proceed east to west. The tower sections will be erected simultaneously.

5. Timetable for Erection

   a. Structural steel erection is anticipated to start February, 1988.

6. Bid Breakdown

The lump sum quotation for the mechanical work is to be divided into categories of work. These categories are to include all associated labor, material, equipment, handling, taxes, overhead and profit.

The breakdown is required for bid evaluation and may serve as the schedule of values for partial payment. The contractor reserves the right to delete portions of the work as categorized if deemed required.

The breakdown is as follows:

   a. HVAC, piping, sheet metal (other than garage) ____________________L.S.

   b. Garage ventilation system complete ____________________L.S.

   c. Breakdown of above

      1) First (1st) floor main lobby $__________________

      2) Atrium area $__________________

H. V. A. C.

REQUEST FOR PROPOSALS (Cont'd.)

6. Bid Breakdown (Cont'd.)

3) Mechanical rooms

Including chillers, chiller water pumps, condenser water pumps complete, chilled water and condenser water piping, fittings, valves, strainers etc., but excluding water treatment. $________________

4) Cooling Towers

Including associated fittings, heaters, etc., and support beams. $________________

5) Water treatment system including chemical storage tanks, chemicals complete. $________________

6) Condenser water system including sleeves, isolators fittings, etc. $________________

c. Breakdown of above

7. Chilled water risers including compression tank $________________

8. Air handling units on all floors except 6, 7 and 8 including all piping, etc. $________________

9. Air handling units on floors 6, 7 and 8. $________________

10. Automatic temperature controls including compressors, dryers, thermostats, etc. $________________

H. V. A. C.

REQUEST FOR PROPOSALS (Cont'd.)

6. Bid Breakdown (Cont'd.)

Cooling Towers (Cont'd.)

11. Insulation

Associated with HVAC complete. $________________

12. Balancing

Water balancing for all systems. $________________

13. Sheet Metal

Including all angles, supports, hanger, vanes, sound attenuators, etc.

a) First (1st) floor lobby $________________

b) Main mechanical/elevator rooms. $________________

c) Riverside office level at third (3rd) floor. $________________

d) Typical floors 4, 5 and 9, from air handling unit to flex connection including elevator lobby. $________________

e) Typical floors 6, 7 and 8, from air handling unit to flex connection including elevator lobby and VAV boxes. $________________

f) Outside air

From outside louver to unit including plenum and from unit to riser, riser and including all horizontal runs. $________________

g) Stairwell pressurization including plenum to outside louver. $________________

H. V. A. C.

REQUEST FOR PROPOSALS (Cont'd.)

6. Bid Breakdown (Cont'd.)

13. Sheet Metal (Cont'd.)

h) Toilet/smoke exhaust system including exhaust for plenum. $______________

i) Miscellaneous $______________

Any remaining duct work not included in above. $______________

14. Work associated with tenant work listed on page 15B-5, paragraph 3. $______________

15. Equipment

a) Air handling units inclusive of labor. $______________

b) Outside air unit. $______________

c) Exhaust fans. $______________

d) Ventilation fans. $______________

e) Heating coils. $______________

f) VAV boxes. $______________

g) Other - specify. $______________

16. Dampers $______________

17. Diffusers

a) Diffusers $______________

18. Flex Duct $______________

a) From spin fitting to perimeter diffusers.

H. V. A. C.

REQUEST FOR PROPOSALS (Cont'd.)

6. Bid Breakdown (Cont'd.)

19. Grilles and registers $________________

20. Balancing

Air balancing system $________________

21. Duct Insulation $________________

22. Miscellaneous $________________

The sum of 1 through 22 above is equal to the lump sum quotation for HVAC, Piping and Sheetmetal.

23. Garage Ventilation

a) Garage ventilation ductwork from outside louvers to fans. $________________

b) Garage fans. $________________

c) Miscellaneous $________________

The sum of 23a, b and c, is equal to the lump sum quotation for garage ventilation system complete.

7. Alternates/Unit Prices

a. Furnish and install architectural louvers shown (See Scope of Work Item #2). $________________

b. Deductive amount for color code piping in mechanical room see page 15A-9, Section 15A, paragraph 16, subparagraph c. $________________

c. Deductive amount to simplify installation period from that specified. $________________

H. V. A. C.

REQUEST FOR PROPOSALS (Cont'd.)

7. Alternates/Unit Prices (Cont'd.)

d. Provide balancing adjusting of mechanical by own forces in lieu of outside service. $__________________

e. Cooling tower field performance test see page 15F-3, Section 15F, paragraph 3, subparagraph u. $__________________

f. Different cooling tower manufactures and cost other than specified see page 15F-3, Section 15F, paragraph 3, subparagraph w. $__________________

g. Furnish seals by other than specified manufacturer, stipulate manufacturer. $__________________

h. Control work by other than companies specified. Bidders are to stipulate company. $__________________

i. Bond cost.

1) Unit Prices

a. Sheet Metal

1. Furnish and install double light troffer to 2 x 2 fluorescent light fixture including 6'-0" of 6" diameter flex complete. $__________________

2. Furnish and install single light troffer to 2 x 2 fluorescent light fixture including 6'-0" of 6" diameter flex complete. $__________________

3. Sheet metal per pound. $__________________

4. Sound liner per square foot of duct surface. $__________________

5. Six inch (6") diameter taps in field. $__________________

H. V. A. C.

REQUEST FOR PROPOSALS (Cont'd.)

7. Alternates/Unit Prices (Cont'd.)

i. Bond cost. (Cont'd.)

1) Unit Prices (Cont'd.)

a. Sheet Metal (Cont'd.)

6. Eight inch (8") diameter taps in field. $__________________

7. Six inch (6") volume dampers. $__________________

8. Eight inch (8") volume dampers. $__________________

9. Miscellaneous volume dampers. $__________________

10. Six inch (6") flex duct insulated. $__________________

11. Eight inch (8") flex duct insulated. $__________________

12. Six inch (6") diameter insulated sheet metal duct. $__________________

13. Receive and install registers. $__________________

14. Receive and install supply grille. $__________________

15. Receive and install troffers. $__________________

16. Unit cost to install additional zone thermostats per specification. $__________________

17. Unit cost for temporary fireman needed during building temporary heating season '88 - '89. $__________________

H. V. A. C.

REQUEST FOR PROPOSALS (Cont'd.)

8. Other

   a. The bidders are encouraged to offer cost savings proposals at the time of bidding.

   b. Any errors, inconsistencies or omissions noted are to be brought to Design/Build Co. attention.

9. Financial Statement

   a. Include with your bid a statement from your bonding company attesting to:

   - Your ability to be bonded.
   - Their ability to furnish a bond for work being performed in Massachusetts.

   b. The successful bidder will be required to furnish their latest audited financial statement for review.

   c. The successful bidder will be required to furnish a bank reference and a bond company reference.

EXHIBIT "C"

H.V.A.C.

CLARIFICATION TO SPECIFICATIONS

1. Page 15A-1, Section 15A, paragraph 3, subparagraph c.

   When the bidders perceive areas of the contract drawings and specifications are in excess of the code, they are encouraged to note said excess on the quotation in the Alternate sections.

2. Page 15A-8, Section 15A, paragraph 12.

   Exclude from work.

3. Page 15A-9, Section 15A, paragraph 16, subparagraph c.

   Color coding of piping is to be carried as a deductive alternate price. Flow arrows and markers specified are part of base quote.

4. Page 15A-9, Item 15A, paragraph 18, subparagraph a).

   Size of concrete pads are to be furnished by mechanical contractor on shop drawings.

5. Page 15A-10, Item 15A, paragraph 19, subparagraph e), f), and g).

   Furnish breakout price for costs associated with noise testing.

6. Page 15A-12, Item 15A, paragraph 24, subparagraphs a, b, and c.

   a) Furnish breakout price for all costs associated with operating instructions as specified.

   b) Furnish deductive alternate price for a more simplified approach to the construction period.

7. Page 15A-13, Item 15A, paragraph 26, subparagraph a.

   The mechanical contractor shall transfer all data from black line drawings to a sepia furnished by the engineer.

8. Page 15B-2, Item 15B, paragraph 2, subparagraph 8.

   Relates to Plumbing work.

EXHIBIT "C"

H.V.A.C.

CLARIFICATION TO SPECIFICATIONS (cont.)

9. Page 15B-3, Item 15B, paragraph 2, subparagraph 9.

   Relates to Fire Protection work.

10. Page 15B-4, Item 15B, paragraph 2.

    Subparagraph 12, by others.
    Subparagraph 13, by others.

11. Page 15B-4, Item 15B, paragraph 2, subparagraph 16.

    Balancing of system shall be by an outside subcontractor skilled to do this work.

    Furnish alternate prices for balancing and adjusting of mechanical systems by own forces in lieu of air balancing service company. Requirements for reports, etc., remain in force.

12. Page 15E-5, Section 15E, paragraph J, subparagraph 6d).

    Temperature control is to be part of the mechanical contract. Reference to temperature control subcontractor, implying a separate contractor, is to be ignored. Temperature control, if subcontracted out, shall be as a sub-sub to the mechanical subcontractor.

13. Page 15F-3, Section 15F, paragraph 3, subparagraph t.

    Scope of the cooling tower support is as follows:

    The mechanical subcontractor shall be responsible to furnish and install all steel supports required for the cooling towers beyond those furnished and shown on the structural drawings (see structural drawings).

14. Page 15F-3, Section 15F, paragraph 3, subparagraph u.

    Field performance tests alternate shall apply as written if the owner accepts the alternate.

15. Page 15F-3, Section 15F, paragraph 3, subparagraph v.

    The mechanical subcontractor shall contract for the cooling towers.

EXHIBIT "C"

H.V.A.C.

CLARIFICATION TO SPECIFICATIONS (cont.)

16. Page 15F-3, Section 15F, paragraph 3, subparagraph w.

    Submit alternates for different cooling tower manufacturers, if possible.

17. Page 15G-2, Section 15G, paragraph 3, subparagraph f).

    Furnish alternate for pump seals.

18. Section 15G

    The mechanical subcontractor shall include pumps required for his work as defined in this section and as further defined herein. Omission, by this scope, of a clear delineation between HVAC pumps and other trade pumps does not relieve the mechanical contractor from responsibility.

19. Page 15G-4, Section 15G, paragraph 3, subparagraph m.

    The mechanical subcontractor shall be responsible to furnish motor controllers for all his pumps if said motors are <u>not</u> part of a motor control center.

20. Page 15G-4, Section 15G, paragraph 4.

    Sewage ejectors are by the plumbing subcontractor. The mechanical contractor shall coordinate and make provision for any alarm that may be required within a mechanical alarm panel.

21. Page 15G-5, Section 15G, paragraph 5.

    Sump pumps are by the plumbing contractor. The mechanical contractor shall coordinate and make provisions for any alarm that may be required within a mechanical alarm panel.

22. Page 15G-6, Section 15G, paragraph 6.

    Diesel fire pump is by the sprinkler subcontractor.

EXHIBIT "C"

H.V.A.C.

CLARIFICATION TO SPECIFICATIONS (cont)

Related work by mechanical subcontractor is to be included as follows:

A. Exhaust piping from factory-installed muffler or manifold to the outside, inclusive of any supports, sleeving, louvers, insulation, and other items necessary for a complete installation.

B. Fuel oil tank and fuel piping to the integral day tank as required. Fuel and piping is to be pipe within a pipe installation from tank to tank.

C. Temperature controls, controls, control gauges, and routing to control center, and other related work from sensing points on the unit to "CAP" panel and as required.

23. Section 15Q

The temperature control work is to be included as part of the mechanical subcontractor's quotation. All requirements, obliged to "the temperature control subcontractor", are to be adhered to by the mechanical subcontractor.

24. Page 15Y-2, Section 15Y, paragraph 5

Electric unit heater is by the electrical subcontractor. Remote controls, if required, are by the mechanical subcontractor.

25. Page 15Y-6, Section 15Y, paragraph 12

Cabinet-type electric unit heaters are by the electrical subcontractor. Remote controls, if required, are by the mechanical subcontractor.

ADDENDA NO. 1

TO

REQUEST FOR PROPOSALS

FOR

H. V. A. C.

The addenda shall be included and made part of the Bid Documents. The bidders shall acknowledge this addenda in their proposal.

1. Documents Section 2, add the Following:

    c. Drawings as listed on Exhibit "D."

2. Scope of Work Section 3, Modify and Add the Following:

    3. Scope of work to include but not limited to:

        (is modified as follows:)

        k) Items a through j.

        l) Reinforced circular holes penetration of structural beams for piping associated with the HVAC is to be performed as part of the HVAC work. Rectangular holes for ductwork are to be by others. The cost of this work is to be shown as breakout price 22. The number and size of holes are to be stated. The holes and method of work require design approval by the structural engineer. Coordination shall be as required by field operations.

        m) Attached is specification Section 10200 Louvers, dated 11/14/87 which shall be used for architectural louvers. (See Alternate 7a).

3. Bid Breakdown Section 6, Modify and Add the Following:

    6. Bid Breakdown

        The following bid breakout prices are required in addition to those stipulated in the original scope of work.

    c) Breakdown of above (revised).

        8a. Air handling units on all floors except 5, 6, 7 and 8 including piping, etc. $________________

ADDENDA NO. 1 (Cont'd.)

H. V. A. C.

3. Bid Breakdown Section 6, Modify and Add the Following: Cont'd.)

c. Breakdown of above (revised) (cont'd.)

8b. Air handling units on floor 5. $________________

13. Sheet Metal

j. Main duct (40 x 12, 34 x 12, etc.) Run out from mechanical room for floors 6, 7 and 8. $________________

22a. Holes in structural beams (see scope Item 1). $________________

b. Number of holes and sizes. $________________

23. Miscellaneous (not included in above) $________________

The sum of 1 through 23 above is equal to the lump sum quotation for HVAC, Piping and Sheetmetal.

4. Alternates/Unit Cost Section 7, Revise and Add as Follows:

c. Deductive amount to simplify instruction period from that specified. $________________

h. 1. Control work by other than companies specified. Bidders are to stipulate company. $________________

h. 2. Furnish alternate cost for a modification to or reduction of control work specified to achieve a reasonable system. $________________

Unit Prices

17. a. VAV box interior zone (floors 6, 7 and 8). $________________

17. b. VAV box perimeter zone (floors 6, 7 and 8). $________________

ADDENDA NO. 1 (Cont'd.)

H. V. A. C.

4. Alternates/Unit Cost Section 7, Revise and Add as Follows: (Cont'd.)

18. Furnish and install perimeter distribution from collar including flex duct and linear diffuser complete for floors 6, 7 and 8. $________________

19. Furnish and install perimeter distribution from collar including flex duct and linear diffuser complete for floors 1, 2, 3, 4, 5 and 9. $________________

20. Unit price for high velocity ductwork (double wall flat oval or rectangular) per linear foot for sizes as follows:

a) 32 x 22 $_____________L.F.
b) 18 x 14 $_____________L.F.
c) 34 x 14 $_____________L.F.
d) 42 x 24 $_____________L.F.

21. High velocity round elbows.

a) 32 x 22 $______________Ea.
b) 18 x 14 $______________Ea.
c) 34 x 14 $______________Ea.
d) 42 x 24 $______________Ea.

22. "T" fittings to 12" round duct. $______________Ea.

23. Fan powered VAV box with pneumatic controls and electric 3 KW coil. $______________Ea.

24a) Unit cost for 10 ton DX (upflow and downflow) computer room units complete with roof top unit and piping to 7th floor. $______________Ea.

b) Unit cost per linear foot for copper piping for 10 ton DX unit. $______________L.F.

ADDENDA NO. 1 (Cont'd.)

H. V. A. C.

4. Alternates/Unit Cost Section 7, Revise and Add as Follows: (Cont'd.)

Unit Prices (Cont'd.)

j. Furnish and install system shown on drawing labeled.

Alternate J, B R and A, dated 11/16/87.

Showing east and west mechanical room for typical floor 6 or 7.

Price one floor. This alternate is not in lieu of the system shown on mechanical. drawing. $__________________

k. Unit cost for temporary firemen needed during building temporary heating season '88 - '89. $__________________

5. Add new sections.

10. A dumpster trash chute has been provided for removal of trash. The successful bidder shall be responsible to clean-up his debris and place same in the containers provided. Provisions of Design/Build Subcontract Article III, Paragraph L, will be strictly adhered to.

11. A material hoist will be available for subcontractor's use. Usage shall be by pre-arranged schedule through the Project Superintendent.

Hoisting necessary using cranes, exclusive of rigging work such as cooling towers, boilers and chillers, are to be stated in estimated hours in the alternate section of the bid breakdown.

Appendix C

AIA Contract Documents

The following documents from the American Institute of Architects (AIA) include copies of those referred to as examples in the text. Also shown are the cover sheets of recently drafted AIA documents written specifically for design/build construction.

AIA Document A111

Standard Form of Agreement between Owner and Contractor – Cost of the Work Plus a Fee

AIA Document A101

Standard Form of Agreement between Owner and Contractor – Stipulated Sum

AIA Document A191

Standard Form of Agreements between Owner and Design/Builder

AIA Document B901 (cover page only)

Standard Form of Agreements between Design/Builder and Architect

AIA Document A491 (cover page only)

Standard Form of Agreements between Design/Builder and Contractor

AIA Document A401 (page 1 only)

Standard Form of Agreement between Contractor and Subcontractor

T H E A M E R I C A N I N S T I T U T E O F A R C H I T E C T S

AIA Document A111

Standard Form of Agreement Between Owner and Contractor

where the basis of payment is the

COST OF THE WORK PLUS A FEE

with or without a Guaranteed Maximum Price

1987 EDITION

THIS DOCUMENT HAS IMPORTANT LEGAL CONSEQUENCES; CONSULTATION WITH AN ATTORNEY IS ENCOURAGED WITH RESPECT TO ITS COMPLETION OR MODIFICATION.

The 1987 Edition of AIA Document A201, General Conditions of the Contract for Construction, is adopted in this document by reference. Do not use with other general conditions unless this document is modified.

This document has been approved and endorsed by The Associated General Contractors of America.

AGREEMENT

made as of the day of in the year of Nineteen Hundred and

BETWEEN the Owner:
(Name and address)

and the Contractor:
(Name and address)

the Project is:
(Name and address)

the Architect is:
(Name and address)

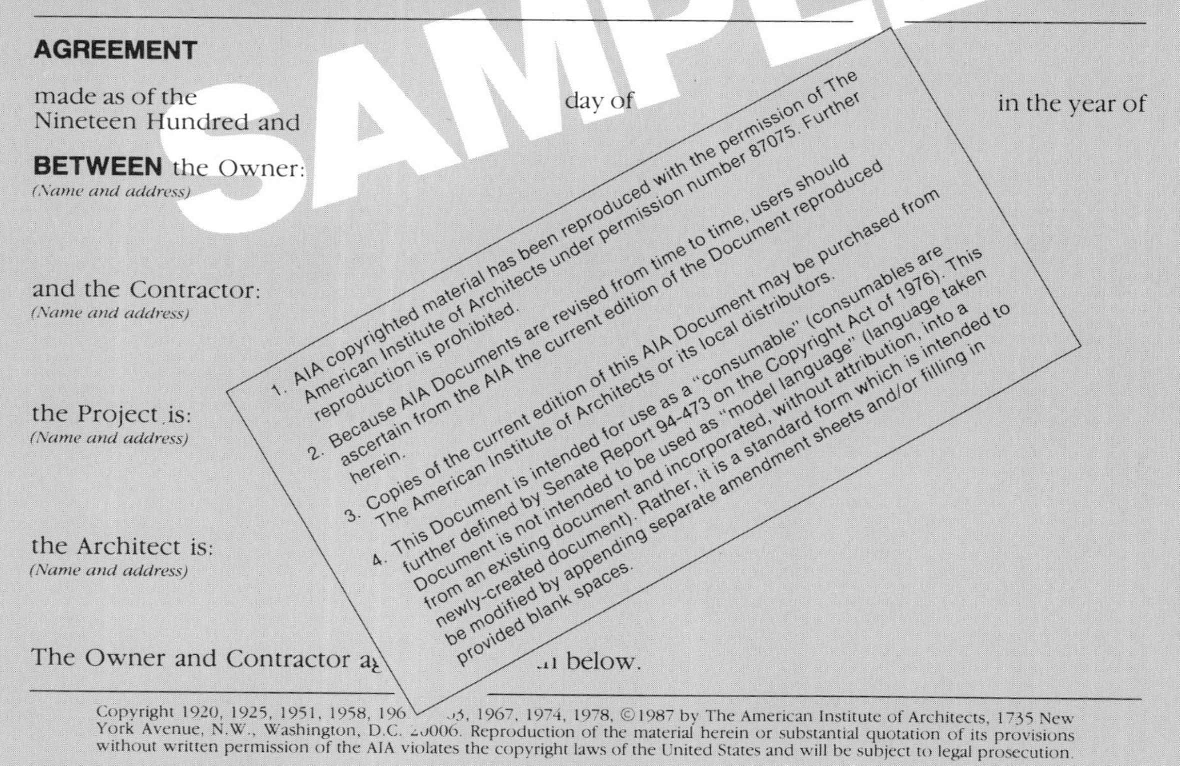

The Owner and Contractor a[illegible] ..n below.

ARTICLE 1
THE CONTRACT DOCUMENTS

1.1 The Contract Documents consist of this Agreement, Conditions of the Contract (General, Supplementary and other Conditions), Drawings, Specifications, Addenda issued prior to execution of this Agreement, other documents listed in this Agreement and Modifications issued after execution of this Agreement; these form the Contract, and are as fully a part of the Contract as if attached to this Agreement or repeated herein. The Contract represents the entire and integrated agreement between the parties hereto and supersedes prior negotiations, representations or agreements, either written or oral. An enumeration of the Contract Documents, other than Modifications, appears in Article 16. If anything in the other Contract Documents is inconsistent with this Agreement, this Agreement shall govern.

ARTICLE 2
THE WORK OF THIS CONTRACT

2.1 The Contractor shall execute the entire Work described in the Contract Documents, except to the extent specifically indicated in the Contract Documents to be the responsibility of others, or as follows:

ARTICLE 3
RELATIONSHIP OF THE PARTIES

3.1 The Contractor accepts the relationship of trust and confidence established by this Agreement and covenants with the Owner to cooperate with the Architect and utilize the Contractor's best skill, efforts and judgment in furthering the interests of the Owner; to furnish efficient business administration and supervision; to make best efforts to furnish at all times an adequate supply of workers and materials; and to perform the Work in the best way and most expeditious and economical manner consistent with the interests of the Owner. The Owner agrees to exercise best efforts to enable the Contractor to perform the Work in the best way and most expeditious manner by furnishing and approving in a timely way information required by the Contractor and making payments to the Contractor in accordance with requirements of the Contract Documents.

ARTICLE 4
DATE OF COMMENCEMENT AND SUBSTANTIAL COMPLETION

4.1 The date of commencement is the date from which the Contract Time of Subparagraph 4.2 is measured; it shall be the date of this Agreement, as first written above, unless a different date is stated below or provision is made for the date to be fixed in a notice to proceed issued by the Owner.

(Insert the date of commencement, if it differs from the date of this Agreement or, if applicable, state that the date will be fixed in a notice to proceed.)

Unless the date of commencement is established by a notice to proceed issued by the Owner, the Contractor shall notify the Owner in writing not less than five days before commencing the Work to permit the timely filing of mortgages, mechanic's liens and other security interests.

4.2 The Contractor shall achieve Substantial Completion of the entire Work not later than

(Insert the calendar date or number of calendar days after the date of commencement. Also insert any requirements for earlier Substantial Completion of certain portions of the Work, if not stated elsewhere in the Contract Documents.)

, subject to adjustments of this Contract Time as provided in the Contract Documents.

(Insert provisions, if any, for liquidated damages relating to failure to complete on time.)

ARTICLE 5
CONTRACT SUM

5.1 The Owner shall pay the Contractor in current funds for the Contractor's performance of the Contract the Contract Sum consisting of the Cost of the Work as defined in Article 7 and the Contractor's Fee determined as follows:

(State a lump sum, percentage of Cost of the Work or other provision for determining the Contractor's Fee, and explain how the Contractor's Fee is to be adjusted for changes in the Work.)

5.2 GUARANTEED MAXIMUM PRICE (IF APPLICABLE)

5.2.1 The sum of the Cost of the Work and the Contractor's Fee is guaranteed by the Contractor not to exceed Dollars (**$**), subject to additions and deductions by Change Order as provided in the Contract Documents. Such maximum sum is referred to in the Contract Documents as the Guaranteed Maximum Price. Costs which would cause the Guaranteed Maximum Price to be exceeded shall be paid by the Contractor without reimbursement by the Owner.

(Insert specific provisions if the Contractor is to participate in any savings.)

5.2.2 The Guaranteed Maximum Price is based upon the following alternates, if any, which are described in the Contract Documents and are hereby accepted by the Owner:

(State the numbers or other identification of accepted alternates, but only if a Guaranteed Maximum Price is inserted in Subparagraph 5.2.1. If decisions on other alternates are to be made by the Owner subsequent to the execution of this Agreement, attach a schedule of such other alternates showing the amount for each and the date until which that amount is valid.)

5.2.3 The amounts agreed to for unit prices, if any, are as follows:

(State unit prices only if a Guaranteed Maximum Price is inserted in Subparagraph 5.2.1.)

ARTICLE 6
CHANGES IN THE WORK

6.1 CONTRACTS WITH A GUARANTEED MAXIMUM PRICE

6.1.1 Adjustments to the Guaranteed Maximum Price on account of changes in the Work may be determined by any of the methods listed in Subparagraph 7.3.3 of the General Conditions.

6.1.2 In calculating adjustments to subcontracts (except those awarded with the Owner's prior consent on the basis of cost plus a fee), the terms "cost" and "fee" as used in Clause 7.3.3.3 of the General Conditions and the terms "costs" and "a reasonable allowance for overhead and profit" as used in Subparagraph 7.3.6 of the General Conditions shall have the meanings assigned to them in the General Conditions and shall not be modified by Articles 5, 7 and 8 of this Agreement. Adjustments to subcontracts awarded with the Owner's prior consent on the basis of cost plus a fee shall be calculated in accordance with the terms of those subcontracts.

6.1.3 In calculating adjustments to this Contract, the terms "cost" and "costs" as used in the above-referenced provisions of the General Conditions shall mean the Cost of the Work as defined in Article 7 of this Agreement and the terms "fee" and "a reasonable allowance for overhead and profit" shall mean the Contractor's Fee as defined in Paragraph 5.1 of this Agreement.

6.2 CONTRACTS WITHOUT A GUARANTEED MAXIMUM PRICE

6.2.1 Increased costs for the items set forth in Article 7 which result from changes in the Work shall become part of the Cost of the Work, and the Contractor's Fee shall be adjusted as provided in Paragraph 5.1.

6.3 ALL CONTRACTS

6.3.1 If no specific provision is made in Paragraph 5.1 for adjustment of the Contractor's Fee in the case of changes in the Work, or if the extent of such changes is such, in the aggregate, that application of the adjustment provisions of Paragraph 5.1 will cause substantial inequity to the Owner or Contractor, the Contractor's Fee shall be equitably adjusted on the basis of the Fee established for the original Work.

ARTICLE 7
COSTS TO BE REIMBURSED

7.1 The term Cost of the Work shall mean costs necessarily incurred by the Contractor in the proper performance of the Work. Such costs shall be at rates not higher than the standard paid at the place of the Project except with prior consent of the Owner. The Cost of the Work shall include only the items set forth in this Article 7.

7.1.1 LABOR COSTS

7.1.1.1 Wages of construction workers directly employed by the Contractor to perform the construction of the Work at the site or, with the Owner's agreement, at off-site workshops.

7.1.1.2 Wages or salaries of the Contractor's supervisory and administrative personnel when stationed at the site with the Owner's agreement.

(If it is intended that the wages or salaries of certain personnel stationed at the Contractor's principal or other offices shall be included in the Cost of the Work, identify in Article 14 the personnel to be included and whether for all or only part of their time.)

7.1.1.3 Wages and salaries of the Contractor's supervisory or administrative personnel engaged, at factories, workshops or on the road, in expediting the production or transportation of materials or equipment required for the Work, but only for that portion of their time required for the Work.

7.1.1.4 Costs paid or incurred by the Contractor for taxes, insurance, contributions, assessments and benefits required by law or collective bargaining agreements and, for personnel not covered by such agreements, customary benefits such as sick leave, medical and health benefits, holidays, vacations and pensions, provided such costs are based on wages and salaries included in the Cost of the Work under Clauses 7.1.1.1 through 7.1.1.3.

7.1.2 SUBCONTRACT COSTS

Payments made by the Contractor to Subcontractors in accordance with the requirements of the subcontracts.

7.1.3 COSTS OF MATERIALS AND EQUIPMENT INCORPORATED IN THE COMPLETED CONSTRUCTION

7.1.3.1 Costs, including transportation, of materials and equipment incorporated or to be incorporated in the completed construction.

7.1.3.2 Costs of materials described in the preceding Clause 7.1.3.1 in excess of those actually installed but required to provide reasonable allowance for waste and for spoilage. Unused excess materials, if any, shall be handed over to the Owner at the completion of the Work or, at the Owner's option, shall be sold by the Contractor; amounts realized, if any, from such sales shall be credited to the Owner as a deduction from the Cost of the Work.

7.1.4 COSTS OF OTHER MATERIALS AND EQUIPMENT, TEMPORARY FACILITIES AND RELATED ITEMS

7.1.4.1 Costs, including transportation, installation, maintenance, dismantling and removal of materials, supplies, temporary facilities, machinery, equipment, and hand tools not customarily owned by the construction workers, which are provided by the Contractor at the site and fully consumed in the performance of the Work; and cost less salvage value on such items if not fully consumed, whether sold to others or retained by the Contractor. Cost for items previously used by the Contractor shall mean fair market value.

7.1.4.2 Rental charges for temporary facilities, machinery, equipment, and hand tools not customarily owned by the construction workers, which are provided by the Contractor at the site, whether rented from the Contractor or others, and costs of transportation, installation, minor repairs and replacements, dismantling and removal thereof. Rates and quantities of equipment rented shall be subject to the Owner's prior approval.

7.1.4.3 Costs of removal of debris from the site.

7.1.4.4 Costs of telegrams and long-distance telephone calls, postage and parcel delivery charges, telephone service at the site and reasonable petty cash expenses of the site office.

7.1.4.5 That portion of the reasonable travel and subsistence expenses of the Contractor's personnel incurred while traveling in discharge of duties connected with the Work.

7.1.5 MISCELLANEOUS COSTS

7.1.5.1 That portion directly attributable to this Contract of premiums for insurance and bonds.

7.1.5.2 Sales, use or similar taxes imposed by a governmental authority which are related to the Work and for which the Contractor is liable.

7.1.5.3 Fees and assessments for the building permit and for other permits, licenses and inspections for which the Contractor is required by the Contract Documents to pay.

7.1.5.4 Fees of testing laboratories for tests required by the Contract Documents, except those related to defective or nonconforming Work for which reimbursement is excluded by Subparagraph 13.5.3 of the General Conditions or other provisions of the Contract Documents and which do not fall within the scope of Subparagraphs 7.2.2 through 7.2.4 below.

7.1.5.5 Royalties and license fees paid for the use of a particular design, process or product required by the Contract Documents; the cost of defending suits or claims for infringement of patent rights arising from such requirement by the Contract Documents; payments made in accordance with legal judgments against the Contractor resulting from such suits or claims and payments of settlements made with the Owner's consent; provided, however, that such costs of legal defenses, judgment and settlements shall not be included in the calculation of the Contractor's Fee or of a Guaranteed Maximum Price, if any, and provided that such royalties, fees and costs are not excluded by the last sentence of Subparagraph 3.17.1 of the General Conditions or other provisions of the Contract Documents.

7.1.5.6 Deposits lost for causes other than the Contractor's fault or negligence.

7.1.6 OTHER COSTS

7.1.6.1 Other costs incurred in the performance of the Work if and to the extent approved in advance in writing by the Owner.

7.2 EMERGENCIES: REPAIRS TO DAMAGED, DEFECTIVE OR NONCONFORMING WORK

The Cost of the Work shall also include costs described in Paragraph 7.1 which are incurred by the Contractor:

7.2.1 In taking action to prevent threatened damage, injury or loss in case of an emergency affecting the safety of persons and property, as provided in Paragraph 10.3 of the General Conditions.

7.2.2 In repairing or correcting Work damaged or improperly executed by construction workers in the employ of the Contractor, provided such damage or improper execution did not result from the fault or negligence of the Contractor or the Contractor's foremen, engineers or superintendents, or other supervisory, administrative or managerial personnel of the Contractor.

7.2.3 In repairing damaged Work other than that described in Subparagraph 7.2.2, provided such damage did not result from the fault or negligence of the Contractor or the Contractor's personnel, and only to the extent that the cost of such repairs is not recoverable by the Contractor from others and the Contractor is not compensated therefor by insurance or otherwise.

7.2.4 In correcting defective or nonconforming Work performed or supplied by a Subcontractor or material supplier and not corrected by them, provided such defective or nonconforming Work did not result from the fault or neglect of the Contractor or the Contractor's personnel adequately to supervise and direct the Work of the Subcontractor or material supplier, and only to the extent that the cost of correcting the defective or nonconforming Work is not recoverable by the Contractor from the Subcontractor or material supplier.

ARTICLE 8
COSTS NOT TO BE REIMBURSED

8.1 The Cost of the Work shall not include:

8.1.1 Salaries and other compensation of the Contractor's personnel stationed at the Contractor's principal office or offices other than the site office, except as specifically provided in Clauses 7.1.1.2 and 7.1.1.3 or as may be provided in Article 14.

8.1.2 Expenses of the Contractor's principal office and offices other than the site office.

8.1.3 Overhead and general expenses, except as may be expressly included in Article 7.

8.1.4 The Contractor's capital expenses, including interest on the Contractor's capital employed for the Work.

8.1.5 Rental costs of machinery and equipment, except as specifically provided in Clause 7.1.4.2.

8.1.6 Except as provided in Subparagraphs 7.2.2 through 7.2.4 and Paragraph 13.5 of this Agreement, costs due to the fault or negligence of the Contractor, Subcontractors, anyone directly or indirectly employed by any of them, or for whose acts any of them may be liable, including but not limited to costs for the correction of damaged, defective or nonconforming Work, disposal and replacement of materials and equipment incorrectly ordered or supplied, and making good damage to property not forming part of the Work.

8.1.7 Any cost not specifically and expressly described in Article 7.

8.1.8 Costs which would cause the Guaranteed Maximum Price, if any, to be exceeded.

ARTICLE 9
DISCOUNTS, REBATES AND REFUNDS

9.1 Cash discounts obtained on payments made by the Contractor shall accrue to the Owner if (1) before making the payment, the Contractor included them in an Application for Payment and received payment therefor from the Owner, or (2) the Owner has deposited funds with the Contractor with which to make payments; otherwise, cash discounts shall accrue to the Contractor. Trade discounts, rebates, refunds and amounts received from sales of surplus materials and equipment shall accrue to the Owner, and the Contractor shall make provisions so that they can be secured.

9.2 Amounts which accrue to the Owner in accordance with the provisions of Paragraph 9.1 shall be credited to the Owner as a deduction from the Cost of the Work.

ARTICLE 10
SUBCONTRACTS AND OTHER AGREEMENTS

10.1 Those portions of the Work that the Contractor does not customarily perform with the Contractor's own personnel shall be performed under subcontracts or by other appropriate agreements with the Contractor. The Contractor shall obtain bids from Subcontractors and from suppliers of materials or equipment fabricated especially for the Work and shall deliver such bids to the Architect. The Owner will then determine, with the advice of the Contractor and subject to the reasonable objection of the Architect, which bids will be accepted. The Owner may designate specific persons or entities from whom the Contractor shall obtain bids; however, if a Guaranteed Maximum Price has been established, the Owner may not prohibit the Contractor from obtaining bids from others. The Contractor shall not be required to contract with anyone to whom the Contractor has reasonable objection.

10.2 If a Guaranteed Maximum Price has been established and a specific bidder among those whose bids are delivered by the Contractor to the Architect (1) is recommended to the Owner by the Contractor; (2) is qualified to perform that portion of the Work; and (3) has submitted a bid which conforms to the requirements of the Contract Documents without reservations or exceptions, but the Owner requires that another bid be accepted; then the Contractor may require that a Change Order be issued to adjust the Guaranteed Maximum Price by the difference between the bid of the person or entity recommended to the Owner by the Contractor and the amount of the subcontract or other agreement actually signed with the person or entity designated by the Owner.

10.3 Subcontracts or other agreements shall conform to the payment provisions of Paragraphs 12.7 and 12.8, and shall not be awarded on the basis of cost plus a fee without the prior consent of the Owner.

ARTICLE 11
ACCOUNTING RECORDS

11.1 The Contractor shall keep full and detailed accounts and exercise such controls as may be necessary for proper financial management under this Contract; the accounting and control systems shall be satisfactory to the Owner. The Owner and the Owner's accountants shall be afforded access to the Contractor's records, books, correspondence, instructions, drawings, receipts, subcontracts, purchase orders, vouchers, memoranda and other data relating to this Contract, and the Contractor shall preserve these for a period of three years after final payment, or for such longer period as may be required by law.

ARTICLE 12
PROGRESS PAYMENTS

12.1 Based upon Applications for Payment submitted to the Architect by the Contractor and Certificates for Payment issued by the Architect, the Owner shall make progress payments on account of the Contract Sum to the Contractor as provided below and elsewhere in the Contract Documents.

12.2 The period covered by each Application for Payment shall be one calendar month ending on the last day of the month, or as follows:

12.3 Provided an Application for Payment is received by the Architect not later than the day of a month, the Owner shall make payment to the Contractor not later than the day of the month. If an Application for Payment is received by the Architect after the application date fixed above, payment shall be made by the Owner not later than days after the Architect receives the Application for Payment.

12.4 With each Application for Payment the Contractor shall submit payrolls, petty cash accounts, receipted invoices or invoices with check vouchers attached, and any other evidence required by the Owner or Architect to demonstrate that cash disbursements already made by the Contractor on account of the Cost of the Work equal or exceed (1) progress payments already received by the Contractor; less (2) that portion of those payments attributable to the Contractor's Fee; plus (3) payrolls for the period covered by the present Application for Payment; plus (4) retainage provided in Subparagraph 12.5.4, if any, applicable to prior progress payments.

 A111-1987 7

12.5 CONTRACTS WITH A GUARANTEED MAXIMUM PRICE

12.5.1 Each Application for Payment shall be based upon the most recent schedule of values submitted by the Contractor in accordance with the Contract Documents. The schedule of values shall allocate the entire Guaranteed Maximum Price among the various portions of the Work, except that the Contractor's Fee shall be shown as a single separate item. The schedule of values shall be prepared in such form and supported by such data to substantiate its accuracy as the Architect may require. This schedule, unless objected to by the Architect, shall be used as a basis for reviewing the Contractor's Applications for Payment.

12.5.2 Applications for Payment shall show the percentage completion of each portion of the Work as of the end of the period covered by the Application for Payment. The percentage completion shall be the lesser of (1) the percentage of that portion of the Work which has actually been completed or (2) the percentage obtained by dividing (a) the expense which has actually been incurred by the Contractor on account of that portion of the Work for which the Contractor has made or intends to make actual payment prior to the next Application for Payment by (b) the share of the Guaranteed Maximum Price allocated to that portion of the Work in the schedule of values.

12.5.3 Subject to other provisions of the Contract Documents, the amount of each progress payment shall be computed as follows:

12.5.3.1 Take that portion of the Guaranteed Maximum Price properly allocable to completed Work as determined by multiplying the percentage completion of each portion of the Work by the share of the Guaranteed Maximum Price allocated to that portion of the Work in the schedule of values. Pending final determination of cost to the Owner of changes in the Work, amounts not in dispute may be included as provided in Subparagraph 7.3.7 of the General Conditions, even though the Guaranteed Maximum Price has not yet been adjusted by Change Order.

12.5.3.2 Add that portion of the Guaranteed Maximum Price properly allocable to materials and equipment delivered and suitably stored at the site for subsequent incorporation in the Work or, if approved in advance by the Owner, suitably stored off the site at a location agreed upon in writing.

12.5.3.3 Add the Contractor's Fee, less retainage of percent (%). The Contractor's Fee shall be computed upon the Cost of the Work described in the two preceding Clauses at the rate stated in Paragraph 5.1 or, if the Contractor's Fee is stated as a fixed sum in that Paragraph, shall be an amount which bears the same ratio to that fixed-sum Fee as the Cost of the Work in the two preceding Clauses bears to a reasonable estimate of the probable Cost of the Work upon its completion.

12.5.3.4 Subtract the aggregate of previous payments made by the Owner.

12.5.3.5 Subtract the shortfall, if any, indicated by the Contractor in the documentation required by Paragraph 12.4 to substantiate prior Applications for Payment, or resulting from errors subsequently discovered by the Owner's accountants in such documentation.

12.5.3.6 Subtract amounts, if any, for which the Architect has withheld or nullified a Certificate for Payment as provided in Paragraph 9.5 of the General Conditions.

12.5.4 Additional retainage, if any, shall be as follows:

(If it is intended to retain additional amounts from progress payments to the Contractor beyond (1) the retainage from the Contractor's Fee provided in Clause 12.5.3.3, (2) the retainage from Subcontractors provided in Paragraph 12.7 below, and (3) the retainage, if any, provided by other provisions of the Contract, insert provision for such additional retainage here. Such provision, if made, should also describe any arrangement for limiting or reducing the amount retained after the Work reaches a certain state of completion.)

12.6 CONTRACTS WITHOUT A GUARANTEED MAXIMUM PRICE

12.6.1 Applications for Payment shall show the Cost of the Work actually incurred by the Contractor through the end of the period covered by the Application for Payment and for which the Contractor has made or intends to make actual payment prior to the next Application for Payment.

12.6.2 Subject to other provisions of the Contract Documents, the amount of each progress payment shall be computed as follows:

12.6.2.1 Take the Cost of the Work as described in Subparagraph 12.6.1.

12.6.2.2 Add the Contractor's Fee, less retainage of percent (%). The Contractor's Fee shall be computed upon the Cost of the Work described in the preceding Clause 12.6.2.1 at the rate stated in Paragraph 5.1 or, if the Contractor's Fee is stated as a fixed sum in that Paragraph, an amount which bears the same ratio to that fixed-sum Fee as the Cost of the Work in the preceding Clause bears to a reasonable estimate of the probable Cost of the Work upon its completion.

12.6.2.3 Subtract the aggregate of previous payments made by the Owner.

12.6.2.4 Subtract the shortfall, if any, indicated by the Contractor in the documentation required by Paragraph 12.4 or to substantiate prior Applications for Payment or resulting from errors subsequently discovered by the Owner's accountants in such documentation.

12.6.2.5 Subtract amounts, if any, for which the Architect has withheld or withdrawn a Certificate for Payment as provided in the Contract Documents.

12.6.3 Additional retainage, if any, shall be as follows:

12.7 Except with the Owner's prior approval, payments to Subcontractors included in the Contractor's Applications for Payment shall not exceed an amount for each Subcontractor calculated as follows:

12.7.1 Take that portion of the Subcontract Sum properly allocable to completed Work as determined by multiplying the percentage completion of each portion of the Subcontractor's Work by the share of the total Subcontract Sum allocated to that portion in the Subcontractor's schedule of values, less retainage of percent (%). Pending final determination of amounts to be paid to the Subcontractor for changes in the Work, amounts not in dispute may be included as provided in Subparagraph 7.3.7 of the General Conditions even though the Subcontract Sum has not yet been adjusted by Change Order.

12.7.2 Add that portion of the Subcontract Sum properly allocable to materials and equipment delivered and suitably stored at the site for subsequent incorporation in the Work or, if approved in advance by the Owner, suitably stored off the site at a location agreed upon in writing, less retainage of percent (%).

12.7.3 Subtract the aggregate of previous payments made by the Contractor to the Subcontractor.

12.7.4 Subtract amounts, if any, for which the Architect has withheld or nullified a Certificate for Payment by the Owner to the Contractor for reasons which are the fault of the Subcontractor.

12.7.5 Add, upon Substantial Completion of the entire Work of the Contractor, a sum sufficient to increase the total payments to the Subcontractor to percent (%) of the Subcontract Sum, less amounts, if any, for incomplete Work and unsettled claims; and, if final completion of the entire Work is thereafter materially delayed through no fault of the Subcontractor, add any additional amounts payable on account of Work of the Subcontractor in accordance with Subparagraph 9.10.3 of the General Conditions.

(If it is intended, prior to Substantial Completion of the entire Work of the Contractor, to reduce or limit the retainage from Subcontractors resulting from the percentages inserted in Subparagraphs 12.7.1 and 12.7.2 above, and this is not explained elsewhere in the Contract Documents, insert here provisions for such reduction or limitation.)

The Subcontract Sum is the total amount stipulated in the subcontract to be paid by the Contractor to the Subcontractor for the Subcontractor's performance of the subcontract.

12.8 Except with the Owner's prior approval, the Contractor shall not make advance payments to suppliers for materials or equipment which have not been delivered and stored at the site.

12.9 In taking action on the Contractor's Applications for Payment, the Architect shall be entitled to rely on the accuracy and completeness of the information furnished by the Contractor and shall not be deemed to represent that the Architect has made a detailed examination, audit or arithmetic verification of the documentation submitted in accordance with Paragraph 12.4 or other supporting data; that the Architect has made exhaustive or continuous on-site inspections or that the Architect has made examinations to ascertain how or for what purposes the Contractor has used amounts previously paid on account of the Contract. Such examinations, audits and verifications, if required by the Owner, will be performed by the Owner's accountants acting in the sole interest of the Owner.

ARTICLE 13
FINAL PAYMENT

13.1 Final payment shall be made by the Owner to the Contractor when (1) the Contract has been fully performed by the Contractor except for the Contractor's responsibility to correct defective or nonconforming Work, as provided in Subparagraph 12.2.2 of the General Conditions, and to satisfy other requirements, if any, which necessarily survive final payment; (2) a final Application for Payment and a final accounting for the Cost of the Work have been submitted by the Contractor and reviewed by the Owner's accountants; and (3) a final Certificate for Payment has then been issued by the Architect; such final payment shall be made by the Owner not more than 30 days after the issuance of the Architect's final Certificate for Payment, or as follows:

13.2 The amount of the final payment shall be calculated as follows:

13.2.1 Take the sum of the Cost of the Work substantiated by the Contractor's final accounting and the Contractor's Fee; but not more than the Guaranteed Maximum Price, if any.

13.2.2 Subtract amounts, if any, for which the Architect withholds, in whole or in part, a final Certificate for Payment as provided in Subparagraph 9.5.1 of the General Conditions or other provisions of the Contract Documents.

13.2.3 Subtract the aggregate of previous payments made by the Owner.

If the aggregate of previous payments made by the Owner exceeds the amount due the Contractor, the Contractor shall reimburse the difference to the Owner.

13.3 The Owner's accountants will review and report in writing on the Contractor's final accounting within 30 days after delivery of the final accounting to the Architect by the Contractor. Based upon such Cost of the Work as the Owner's accountants report to be substantiated by the Contractor's final accounting, and provided the other conditions of Paragraph 13.1 have been met, the Architect will, within seven days after receipt of the written report of the Owner's accountants, either issue to the Owner a final Certificate for Payment with a copy to the Contractor, or notify the Contractor and Owner in writing of the Architect's reasons for withholding a certificate as provided in Subparagraph 9.5.1 of the General Conditions. The time periods stated in this Paragraph 13.3 supersede those stated in Subparagraph 9.4.1 of the General Conditions.

13.4 If the Owner's accountants report the Cost of the Work as substantiated by the Contractor's final accounting to be less than claimed by the Contractor, the Contractor shall be entitled to demand arbitration of the disputed amount without a further decision of the Architect. Such demand for arbitration shall be made by the Contractor within 30 days after the Contractor's receipt of a copy of the Architect's final Certificate for Payment; failure to demand arbitration within this 30-day period shall result in the substantiated amount reported by the Owner's accountants becoming binding on the Contractor. Pending a final resolution by arbitration, the Owner shall pay the Contractor the amount certified in the Architect's final Certificate for Payment.

13.5 If, subsequent to final payment and at the Owner's request, the Contractor incurs costs described in Article 7 and not excluded by Article 8 to correct defective or nonconforming Work, the Owner shall reimburse the Contractor such costs and the Contractor's Fee applicable thereto on the same basis as if such costs had been incurred prior to final payment, but not in excess of the Guaranteed Maximum Price, if any. If the Contractor has participated in savings as provided in Paragraph 5.2, the amount of such savings shall be recalculated and appropriate credit given to the Owner in determining the net amount to be paid by the Owner to the Contractor.

ARTICLE 14
MISCELLANEOUS PROVISIONS

14.1 Where reference is made in this Agreement to a provision of the General Conditions or another Contract Document, the reference refers to that provision as amended or supplemented by other provisions of the Contract Documents.

14.2 Payments due and unpaid under the Contract shall bear interest from the date payment is due at the rate stated below, or in the absence thereof, at the legal rate prevailing from time to time at the place where the Project is located.

(Insert rate of interest agreed upon, if any.)

(Usury laws and requirements under the Federal Truth in Lending Act, similar state and local consumer credit laws and other regulations at the Owner's and Contractor's principal places of business, the location of the Project and elsewhere may affect the validity of this provision. Legal advice should be obtained with respect to deletions or modifications, and also regarding requirements such as written disclosures or waivers.)

14.3 Other provisions:

ARTICLE 15
TERMINATION OR SUSPENSION

15.1 The Contract may be terminated by the Contractor as provided in Article 14 of the General Conditions; however, the amount to be paid to the Contractor under Subparagraph 14.1.2 of the General Conditions shall not exceed the amount the Contractor would be entitled to receive under Paragraph 15.3 below, except that the Contractor's Fee shall be calculated as if the Work had been fully completed by the Contractor, including a reasonable estimate of the Cost of the Work for Work not actually completed.

15.2 If a Guaranteed Maximum Price is established in Article 5, the Contract may be terminated by the Owner for cause as provided in Article 14 of the General Conditions; however, the amount, if any, to be paid to the Contractor under Subparagraph 14.2.4 of the General Conditions shall not cause the Guaranteed Maximum Price to be exceeded, nor shall it exceed the amount the Contractor would be entitled to receive under Paragraph 15.3 below.

15.3 If no Guaranteed Maximum Price is established in Article 5, the Contract may be terminated by the Owner for cause as provided in Article 14 of the General Conditions; however, the Owner shall then pay the Contractor an amount calculated as follows:

15.3.1 Take the Cost of the Work incurred by the Contractor to the date of termination.

15.3.2 Add the Contractor's Fee computed upon the Cost of the Work to the date of termination at the rate stated in Paragraph 5.1 or, if the Contractor's Fee is stated as a fixed sum in that Paragraph, an amount which bears the same ratio to that fixed-sum Fee as the Cost of the Work at the time of termination bears to a reasonable estimate of the probable Cost of the Work upon its completion.

15.3.3 Subtract the aggregate of previous payments made by the Owner.

The Owner shall also pay the Contractor fair compensation, either by purchase or rental at the election of the Owner, for any equipment owned by the Contractor which the Owner elects to retain and which is not otherwise included in the Cost of the Work under Subparagraph 15.3.1. To the extent that the Owner elects to take legal assignment of subcontracts and purchase orders (including rental agreements), the Contractor shall, as a condition of receiving the payments referred to in this Article 15, execute and deliver all such papers and take all such steps, including the legal assignment of such subcontracts and other contractual rights of the Contractor, as the Owner may require for the purpose of fully vesting in the Owner the rights and benefits of the Contractor under such subcontracts or purchase orders.

15.4 The Work may be suspended by the Owner as provided in Article 14 of the General Conditions; in such case, the Guaranteed Maximum Price, if any, shall be increased as provided in Subparagraph 14.3.2 of the General Conditions except that the term "cost of performance of the Contract" in that Subparagraph shall be understood to mean the Cost of the Work and the term "profit" shall be understood to mean the Contractor's Fee as described in Paragraphs 5.1 and 6.3 of this Agreement.

ARTICLE 16
ENUMERATION OF CONTRACT DOCUMENTS

16.1 The Contract Documents, except for Modifications issued after execution of this Agreement, are enumerated as follows:

16.1.1 The Agreement is this executed Standard Form of Agreement Between Owner and Contractor, AIA Document A111, 1987 Edition.

16.1.2 The General Conditions are the General Conditions of the Contract for Construction, AIA Document A201, 1987 Edition.

16.1.3 The Supplementary and other Conditions of the Contract are those contained in the Project Manual dated , and are as follows:

Document	Title	Pages

16.1.4 The Specifications are those contained in the Project Manual dated as in Paragraph 16.1.3, and are as follows:
(Either list the Specifications here or refer to an exhibit attached to this Agreement.)

Section	Title	Pages

16.1.5 The Drawings are as follows, and are dated unless a different date is shown below:

(Either list the Drawings here or refer to an exhibit attached to this Agreement.)

Number	**Title**	**Date**

16.1.6 The Addenda, if any, are as follows:

Number	**Date**	**Pages**

Portions of Addenda relating to bidding requirements are not part of the Contract Documents unless the bidding requirements are also enumerated in this Article 16.

16.1.7 Other Documents, if any, forming part of the Contract Documents are as follows:

(List here any additional documents which are intended to form part of the Contract Documents. The General Conditions provide that bidding requirements such as advertisement or invitation to bid, Instructions to Bidders, sample forms and the Contractor's bid are not part of the Contract Documents unless enumerated in this Agreement. They should be listed here only if intended to be part of the Contract Documents.)

This Agreement is entered into as of the day and year first written above and is executed in at least three original copies of which one is to be delivered to the Contractor, one to the Architect for use in the administration of the Contract, and the remainder to the Owner.

OWNER	CONTRACTOR
(Signature)	*(Signature)*
(Printed name and title)	*(Printed name and title)*

THE AMERICAN INSTITUTE OF ARCHITECTS

AIA Document A101

Standard Form of Agreement Between Owner and Contractor

where the basis of payment is a

STIPULATED SUM

1987 EDITION

THIS DOCUMENT HAS IMPORTANT LEGAL CONSEQUENCES; CONSULTATION WITH AN ATTORNEY IS ENCOURAGED WITH RESPECT TO ITS COMPLETION OR MODIFICATION

The 1987 Edition of AIA Document A201, General Conditions of the Contract for Construction, is adopted in this document by reference. Do not use with other general conditions unless this document is modified.

This document has been approved and endorsed by The Associated General Contractors of America.

AGREEMENT

made as of the day of in the year of Nineteen Hundred and

BETWEEN the Owner:
(Name and address)

and the Contractor:
(Name and address)

The Project is:
(Name and location)

The Architect is:
(Name and address)

The Owner and Contractor ag set forth below.

ARTICLE 1
THE CONTRACT DOCUMENTS

The Contract Documents consist of this Agreement, Conditions of the Contract (General, Supplementary and other Conditions), Drawings, Specifications, Addenda issued prior to execution of this Agreement, other documents listed in this Agreement and Modifications issued after execution of this Agreement; these form the Contract, and are as fully a part of the Contract as if attached to this Agreement or repeated herein. The Contract represents the entire and integrated agreement between the parties hereto and supersedes prior negotiations, representations or agreements, either written or oral. An enumeration of the Contract Documents, other than Modifications, appears in Article 9.

ARTICLE 2
THE WORK OF THIS CONTRACT

The Contractor shall execute the entire Work described in the Contract Documents, except to the extent specifically indicated in the Contract Documents to be the responsibility of others, or as follows:

ARTICLE 3
DATE OF COMMENCEMENT AND SUBSTANTIAL COMPLETION

3.1 The date of commencement is the date from which the Contract Time of Paragraph 3.2 is measured, and shall be the date of this Agreement, as first written above, unless a different date is stated below or provision is made for the date to be fixed in a notice to proceed issued by the Owner.

(Insert the date of commencement, if it differs from the date of this Agreement or, if applicable, state that the date will be fixed in a notice to proceed.)

Unless the date of commencement is established by a notice to proceed issued by the Owner, the Contractor shall notify the Owner in writing not less than five days before commencing the Work to permit the timely filing of mortgages, mechanic's liens and other security interests.

3.2 The Contractor shall achieve Substantial Completion of the entire Work not later than

(Insert the calendar date or number of calendar days after the date of commencement. Also insert any requirements for earlier Substantial Completion of certain portions of the Work, if not stated elsewhere in the Contract Documents.)

, subject to adjustments of this Contract Time as provided in the Contract Documents.

(Insert provisions, if any, for liquidated damages relating to failure to complete on time.)

ARTICLE 4
CONTRACT SUM

4.1 The Owner shall pay the Contractor in current funds for the Contractor's performance of the Contract the Contract Sum of Dollars ($), subject to additions and deductions as provided in the Contract Documents.

4.2 The Contract Sum is based upon the following alternates, if any, which are described in the Contract Documents and are hereby accepted by the Owner:

(State the numbers or other identification of accepted alternates. If decisions on other alternates are to be made by the Owner subsequent to the execution of this Agreement, attach a schedule of such other alternates showing the amount for each and the date until which that amount is valid.)

4.3 Unit prices, if any, are as follows:

ARTICLE 5
PROGRESS PAYMENTS

5.1 Based upon Applications for Payment submitted to the Architect by the Contractor and Certificates for Payment issued by the Architect, the Owner shall make progress payments on account of the Contract Sum to the Contractor as provided below and elsewhere in the Contract Documents.

5.2 The period covered by each Application for Payment shall be one calendar month ending on the last day of the month, or as follows:

5.3 Provided an Application for Payment is received by the Architect not later than the day of a month, the Owner shall make payment to the Contractor not later than the day of the month. If an Application for Payment is received by the Architect after the application date fixed above, payment shall be made by the Owner not later than days after the Architect receives the Application for Payment.

5.4 Each Application for Payment shall be based upon the Schedule of Values submitted by the Contractor in accordance with the Contract Documents. The Schedule of Values shall allocate the entire Contract Sum among the various portions of the Work and be prepared in such form and supported by such data to substantiate its accuracy as the Architect may require. This Schedule, unless objected to by the Architect, shall be used as a basis for reviewing the Contractor's Applications for Payment.

5.5 Applications for Payment shall indicate the percentage of completion of each portion of the Work as of the end of the period covered by the Application for Payment.

5.6 Subject to the provisions of the Contract Documents, the amount of each progress payment shall be computed as follows:

5.6.1 Take that portion of the Contract Sum properly allocable to completed Work as determined by multiplying the percentage completion of each portion of the Work by the share of the total Contract Sum allocated to that portion of the Work in the Schedule of Values, less retainage of percent (%). Pending final determination of cost to the Owner of changes in the Work, amounts not in dispute may be included as provided in Subparagraph 7.3.7 of the General Conditions even though the Contract Sum has not yet been adjusted by Change Order;

5.6.2 Add that portion of the Contract Sum properly allocable to materials and equipment delivered and suitably stored at the site for subsequent incorporation in the completed construction (or, if approved in advance by the Owner, suitably stored off the site at a location agreed upon in writing), less retainage of percent (%);

5.6.3 Subtract the aggregate of previous payments made by the Owner; and

5.6.4 Subtract amounts, if any, for which the Architect has withheld or nullified a Certificate for Payment as provided in Paragraph 9.5 of the General Conditions.

5.7 The progress payment amount determined in accordance with Paragraph 5.6 shall be further modified under the following circumstances:

5.7.1 Add, upon Substantial Completion of the Work, a sum sufficient to increase the total payments to percent (%) of the Contract Sum, less such amounts as the Architect shall determine for incomplete Work and unsettled claims; and

5.7.2 Add, if final completion of the Work is thereafter materially delayed through no fault of the Contractor, any additional amounts payable in accordance with Subparagraph 9.10.3 of the General Conditions.

5.8 Reduction or limitation of retainage, if any, shall be as follows:

(If it is intended, prior to Substantial Completion of the entire Work, to reduce or limit the retainage resulting from the percentages inserted in Subparagraphs 5.6.1 and 5.6.2 above, and this is not explained elsewhere in the Contract Documents, insert here provisions for such reduction or limitation.)

ARTICLE 6
FINAL PAYMENT

Final payment, constituting the entire unpaid balance of the Contract Sum, shall be made by the Owner to the Contractor when (1) the Contract has been fully performed by the Contractor except for the Contractor's responsibility to correct nonconforming Work as provided in Subparagraph 12.2.2 of the General Conditions and to satisfy other requirements, if any, which necessarily survive final payment; and (2) a final Certificate for Payment has been issued by the Architect; such final payment shall be made by the Owner not more than 30 days after the issuance of the Architect's final Certificate for Payment, or as follows:

ARTICLE 7
MISCELLANEOUS PROVISIONS

7.1 Where reference is made in this Agreement to a provision of the General Conditions or another Contract Document, the reference refers to that provision as amended or supplemented by other provisions of the Contract Documents.

7.2 Payments due and unpaid under the Contract shall bear interest from the date payment is due at the rate stated below, or in the absence thereof, at the legal rate prevailing from time to time at the place where the Project is located.

(Insert rate of interest agreed upon, if any.)

(Usury laws and requirements under the Federal Truth in Lending Act, similar state and local consumer credit laws and other regulations at the Owner's and Contractor's principal places of business, the location of the Project and elsewhere may affect the validity of this provision. Legal advice should be obtained with respect to deletions or modifications, and also regarding requirements such as written disclosures or waivers.)

7.3 Other provisions:

ARTICLE 8
TERMINATION OR SUSPENSION

8.1 The Contract may be terminated by the Owner or the Contractor as provided in Article 14 of the General Conditions.

8.2 The Work may be suspended by the Owner as provided in Article 14 of the General Conditions.

ARTICLE 9
ENUMERATION OF CONTRACT DOCUMENTS

9.1 The Contract Documents, except for Modifications issued after execution of this Agreement, are enumerated as follows:

9.1.1 The Agreement is this executed Standard Form of Agreement Between Owner and Contractor, AIA Document A101, 1987 Edition.

9.1.2 The General Conditions are the General Conditions of the Contract for Construction, AIA Document A201, 1987 Edition.

9.1.3 The Supplementary and other Conditions of the Contract are those contained in the Project Manual dated , and are as follows:

Document	Title	Pages

9.1.4 The Specifications are those contained in the Project Manual dated as in Subparagraph 9.1.3, and are as follows:
(Either list the Specifications here or refer to an exhibit attached to this Agreement.)

Section	Title	Pages

9.1.5 The Drawings are as follows, and are dated unless a different date is shown below:
(Either list the Drawings here or refer to an exhibit attached to this Agreement.)

Number	Title	Date

9.1.6 The Addenda, if any, are as follows:

Number	Date	Pages

Portions of Addenda relating to bidding requirements are not part of the Contract Documents unless the bidding requirements are also enumerated in this Article 9.

9.1.7 Other documents, if any, forming part of the Contract Documents are as follows:

(List here any additional documents which are intended to form part of the Contract Documents. The General Conditions provide that bidding requirements such as advertisement or invitation to bid, Instructions to Bidders, sample forms and the Contractor's bid are not part of the Contract Documents unless enumerated in this Agreement. They should be listed here only if intended to be part of the Contract Documents.)

This Agreement is entered into as of the day and year first written above and is executed in at least three original copies of which one is to be delivered to the Contractor, one to the Architect for use in the administration of the Contract, and the remainder to the Owner.

OWNER

CONTRACTOR

(Signature)

(Signature)

(Printed name and title)

(Printed name and title)

THE AMERICAN INSTITUTE OF ARCHITECTS

AIA Document A191

Standard Form of Agreements Between Owner and Design/Builder

1985 EDITION

THIS DOCUMENT HAS IMPORTANT LEGAL CONSEQUENCES; CONSULTATION WITH AN ATTORNEY IS ENCOURAGED.

TABLE OF ARTICLES

PART 1 AGREEMENT—PRELIMINARY DESI... BUDGETING

PA... ...CTION

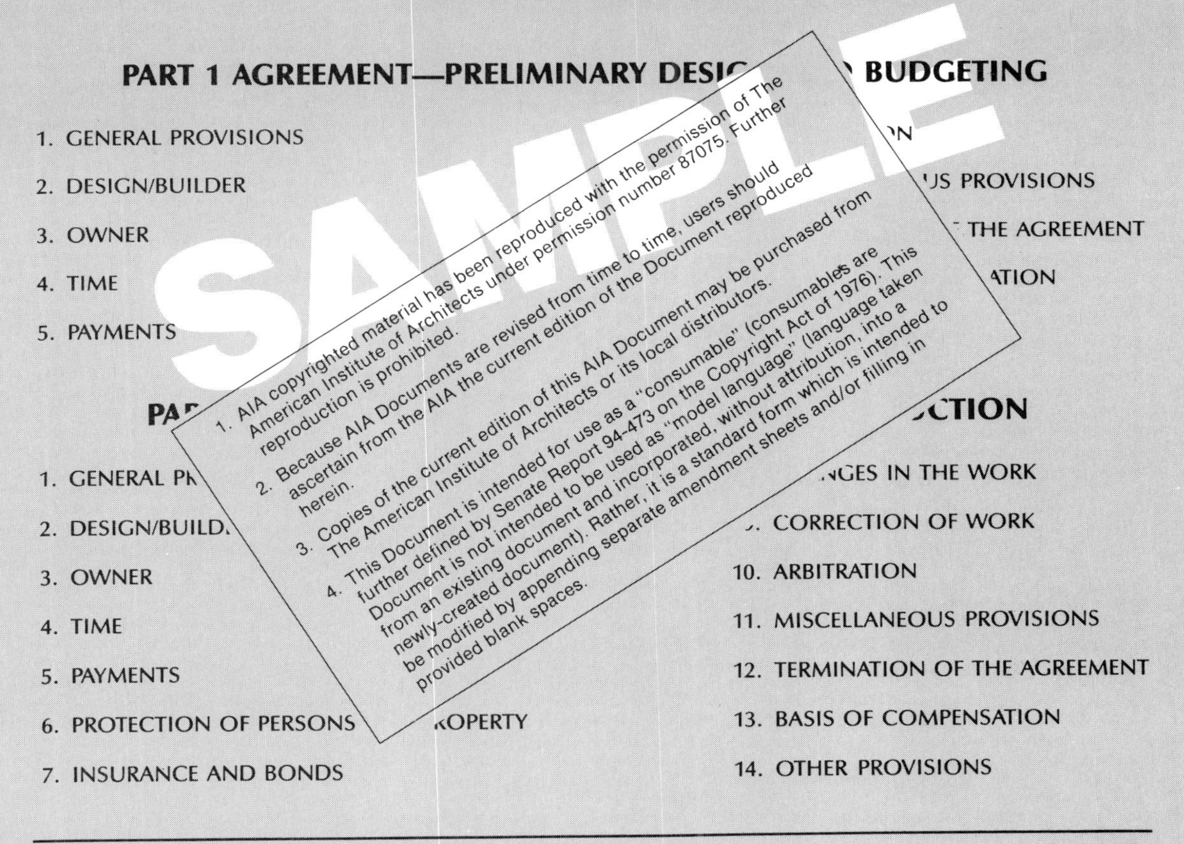

AIA DOCUMENT A191 • OWNER-DESIGN/BUILDER AGREEMENT • FIRST EDITION • AIA® • ©1985 • THE AMERICAN INSTITUTE OF ARCHITECTS, 1735 NEW YORK AVENUE, N.W., WASHINGTON, D.C. 20006

A191-1985
COVER PAGE

THE AMERICAN INSTITUTE OF ARCHITECTS

AIA Document A191

Standard Form of Agreement Between Owner and Design/Builder

1985 EDITION

THIS DOCUMENT HAS IMPORTANT LEGAL CONSEQUENCES; CONSULTATION WITH AN ATTORNEY IS ENCOURAGED.

This Document comprises two separate Agreements: Part 1 Agreement—Preliminary Design and Budgeting and Part 2 Agreement—Final Design and Construction. Hereinafter, the Part 1 Agreement is referred to as Part 1 and the Part 2 Agreement is referred to as Part 2. Before executing Part 1, the parties should reach substantial agreement on Part 2.

PART 1 AGREEMENT—PRELIMINARY DESIGN AND BUDGETING

AGREEMENT

made as of the day of in the year of Nineteen Hundred and

BETWEEN the Owner:
(Name and address)

and the Design/Builder:
(Name and address)

For the following Project:
(Include Project name, location and detailed description of scope.)

The architectural services described in Article 2 will be provided by the following person or entity who is lawfully licensed to practice architecture:
(Name and address)

The Owner and the Design/Builder agree as set forth below.

Terms and Conditions—Part 1 Agreement

ARTICLE 1
GENERAL PROVISIONS

1.1 BASIC DEFINITIONS

1.1.1 The Project is the total design and construction for which the Design/Builder is responsible under Part 1, including all professional design services and all labor, materials and equipment used or incorporated in such design and construction.

1.1.2 The Work comprises the completed construction designed under the Project and includes labor necessary to produce such construction, and materials and equipment incorporated or to be incorporated in such construction.

1.2 EXECUTION, CORRELATION AND INTENT

1.2.1 This Part 1 shall be signed in not less than duplicate by the Owner and Design/Builder.

1.2.2 Nothing contained in the Design/Builder Contract Documents shall create a professional obligation or contractual relationship between the Owner and any third party.

1.3 OWNERSHIP AND USE OF DOCUMENTS

1.3.1 The drawings, specifications and other documents furnished by the Design/Builder are instruments of service and shall not become the property of the Owner whether or not the Project for which they are made is commenced. Drawings, specifications and other documents furnished by the Design/Builder shall not be used by the Owner on other projects, for additions to this Project or, unless the Design/Builder is in default under Part 1, for completion of this Project by others, except by written agreement relating to use, liability and compensation.

1.3.2 Submission or distribution of documents to meet official regulatory requirements or for other purposes in connection with the Project is not to be construed as publication in derogation of the Design/Builder's or the Architect's common law copyrights or other reserved rights. The Owner shall own neither the documents nor the copyrights.

ARTICLE 2
DESIGN/BUILDER

2.1 SERVICES AND RESPONSIBILITIES

2.1.1 Design services shall be performed by qualified architects, engineers and other professionals selected and paid by the Design/Builder. The professional obligations of such persons shall be undertaken and performed in the interest of the Design/Builder. Construction services shall be performed by qualified construction contractors and suppliers, selected and paid by the Design/Builder and acting in the interest of the Design/Builder. Nothing contained in Part 1 shall create any professional obligation or contractual relationship between such persons and the Owner.

2.1.2 The Design/Builder shall be responsible to the Owner for acts and omissions of the Design/Builder's employees and parties in privity of contract with the Design/Builder to perform a portion of the Work, including their agents and employees.

2.2 BASIC SERVICES

2.2.1 The Design/Builder's Basic Services are as described below and in Article 10.

2.2.2 The Design/Builder shall review the Owner's program to ascertain requirements of the Project and shall review such requirements with the Owner.

2.2.3 The Design/Builder shall provide, after consultation with the Owner, a preliminary evaluation of the program and Project budget, each in terms of the other.

2.2.4 The Design/Builder shall review with the Owner alternative approaches to design and construction of the Project.

2.2.5 The Design/Builder shall submit to the Owner a Proposal including the completed Preliminary Design Documents, a statement of the proposed contract sum, a proposed schedule for completion of the Work under Part 2 and all other information necessary to complete Part 2. Preliminary Design Documents shall consist of preliminary design drawings, outline specifications and other documents to fix and describe the size, quality and character of the entire Project, its architectural, structural, mechanical and electrical systems, and the materials and such other elements of the Project as may be appropriate. If the Proposal is accepted by the Owner, they shall then execute Part 2. Modifications to the Proposal before execution of Part 2 shall be recorded in writing as an addendum and be identified in the Contract Documents of Part 2.

2.3 ADDITIONAL SERVICES

All other services requested by the Owner and mutually agreed to in writing by the Owner and Design/Builder in Part 1, including, among others, changes in Project scope and program, shall constitute Additional Services and shall be paid for by the Owner as provided in Part 1.

ARTICLE 3
OWNER

3.1 The Owner shall provide information regarding requirements for the Project, including but not limited to the Owner's design objectives, constraints and criteria.

3.2 If the Owner provides a budget for the Project, it shall explicitly include as separate line items contingencies for changes in the design and construction, and other costs which are the responsibility of the Owner. No budget shall constitute a fixed limit of construction cost unless such limit has been agreed to in writing by the Design/Builder.

3.3 The Owner shall designate a representative authorized to act on the Owner's behalf with respect to the Project. The Owner or such authorized representative shall examine the documents submitted by the Design/Builder and shall promptly render decisions pertaining thereto to avoid delay in the orderly progress of design and construction.

3.4 The Owner shall cooperate with the Design/Builder in identifying required permits, licenses and inspections, and shall take appropriate action with reasonable promptness.

3.5 Prior to commencement of Basic Services, the Owner shall furnish a legal description and a certified land survey of the site, giving, as applicable, grades and lines of streets, alleys, pavements and adjoining property; rights-of-way, restrictions, easements, encroachments, zoning, deed restrictions, elevations and contours of the site; locations, dimensions and complete data pertaining to existing buildings, other improvements and trees; and full information concerning available services and utility lines, both public and private, above and below grade, including inverts and depths.

3.6 The Owner shall furnish services of geotechnical engineers and other consultants when such services are deemed necessary by the Design/Builder. Geotechnical engineers or other consultants shall be selected by mutual agreement. Such services shall include, as required, applicable test borings, test pits, soil bearing values, percolation tests, air and water pollution tests, and other necessary operations for determining subsoil, air and water conditions, with reports and appropriate professional recommendations.

3.7 The services, information, surveys and reports required by Paragraphs 3.5 and 3.6 shall be furnished at the Owner's expense, and the Design/Builder shall be entitled to rely upon their accuracy and completeness.

3.8 If the Owner observes or otherwise becomes aware of any fault or defect in the Work, the Owner shall give prompt written notice thereof to the Design/Builder.

3.9 The Owner shall furnish required information and services and shall render decisions with reasonable promptness to avoid delay in the orderly progress of the Design/Builder's services.

3.10 The Owner shall communicate with contractors only through the Design/Builder.

ARTICLE 4
TIME

4.1 The Design/Builder shall provide the Basic and Additional Services as expeditiously as is consistent with reasonable skill and care and shall complete the services in the time provided in Article 10.

ARTICLE 5
PAYMENTS

5.1 The initial payment provided in Article 9 shall be made upon execution of this Part 1 and credited to the Owner's account as provided in Subparagraph 9.1.2.

5.2 Subsequent payments for Part 1 Basic Services, Additional Services and Reimbursable Expenses shall be made monthly on the basis set forth in Article 9.

5.3 Within ten days of the Owner's receipt of a properly submitted Application for Payment, the Owner shall make payment to the Design/Builder.

5.4 Payments due the Design/Builder under Part 1 which are not paid when due shall bear interest from the date due at the rate specified in Paragraph 9.5, or in the absence of a specified rate, at the legal rate prevailing where the principal office of the Design/Builder is located.

ARTICLE 6
ARBITRATION

6.1 Claims, disputes and other matters in question between the parties to Part 1 arising out of or relating to Part 1 shall be decided by arbitration in accordance with the Construction Industry Arbitration Rules of the American Arbitration Association then in effect unless the parties agree otherwise. No arbitration arising out of or relating to Part 1 shall include, by consolidation or joinder or in any other manner, an additional person not a party to Part 1 except by written consent containing specific reference to Part 1 and signed by the Owner, Design/Builder and any other person sought to be joined. Consent to arbitration involving an additional person or persons shall not constitute consent to arbitration of a dispute not described therein or with a person not named therein. This provision shall be specifically enforceable in any court of competent jurisdiction.

6.2 Notice of demand for arbitration shall be filed in writing with the other party to Part 1 and with the American Arbitration Association. The demand shall be made within a reasonable time after the claim, dispute or other matter in question has arisen. In no event shall the demand for arbitration be made after the date when the applicable statute of limitations would bar institution of a legal or equitable proceeding based on such claim, dispute or other matter in question.

6.3 The award rendered by arbitrators shall be final, and judgment may be entered upon it in accordance with applicable law in any court having jurisdiction.

6.4 Unless otherwise agreed in writing, the Design/Builder shall carry on the services and maintain progress during any arbitration proceedings, and the Owner shall continue to make payments to the Design/Builder in accordance with Part 1.

6.5 This Article 6 shall survive completion or termination of Part 1.

ARTICLE 7
MISCELLANEOUS PROVISIONS

7.1 This Part 1 shall be governed by the law of the place where the principal office of the Design/Builder is located.

7.2 The table of contents and the headings of articles and paragraphs are for convenience only and shall not modify rights and obligations created by Part 1.

7.3 In case a provision of Part 1 is held to be invalid, illegal or unenforceable, the validity, legality and enforceability of the remaining provisions shall not be affected.

7.4 SUCCESSORS AND ASSIGNS

7.4.1 This Part 1 shall be binding on successors, assigns and legal representatives of and persons in privity of contract with the Owner or Design/Builder. Neither party shall assign, sublet or transfer an interest in Part 1 without written consent of the other.

7.4.2 This Paragraph 7.4 shall survive completion or termination of Part 1.

7.5 EXTENT OF AGREEMENT

7.5.1 Part 1 represents the entire agreement for Preliminary Design and Budgeting and supersedes all prior negotiations, representations or agreements. Part 1 may be amended only by written instrument signed by both Owner and Design/Builder.

7.6 In case of termination of the Architect, the Design/Builder shall provide the services of another lawfully licensed person or entity against whom the Owner makes no reasonable objection.

ARTICLE 8
TERMINATION OF THE AGREEMENT

8.1 Part 1 may be terminated by either party upon seven days' written notice should the other party fail substantially to perform in accordance with its terms through no fault of the party initiating the termination.

8.2 Part 1 may be terminated by the Owner upon at least seven days' written notice to the Design/Builder in the event that the Project is permanently abandoned.

8.3 In the event of termination not the fault of the Design/Builder, the Design/Builder shall be compensated for services performed to termination date, together with Reimbursable Expenses then due and Termination Expenses. Termination Expenses are expenses directly attributable to termination, including a reasonable amount for overhead and profit, for which the Design/Builder is not otherwise compensated under Part 1.

ARTICLE 9
BASIS OF COMPENSATION

The Owner shall compensate the Design/Builder in accordance with Article 5, Payments, and the other provisions of Part 1 as described below.

9.1 COMPENSATION FOR BASIC SERVICES

9.1.1 FOR BASIC SERVICES, compensation shall be as follows:

9.1.2 AN INITIAL PAYMENT of dollars ($) shall be made upon execution of Part 1 and credited to the Owner's account as follows:

9.1.3 SUBSEQUENT PAYMENTS shall be as follows:

9.2 COMPENSATION FOR ADDITIONAL SERVICES

9.2.1 FOR ADDITIONAL SERVICES, compensation shall be as follows:

9.3 REIMBURSABLE EXPENSES

9.3.1 Reimbursable Expenses are in addition to Compensation for Basic and Additional Services and include actual expenditures made by the Design/Builder and the Design/Builder's employees and contractors in the interest of the Project for the expenses listed as follows:

9.3.2 FOR REIMBURSABLE EXPENSES, compensation shall be a multiple of ( ) times the amounts expended.

9.4 DIRECT PERSONNEL EXPENSE is defined as the direct salaries of personnel engaged on the Project, and the portion of the cost of their mandatory and customary contributions and benefits related thereto, such as employment taxes and other statutory employee benefits, insurance, sick leave, holidays, vacations, pensions and similar contributions and benefits.

.5 INTEREST PAYMENTS

·.5.1 The rate of interest for past due payments shall be as follows:

(Usury laws and requirements under the Federal Truth in Lending Act, similar state and local consumer credit laws and other regulations at the Owner's and Design/Builder's principal places of business, at the location of the Project and elsewhere may affect the validity of this provision. Specific legal advice should be obtained with respect to deletion, modification or other requirements, such as written disclosures or waivers.)

9.6 IF THE SCOPE of the Project is changed materially, the amount of compensation shall be equitably adjusted.

9.7 The compensation set forth herein shall be equitably adjusted if through no fault of the Design/Builder the services have not been completed within ( ) months of the date of Part 1.

ARTICLE 10
OTHER PROVISIONS

10.1 The Basic Services to be performed shall be commenced on   and, subject to authorized adjustments and to delays not caused by the Design/Builder, shall be completed in   (   ) calendar days.

10.2 The Basic Services beyond those described in Article 2 are:

This Part 1 entered into as of the day and year first written above.

OWNER	DESIGN/BUILDER
______________________	______________________
______________________	______________________
______________________	______________________
BY ______________________	BY ______________________

THE AMERICAN INSTITUTE OF ARCHITECTS

AIA Document A191

Standard Form of Agreement Between Owner and Design/Builder

1985 EDITION

THIS DOCUMENT HAS IMPORTANT LEGAL CONSEQUENCES; CONSULTATION WITH AN ATTORNEY IS ENCOURAGED.

This Document comprises two separate Agreements: Part 1 Agreement—Preliminary Design and Budgeting and Part 2 Agreement—Final Design and Construction. Hereinafter, the Part 1 Agreement is referred to as Part 1 and the Part 2 Agreement is referred to as Part 2.

PART 2 AGREEMENT—FINAL DESIGN AND CONSTRUCTION

AGREEMENT

made as of the day of in the year of Nineteen Hundred and

BETWEEN the Owner:
(Name and address)

and the Design/Builder:
(Name and address)

For the following Project:
(Include Project name, location and detailed description of scope.)

The architectural services described in Article 2 will be provided by the following person or entity who is lawfully licensed to practice architecture:
(Name and address)

The Owner and the Design/Builder agree as set forth below.

Terms and Conditions—Part 2 Agreement

ARTICLE 1
GENERAL PROVISIONS

1.1 BASIC DEFINITIONS

1.1.1 The Contract Documents consist of the Design/Builder's Proposal identified in Article 14, this Part 2, the Construction Documents approved by the Owner in accordance with Subparagraph 2.2.2 below and Modifications issued after execution of Part 2. A Modification is a Change Order or a written amendment to Part 2 signed by both parties. These form the Contract, and are as fully a part of the Contract as if attached to this Part 2 or repeated herein.

1.1.2 The Project is the total design and construction for which the Design/Builder is responsible under Part 2, including all professional design services and all labor, materials and equipment used or incorporated in such design and construction.

1.1.3 The Work comprises the completed construction designed under the Project and includes labor necessary to produce such construction, and materials and equipment incorporated or to be incorporated in such construction.

1.2 EXECUTION, CORRELATION AND INTENT

1.2.1 This Part 2 shall be signed in not less than duplicate by the Owner and Design/Builder.

1.2.2 It is the intent of the Owner and Design/Builder that the Contract Documents include all items necessary for proper execution and completion of the Work. The Contract Documents are complementary, and what is required by any one shall be as binding as if required by all. Work not covered in the Contract Documents will not be required unless it is consistent with and is reasonably inferable from the Contract Documents as being necessary to produce the intended results. Words and abbreviations which have well-known technical or trade meanings are used in the Contract Documents in accordance with such recognized meanings.

1.3 OWNERSHIP AND USE OF DOCUMENTS

1.3.1 The drawings, specifications and other documents furnished by the Design/Builder are instruments of service and shall not become the property of the Owner whether or not the Project for which they are made is commenced. Drawings, specifications and other documents furnished by the Design/Builder shall not be used by the Owner on other projects, for additions to this Project or, unless the Design/Builder is in default under Part 2, for completion of this Project by others, except by written agreement relating to use, liability and compensation.

1.3.2 Submission or distribution of documents to meet official regulatory requirements or for other purposes in connection with the Project is not to be construed as publication in derogation of the Design/Builder's or the Architect's common law copyrights or other reserved rights. The Owner shall own neither the documents nor the copyrights.

ARTICLE 2
DESIGN/BUILDER

2.1 SERVICES AND RESPONSIBILITIES

2.1.1 Design services shall be performed by qualified architects, engineers and other professionals selected and paid by the Design/Builder. The professional obligations of such persons shall be undertaken and performed in the interest of the Design/Builder. Construction services shall be performed by qualified construction contractors and suppliers, selected and paid by the Design/Builder and acting in the interest of the Design/Builder. Nothing contained in Part 2 shall create any professional obligation or contractual relationship between such persons and the Owner.

2.2 BASIC SERVICES

2.2.1 The Design/Builder's Basic Services are described below and in Article 14.

2.2.2 Based on the Design/Builder's Proposal, the Design/Builder shall submit Construction Documents for review and approval by the Owner. Construction Documents shall include technical drawings, schedules, diagrams and specifications, setting forth in detail the requirements for construction of the Work, and shall:

.1 develop the intent of the Design/Builder's Proposal in greater detail;
.2 provide information customarily necessary for the use of those in the building trades; and
.3 include documents customarily required for regulatory agency approvals.

2.2.3 The Design/Builder shall assist the Owner in filing documents required to obtain necessary approvals of governmental authorities having jurisdiction over the Project.

2.2.4 Unless otherwise provided in the Contract Documents, the Design/Builder shall provide or cause to be provided and shall pay for design services, labor, materials, equipment, tools, construction equipment and machinery, water, heat, utilities, transportation and other facilities and services necessary for proper execution and completion of the Work, whether temporary or permanent and whether or not incorporated or to be incorporated in the Work.

2.2.5 The Design/Builder shall be responsible for and shall coordinate all construction means, methods, techniques, sequences and procedures.

2.2.6 The Design/Builder shall keep the Owner informed of the progress and quality of the Work.

2.2.7 If requested in writing by the Owner, the Design/Builder, with reasonable promptness and in accordance with time limits agreed upon, shall interpret the requirements of the Contract Documents and initially shall decide, subject to demand for arbitration, claims, disputes and other matters in question relating to performance thereunder by both Owner and Design/Builder. Such interpretations and decisions shall be in writing, shall not be presumed to be correct and shall be given such weight as the arbitrator(s) or the court shall determine.

2.2.8 The Design/Builder shall correct Work which does not conform to the Construction Documents.

2.2.9 The Design/Builder warrants to the Owner that materials and equipment incorporated in the Work will be new unless otherwise specified, and that the Work will be of good quality, free from faults and defects, and in conformance with the Contract Documents. Work not conforming to these requirements shall be corrected in accordance with Article 9.

2.2.10 The Design/Builder shall pay all sales, consumer, use and similar taxes which were in effect at the time the Design/Builder's Proposal was first submitted to the Owner, and shall secure and pay for building and other permits and governmental fees, licenses and inspections necessary for the proper execution and completion of the Work which are either customarily secured after execution of Part 2 or are legally required at the time the Design/Builder's Proposal was first submitted to the Owner.

2.2.11 The Design/Builder shall give notices and comply with laws, ordinances, rules, regulations and lawful orders of public authorities relating to the Project.

2.2.12 The Design/Builder shall pay royalties and license fees. The Design/Builder shall defend suits or claims for infringement of patent rights and shall save the Owner harmless from loss on account thereof, except that the Owner shall be responsible for such loss when a particular design, process or product of a particular manufacturer is required by the Owner. However, if the Design/Builder has reason to believe the use of a required design, process or product is an infringement of a patent, the Design/Builder shall be responsible for such loss unless such information is promptly given to the Owner.

2.2.13 The Design/Builder shall be responsible to the Owner for acts and omissions of the Design/Builder's employees and parties in privity of contract with the Design/Builder to perform a portion of the Work, including their agents and employees.

2.2.14 The Design/Builder shall keep the premises free from accumulation of waste materials or rubbish caused by the Design/Builder's operations. At the completion of the Work, the Design/Builder shall remove from and about the Project the Design/Builder's tools, construction equipment, machinery, surplus materials, waste materials and rubbish.

2.2.15 The Design/Builder shall prepare Change Orders for the Owner's approval and execution in accordance with Part 2 and shall have authority to make minor changes in the design and construction consistent with the intent of Part 2 not involving an adjustment in the contract sum or an extension of the contract time. The Design/Builder shall promptly inform the Owner, in writing, of minor changes in the design and construction.

2.2.16 The Design/Builder shall notify the Owner when the Work or an agreed upon portion thereof is substantially completed by issuing a Certificate of Substantial Completion which shall establish the Date of Substantial Completion, shall state the responsibility of each party for security, maintenance, heat, utilities, damage to the Work and insurance, shall include a list of items to be completed or corrected and shall fix the time within which the Design/Builder shall complete items listed therein. Disputes between the Owner and Design/Builder regarding the Certificate of Substantial Completion shall be resolved by arbitration.

2.2.17 The Design/Builder shall maintain in good order at the site one record copy of the drawings, specifications, product data, samples, shop drawings, Change Orders and other Modifications, marked currently to record changes made during construction. These shall be delivered to the Owner upon completion of the design and construction and prior to final payment.

ARTICLE 3
OWNER

3.1 The Owner shall designate a representative authorized to act on the Owner's behalf with respect to the Project. The Owner or such authorized representative shall examine documents submitted by the Design/Builder and shall promptly render decisions pertaining thereto to avoid delay in the orderly progress of the Work.

3.2 The Owner may appoint an on-site project representative to observe the Work and to have such other responsibilities as the Owner and Design/Builder agree in writing prior to execution of Part 2.

3.3 The Owner shall cooperate with the Design/Builder in securing building and other permits, licenses and inspections, and shall pay the fees for such permits, licenses and inspections if the cost of such fees is not identified as being included in the Design/Builder's Proposal.

3.4 The Owner shall furnish services by land surveyors, geotechnical engineers and other consultants for subsoil, air and water conditions, in addition to those provided under Part 1 when such services are deemed necessary by the Design/Builder to carry out properly the design services under this Part 2.

3.5 The Owner shall furnish structural, mechanical, chemical, geotechnical and other laboratory or on-site tests, inspections and reports as required by law or the Contract Documents.

3.6 The services, information, surveys and reports required by Paragraphs 3.4 and 3.5 shall be furnished at the Owner's expense, and the Design/Builder shall be entitled to rely upon their accuracy and completeness.

3.7 If the Owner observes or otherwise becomes aware of a fault or defect in the Work or nonconformity with the Design or Construction Documents, the Owner shall give prompt written notice thereof to the Design/Builder.

3.8 The Owner shall furnish required information and services and shall promptly render decisions pertaining thereto to avoid delay in the orderly progress of the design and construction.

3.9 The Owner shall, at the request of the Design/Builder and upon execution of Part 2, provide a certified or notarized statement of funds available for the Project and their source.

3.10 The Owner shall communicate with contractors only through the Design/Builder.

ARTICLE 4
TIME

4.1 The Design/Builder shall provide services as expeditiously as is consistent with reasonable skill and care and the orderly progress of design and construction.

4.2 Time limits stated in the Contract Documents are of the essence of Part 2. The Work to be performed under Part

2 shall commence upon execution of a notice to proceed unless otherwise agreed and, subject to authorized Modifications, Substantial Completion shall be achieved as indicated in Article 14.

4.3 The Date of Substantial Completion of the Work or an agreed upon portion thereof is the date when construction or an agreed upon portion thereof is sufficiently complete so the Owner can occupy and utilize the Work or agreed upon portion thereof for its intended use.

4.4 The schedule provided in the Design/Builder's Proposal shall include a construction schedule consistent with Paragraph 4.2 above.

4.5 If the Design/Builder is delayed in the progress of the Project by acts or neglect of the Owner, Owner's employees, separate contractors employed by the Owner, changes ordered in the Work not caused by the fault of the Design/Builder, labor disputes, fire, unusual delay in transportation, adverse weather conditions not reasonably anticipatable, unavoidable casualties, or other causes beyond the Design/Builder's control, or by delay authorized by the Owner's pending arbitration or another cause which the Owner and Design/Builder agree is justifiable, the contract time shall be reasonably extended by Change Order.

ARTICLE 5
PAYMENTS

5.1 PROGRESS PAYMENTS

5.1.1 The Design/Builder shall deliver to the Owner itemized Applications for Payment in such detail as indicated in Article 14.

5.1.2 Within ten days of the Owner's receipt of a properly submitted and correct Application for Payment, the Owner shall make payment to the Design/Builder.

5.1.3 The Application for Payment shall constitute a representation by the Design/Builder to the Owner that, to the best of the Design/Builder's knowledge, information and belief, the design and construction have progressed to the point indicated; the quality of the Work covered by the application is in accordance with the Contract Documents; and the Design/Builder is entitled to payment in the amount requested.

5.1.4 The Design/Builder shall pay each contractor, upon receipt of payment from the Owner, out of the amount paid to the Design/Builder on account of such contractor's work, the amount to which said contractor is entitled in accordance with the terms of the Design/Builder's contract with such contractor. The Design/Builder shall, by appropriate agreement with each contractor, require each contractor to make payments to subcontractors in similar manner.

5.1.5 The Owner shall have no obligation to pay or to be responsible in any way for payment to a contractor of the Design/Builder except as may otherwise be required by law.

5.1.6 No progress payment or partial or entire use or occupancy of the Project by the Owner shall constitute an acceptance of Work not in accordance with the Contract Documents.

5.1.7 The Design/Builder warrants that: (1) title to Work, materials and equipment covered by an Application for Payment will pass to the Owner either by incorporation in construction or upon receipt of payment by the Design/Builder, whichever occurs first; (2) Work, materials and equipment covered by previous Applications for Payment are free and clear of liens, claims, security interests or encumbrances, hereinafter referred to as "liens"; and (3) no Work, materials or equipment covered by an Application for Payment will have been acquired by the Design/Builder, or any other person performing work at the site or furnishing materials or equipment for the Project, subject to an agreement under which an interest therein or an encumbrance thereon is retained by the seller or otherwise imposed by the Design/Builder or such other person.

5.1.8 If the Contract provides for retainage, then at the date of Substantial Completion or occupancy of the Work or any agreed upon portion thereof by the Owner, whichever occurs first, the Design/Builder may apply for and the Owner, if the Design/Builder has satisfied the requirements of Paragraph 5.2.1 and any other requirements of the Contract relating to retainage, shall pay the Design/Builder the amount retained, if any, for the Work or for the portion completed or occupied, less the reasonable value of incorrect or incomplete Work. Final payment of such withheld sum shall be made upon correction or completion of such Work.

5.2 FINAL PAYMENT

5.2.1 Neither final payment nor amounts retained, if any, shall become due until the Design/Builder submits to the Owner (1) an affidavit that payrolls, bills for materials and equipment, and other indebtedness connected with the Project for which the Owner or Owner's property might be liable have been paid or otherwise satisfied, (2) consent of surety, if any, to final payment, (3) a certificate that insurance required by the Contract Documents is in force following completion of the Work, and (4) if required by the Owner, other data establishing payment or satisfaction of obligations, such as receipts, releases and waivers of liens arising out of Part 2, to the extent and in such form as may be designated by the Owner. If a contractor refuses to furnish a release or waiver required by the Owner, the Design/Builder may furnish a bond satisfactory to the Owner to indemnify the Owner against such lien. If such lien remains unsatisfied after payments are made, the Design/Builder shall reimburse the Owner for moneys the latter may be compelled to pay in discharging such lien, including all costs and reasonable attorneys' fees.

5.2.2 Final payment constituting the entire unpaid balance due shall be paid by the Owner to the Design/Builder upon the Owner's receipt of the Design/Builder's final Application for Payment when the Work has been completed and the Contract fully performed except for those responsibilities of the Design/Builder which survive final payment.

5.2.3 The making of final payment shall constitute a waiver of all claims by the Owner except those arising from:

- **.1** unsettled liens;
- **.2** faulty or defective Work appearing after Substantial Completion;
- **.3** failure of the Work to comply with requirements of the Contract Documents; or
- **.4** terms of special warranties required by the Contract Documents.

5.2.4 Acceptance of final payment shall constitute a waiver of all claims by the Design/Builder except those previously made in writing and identified by the Design/Builder as unsettled at the time of final Application for Payment.

5.3 INTEREST PAYMENTS

5.3.1 Payments due the Design/Builder under Part 2 which are not paid when due shall bear interest from the date due at the rate specified in Article 13, or in the absence of a specified rate, at the legal rate prevailing where the principal improvements are to be located.

ARTICLE 6
PROTECTION OF PERSONS AND PROPERTY

6.1 The Design/Builder shall be responsible for initiating, maintaining and providing supervision of safety precautions and programs in connection with the Work.

6.2 The Design/Builder shall take reasonable precautions for safety of, and shall provide reasonable protection to prevent damage, injury or loss to: (1) employees on the Work and other persons who may be affected thereby; (2) the Work and materials and equipment to be incorporated therein; and (3) other property at or adjacent to the site.

6.3 The Design/Builder shall give notices and comply with applicable laws, ordinances, rules, regulations and orders of public authorities bearing on the safety of persons and property and their protection from damage, injury or loss.

6.4 The Design/Builder shall be liable for damage or loss (other than damage or loss to property insured under the property insurance provided or required by the Contract Documents to be provided by the Owner) to property at the site caused in whole or in part by the Design/Builder, a contractor of the Design/Builder or anyone directly or indirectly employed by either of them, or by anyone for whose acts they may be liable, except damage or loss attributable to the acts or omissions of the Owner, the Owner's separate contractors or anyone directly or indirectly employed by them or by anyone for whose acts they may be liable and not attributable to the fault or negligence of the Design/Builder.

ARTICLE 7
INSURANCE AND BONDS

7.1 DESIGN/BUILDER'S LIABILITY INSURANCE

7.1.1 The Design/Builder shall purchase and maintain in a company or companies authorized to do business in the state in which the Work is located such insurance as will protect the Design/Builder from claims set forth below which may arise out of or result from operations under the Contract by the Design/Builder or by a contractor of the Design/Builder, or by anyone directly or indirectly employed by any of them, or by anyone for whose acts they may be liable:

.1 claims under workers' or workmen's compensation, disability benefit and other similar employee benefit laws which are applicable to the Work to be performed;

.2 claims for damages because of bodily injury, occupational sickness or disease, or death of the Design/Builder's employees under any applicable employer's liability law;

.3 claims for damages because of bodily injury, sickness or disease, or death of persons other than the Design/Builder's employees;

.4 claims for damages covered by usual personal injury liability coverage which are sustained (1) by a person as a result of an offense directly or indirectly related to employment of such person by the Design/Builder or (2) by another person;

.5 claims for damages, other than to the Work at the site, because of injury to or destruction of tangible property, including loss of use; and

.6 claims for damages for bodily injury or death of a person or property damage arising out of ownership, maintenance or use of a motor vehicle.

7.1.2 The insurance required by the above Subparagraph 7.1.1 shall be written for not less than limits of liability specified in the Contract Documents or required by law, whichever are greater.

7.1.3 The Design/Builder's liability insurance shall include contractual liability insurance applicable to the Design/Builder's obligations under Paragraph 11.7.

7.1.4 Certificates of Insurance, and copies of policies if requested, acceptable to the Owner shall be delivered to the Owner prior to commencement of design and construction. These Certificates as well as insurance policies required by this Paragraph shall contain a provision that coverage will not be cancelled or allowed to expire until at least thirty days' prior written notice has been given to the Owner. If any of the foregoing insurance coverages are required to remain in force after final payment, an additional certificate evidencing continuation of such coverage shall be submitted along with the application for final payment.

7.2 OWNER'S LIABILITY INSURANCE

7.2.1 The Owner shall be responsible for purchasing and maintaining, in a company or companies authorized to do business in the state in which the principal improvements are to be located, Owner's liability insurance to protect the Owner against claims which may arise from operations under this Project.

7.3 PROPERTY INSURANCE

7.3.1 Unless otherwise provided under this Part 2, the Owner shall purchase and maintain, in a company or companies authorized to do business in the state in which the principal improvements are to be located, property insurance upon the Work at the site to the full insurable value thereof. Property insurance shall include interests of the Owner, the Design/Builder, and their respective contractors and subcontractors in the Work. It shall insure against perils of fire and extended coverage and shall include all risk insurance for physical loss or damage including, without duplication of coverage, theft, vandalism and malicious mischief. If the Owner does not intend to purchase such insurance for the full insurable value of the entire Work, the Owner shall inform the Design/Builder in writing prior to commencement of the Work. The Design/Builder may then effect insurance for the Work at the site which will protect the interests of the Design/Builder and the Design/Builder's contractors and subcontractors, and by appropriate Change Order the cost thereof shall be charged to the Owner. If the Design/Builder is damaged by failure of the Owner to purchase or maintain such insurance without notice to the Design/Builder, then the Owner shall bear all reasonable costs properly attributable thereto. If not covered under the all risk insurance or not otherwise provided in the Contract Documents, the Design/Builder shall effect and maintain similar property insurance on portions of the Work stored off-site or in transit when such portions of the Work are to be included in an Application for Payment.

7.3.2 Unless otherwise provided under this Part 2, the Owner shall purchase and maintain such boiler and machinery insurance as may be required by the Contract Documents or by law and which shall specifically cover such insured objects during installation and until final acceptance by the Owner. This insurance shall cover interests of the Owner, the Design/Builder, and the Design/Builder's contractors and subcontractors in the Work.

7.3.3 A loss insured under Owner's property insurance is to be adjusted with the Owner and made payable to the Owner as trustee for the insureds, as their interests may appear, subject to requirements of any applicable mortgagee clause and of Subparagraph 7.3.8. The Design/Builder shall pay contractors their shares of insurance proceeds received by the Design/Builder, and by appropriate agreement, written where legally required for validity, shall require contractors to make payments to their subcontractors in similar manner.

7.3.4 Before an exposure to loss may occur, the Owner shall file with the Design/Builder a copy of each policy required by this Paragraph 7.3. Each policy shall contain only those endorsements specifically related to this Project. Each policy shall contain a provision that the policy will not be cancelled or allowed to expire until at least thirty days' prior written notice has been given the Design/Builder.

7.3.5 If the Design/Builder requests in writing that insurance for risks other than those described herein or for other special hazards be included in the property insurance policy, the Owner shall, if possible, obtain such insurance, and the cost thereof shall be charged to the Design/Builder by appropriate Change Order.

7.3.6 The Owner and Design/Builder waive all rights against each other and the contractors, subcontractors, agents and employees, each of the other, for damages caused by fire or other perils to the extent covered by property insurance obtained pursuant to this Paragraph 7.3 or other property insurance applicable to the Work, except such rights as they may have to proceeds of such insurance held by the Owner as trustee. The Owner or Design/Builder, as appropriate, shall require from contractors and subcontractors by appropriate agreements, written where legally required for validity, similar waivers each in favor of other parties enumerated in this Paragraph 7.3. The policies shall be endorsed to include such waivers of subrogation.

7.3.7 If required in writing by a party in interest, the Owner as trustee shall provide, upon occurrence of an insured loss, a bond for proper performance of the Owner's duties. The cost of required bonds shall be charged against proceeds received as trustee. The Owner shall deposit proceeds so received in a separate account and shall distribute them in accordance with such agreement as the parties in interest may reach, or in accordance with an arbitration award in which case the procedure shall be as provided in Article 10. If after such loss no other special agreement is made, replacement of damaged Work shall be covered by appropriate Change Order.

7.3.8 The Owner, as trustee, shall have power to adjust and settle a loss with insurers unless one of the parties in interest shall object, in writing, within ten days after occurrence of loss, to the Owner's exercise of this power. If such objection be made, the Owner as trustee shall make settlement with the insurers in accordance with the decision of arbitration as provided in Article 10. If distribution of insurance proceeds by arbitration is required, the arbitrators will direct such distribution.

7.3.9 If the Owner finds it necessary to occupy or use a portion or portions of the Work before Substantial Completion, such occupancy or use shall not commence prior to a time agreed to by the Owner and Design/Builder and to which the insurance company or companies providing property insurance have consented by endorsement to the policy or policies. The property insurance shall not lapse or be cancelled on account of such partial occupancy or use. Consent of the Design/Builder and of the insurance company or companies to such occupancy or use shall not be unreasonably withheld.

7.4 LOSS OF USE INSURANCE

7.4.1 The Owner, at the Owner's option, may purchase and maintain such insurance as will insure the Owner against loss of use of the Owner's property due to fire or other hazards, however caused. The Owner waives all rights of action against the Design/Builder, and its contractors and their agents and employees, for loss of use of the Owner's property, including consequential losses due to fire or other hazards, however caused, to the extent covered by insurance under this Paragraph 7.4.

7.5 PERFORMANCE BOND AND PAYMENT BOND

7.5.1 The Owner shall have the right to require the Design/Builder to furnish bonds covering the faithful performance of the Contract and the payment of all obligations arising thereunder if and as required in the Contract Documents or in Article 14.

ARTICLE 8
CHANGES IN THE WORK

8.1 CHANGE ORDERS

8.1.1 A Change Order is a written order signed by the Owner and Design/Builder, and issued after execution of Part 2, authorizing a change in the Work or adjustment in the contract sum or contract time. The contract sum and contract time may be changed only by Change Order.

8.1.2 The Owner, without invalidating Part 2, may order changes in the Work within the general scope of Part 2 consisting of additions, deletions or other revisions, and the contract sum and contract time shall be adjusted accordingly. Such changes in the Work shall be authorized by Change Order, and shall be performed under applicable conditions of the Contract Documents.

8.1.3 If the Owner requests the Design/Builder to submit a proposal for a change in the Work and then elects not to proceed with the change, a Change Order shall be issued to reimburse the Design/Builder for any costs incurred for Design Services or proposed revisions to the Contract Documents.

8.1.4 Cost or credit to the Owner resulting from a change in the Work shall be determined in one or more of the following ways:

- **.1** by mutual acceptance of a lump sum properly itemized and supported by sufficient substantiating data to permit evaluation;
- **.2** by unit prices stated in the Contract Documents or subsequently agreed upon;

.3 by cost to be determined in a manner agreed upon by the parties and a mutually acceptable fixed or percentage fee; or
.4 by the method provided below.

8.1.5 If none of the methods set forth in Clauses 8.1.4.1, 8.1.4.2 or 8.1.4.3 is agreed upon, the Design/Builder, provided a written order signed by the Owner is received, shall promptly proceed with the Work involved. The cost of such Work shall then be determined on the basis of reasonable expenditures and savings of those performing the Work attributable to the change, including the expenditures for design services and revisions to the Contract Documents. In case of an increase in the contract sum, the cost shall include a reasonable allowance for overhead and profit. In case of the methods set forth in Clauses 8.1.4.3 and 8.1.4.4, the Design/Builder shall keep and present an itemized accounting together with appropriate supporting data for inclusion in a Change Order. Unless otherwise provided in the Contract Documents, cost shall be limited to the following: cost of materials, including sales tax and cost of delivery; cost of labor, including social security, old age and unemployment insurance, and fringe benefits required by agreement or custom; workers' or workmen's compensation insurance; bond premiums; rental value of equipment and machinery; additional costs of supervision and field office personnel directly attributable to the change; and fees paid to architects, engineers and other professionals. Pending final determination of cost to the Owner, payments on account shall be made on the Application for Payment. The amount of credit to be allowed by the Design/Builder to the Owner for deletion or change which results in a net decrease in the contract sum will be actual net cost. When both additions and credits covering related Work or substitutions are involved in a change, the allowance for overhead and profit shall be figured on the basis of the net increase, if any, with respect to that change.

8.1.6 If unit prices are stated in the Contract Documents or subsequently agreed upon, and if quantities originally contemplated are so changed in a proposed Change Order that application of agreed unit prices to quantities proposed will cause substantial inequity to the Owner or Design/Builder, applicable unit prices shall be equitably adjusted.

8.2 CONCEALED CONDITIONS

8.2.1 If concealed or unknown conditions of an unusual nature that affect the performance of the Work and vary from those indicated by the Contract Documents are encountered below ground or in an existing structure other than the Work, which conditions are not ordinarily found to exist or which differ materially from those generally recognized as inherent in work of the character provided for in this Part 2, notice by the observing party shall be given promptly to the other party and, if possible, before conditions are disturbed and in no event later than twenty-one days after first observance of the conditions. The contract sum shall be equitably adjusted for such concealed or unknown conditions by Change Order upon claim by either party made within twenty-one days after the claimant becomes aware of the conditions.

8.3 REGULATORY CHANGES

8.3.1 The Design/Builder shall be compensated for changes in the Work necessitated by the enactment or revision of codes, laws or regulations subsequent to the submission of the Design/Builder's Proposal under Part 1.

ARTICLE 9
CORRECTION OF WORK

9.1 The Design/Builder shall promptly correct Work rejected by the Owner or known by the Design/Builder to be defective or failing to conform to the Construction Documents, whether observed before or after Substantial Completion and whether or not fabricated, installed or completed, and shall correct Work under this Part 2 found to be defective or nonconforming within a period of one year from the date of Substantial Completion of the Work or designated portion thereof, or within such longer period provided by any applicable special warranty in the Contract Documents.

9.2 Nothing contained in this Article 9 shall be construed to establish a period of limitation with respect to other obligations of the Design/Builder under this Part 2. Paragraph 9.1 relates only to the specific obligation of the Design/Builder to correct the Work, and has no relationship to the time within which the obligation to comply with the Contract Documents may be sought to be enforced, nor to the time within which proceedings may be commenced to establish the Design/Builder's liability with respect to the Design/Builder's obligations other than correction of the Work.

9.3 If the Design/Builder fails to correct defective Work as required or persistently fails to carry out Work in accordance with the Contract Documents, the Owner, by written order signed personally or by an agent specifically so empowered by the Owner in writing, may order the Design/Builder to stop the Work, or any portion thereof, until the cause for such order has been eliminated; however, the Owner's right to stop the Work shall not give rise to a duty on the part of the Owner to exercise the right for benefit of the Design/Builder or other persons or entities.

9.4 If the Design/Builder defaults or neglects to carry out the Work in accordance with the Contract Documents and fails within seven days after receipt of written notice from the Owner to commence and continue correction of such default or neglect with diligence and promptness, the Owner may give a second written notice to the Design/Builder and, seven days following receipt by the Design/Builder of that second written notice and without prejudice to other remedies the Owner may have, correct such deficiencies. In such case an appropriate Change Order shall be issued deducting from payments then or thereafter due the Design/Builder costs of correcting such deficiencies. If the payments then or thereafter due the Design/Builder are not sufficient to cover the amount of the deduction, the Design/Builder shall pay the difference to the Owner. Such action by the Owner shall be subject to arbitration.

ARTICLE 10
ARBITRATION

10.1 Claims, disputes and other matters in question between the parties to this Part 2 arising out of or relating to Part 2 shall be decided by arbitration in accordance with the Construction Industry Arbitration Rules of the American Arbitration Association then in effect unless the parties agree otherwise. No arbitration arising out of or relating to this Part 2 shall include, by consolidation or joinder or in any other manner, an additional person not a party to Part 2 except by written consent containing specific reference to Part 2 and signed by the Owner, Design/Builder and any other person sought to be joined. Consent to arbitration involving an additional person or persons shall not constitute consent to arbitration of a dispute not described or with a person not named therein. This provision shall be specifically enforceable in any court of competent jurisdiction.

10.2 Notice of demand for arbitration shall be filed in writing with the other party to this Part 2 and with the American Arbitration Association. The demand shall be made within a reasonable time after the claim, dispute or other matter in question has arisen. In no event shall the demand for arbitration be made after the date when the applicable statute of limitations would bar institution of a legal or equitable proceeding based on such claim, dispute or other matter in question.

10.3 The award rendered by arbitrators shall be final, and judgment may be entered upon it in accordance with applicable law in any court having jurisdiction.

10.4 Unless otherwise agreed in writing, the Design/Builder shall carry on the Work and maintain its progress during any arbitration proceedings, and the Owner shall continue to make payments to the Design/Builder in accordance with the Contract Documents.

10.5 This Article 10 shall survive completion or termination of Part 2.

ARTICLE 11
MISCELLANEOUS PROVISIONS

11.1 This Part 2 shall be governed by the law of the place where the Work is located.

11.2 The table of contents and the headings of articles and paragraphs are for convenience only and shall not modify rights and obligations created by this Part 2.

11.3 In case a provision of Part 2 is held to be invalid, illegal or unenforceable, the validity, legality and enforceability of the remaining provisions shall not be affected.

11.4 SUBCONTRACTS

11.4.1 The Design/Builder, as soon as practicable after execution of Part 2, shall furnish to the Owner in writing the names of the persons or entities the Design/Builder will engage as contractors for the Project.

11.4.2 Nothing contained in the Design/Builder Contract Documents shall create a professional obligation or contractual relationship between the Owner and any third party.

11.5 WORK BY OWNER OR OWNER'S CONTRACTORS

11.5.1 The Owner reserves the right to perform work related to, but not part of, the Project and to award separate contracts in connection with other work at the site. If the Design/Builder claims that delay or additional cost is involved because of such action by the Owner, the Design/Builder shall make such claims as provided in Subparagraph 11.6.

11.5.2 The Design/Builder shall afford the Owner's separate contractors reasonable opportunity for introduction and storage of their materials and equipment for execution of their work. The Design/Builder shall incorporate and coordinate the Design/Builder's Work with work of the Owner's separate contractors as required by the Contract Documents.

11.5.3 Costs caused by defective or ill-timed work shall be borne by the party responsible.

11.6 CLAIMS FOR DAMAGES

11.6.1 Should either party to Part 2 suffer injury or damage to person or property because of an act or omission of the other party, the other party's employees or agents, or another for whose acts the other party is legally liable, claim shall be made in writing to the other party within a reasonable time after such injury or damage is or should have been first observed.

11.7 INDEMNIFICATION

11.7.1 To the fullest extent permitted by law, the Design/Builder shall indemnify and hold harmless the Owner and the Owner's consultants and separate contractors, any of their subcontractors, sub-subcontractors, agents and employees from and against claims, damages, losses and expenses, including but not limited to attorneys' fees, arising out of or resulting from performance of the Work. These indemnification obligations shall be limited to claims, damages, losses or expenses (1) that are attributable to bodily injury, sickness, disease or death, or to injury to or destruction of tangible property (other than the Work itself) including loss of use resulting therefrom, and (2) to the extent such claims, damages, losses or expenses are caused in whole or in part by negligent acts or omissions of the Design/Builder, the Design/Builder's contractors, anyone directly or indirectly employed by either or anyone for whose acts either may be liable, regardless of whether or not they are caused in part by a party indemnified hereunder. Such obligation shall not be construed to negate, abridge or otherwise reduce other rights or obligations of indemnity which would otherwise exist as to a party or person described in this Paragraph 11.7.

11.7.2 In claims against the Owner or its consultants and its contractors, any of their subcontractors, sub-subcontractors, agents or employees by an employee of the Design/Builder, its contractors, anyone directly or indirectly employed by them or anyone for whose acts they may be liable, the indemnification obligation under this Paragraph 11.7 shall not be limited by a limitation on amount or type of damages, compensation or benefits payable by or for the Design/Builder, or a Design/Builder's contractor, under workers' or workmen's compensation acts, disability benefit acts or other employee benefit acts.

11.8 SUCCESSORS AND ASSIGNS

11.8.1 This Part 2 shall be binding on successors, assigns, and legal representatives of and persons in privity of contract with the Owner or Design/Builder. Neither party shall assign, sublet or transfer an interest in Part 2 without the written consent of the other.

11.8.2 This Paragraph 11.8 shall survive completion or termination of Part 2.

11.9 In case of termination of the Architect, the Design/Builder shall provide the services of another lawfully licensed person or entity against whom the Owner makes no reasonable objection.

11.10 EXTENT OF AGREEMENT

11.10.1 Part 2 represents the entire agreement between the Owner and Design/Builder and supersedes Part 1 and prior negotiations, representations or agreements. Part 2 may be amended only by written instrument signed by both Owner and Design/Builder.

ARTICLE 12
TERMINATION OF THE AGREEMENT

12.1 TERMINATION BY THE OWNER

12.1.1 This Part 2 may be terminated by the Owner upon fourteen days' written notice to the Design/Builder in the event that the Project is abandoned. If such termination occurs, the Owner shall pay the Design/Builder for Work completed and for proven loss sustained upon materials, equipment, tools, and construction equipment and machinery, including reasonable profit and applicable damages.

12.1.2 If the Design/Builder defaults or persistently fails or neglects to carry out the Work in accordance with the Contract Documents or fails to perform the provisions of Part 2, the Owner may give written notice that the Owner intends to terminate Part 2. If the Design/Builder fails to correct the defaults, failure or neglect within seven days after being given notice, the Owner may then give a second written notice and, after an additional seven days, the Owner may without prejudice to any other remedy make good such deficiencies and may deduct the cost thereof from the payment due the Design/Builder or, at the Owner's option, may terminate the employment of the Design/Builder and take possession of the site and of all materials, equipment, tools and construction equipment and machinery thereon owned by the Design/Builder and finish the Work by whatever method the Owner may deem expedient. If the unpaid balance of the contract sum exceeds the expense of finishing the Work, the excess shall be paid to the Design/Builder, but if the expense exceeds the unpaid balance, the Design/Builder shall pay the difference to the Owner.

12.2 TERMINATION BY THE DESIGN/BUILDER

12.2.1 If the Owner fails to make payment when due, the Design/Builder may give written notice of the Design/Builder's intention to terminate Part 2. If the Design/Builder fails to receive payment within seven days after receipt of such notice by the Owner, the Design/Builder may give a second written notice and, seven days after receipt of such second written notice by the Owner, may terminate Part 2 and recover from the Owner payment for Work executed and for proven losses sustained upon materials, equipment, tools, and construction equipment and machinery, including reasonable profit and applicable damages.

ARTICLE 13
BASIS OF COMPENSATION

The Owner shall compensate the Design/Builder in accordance with Article 5, Payments, and the other provisions of this Part 2 as described below.

13.1 COMPENSATION

13.1.1 FOR BASIC SERVICES, as described in Paragraphs 2.2.2 through 2.2.17, and for any other services included in Article 14 as part of Basic Services, Basic Compensation shall be as follows:

13.2 REIMBURSABLE EXPENSES

13.2.1 Reimbursable Expenses are in addition to the compensation for Basic and Additional Services and include actual expenditures made by the Design/Builder in the interest of the Project for the expenses listed as follows:

13.2.2 FOR REIMBURSABLE EXPENSES, compensation shall be a multiple of ( ) times the amounts expended.

13.3 INTEREST PAYMENTS

13.3.1 The rate of interest for past due payments shall be as follows:

(Usury laws and requirements under the Federal Truth in Lending Act, similar state and local consumer credit laws and other regulations at the Owner's and Design/Builder's principal places of business, at the location of the Project and elsewhere may affect the validity of this provision. Specific legal advice should be obtained with respect to deletion, modification or other requirements, such as written disclosures or waivers.)

ARTICLE 14
OTHER PROVISIONS

14.1 The Basic Services to be performed shall be commenced on and, subject to authorized adjustments and to delays not caused by the Design/Builder, Substantial Completion shall be achieved in ( ) calendar days.

14.2 The Basic Services beyond those described in Article 2 are:

14.3 The Design/Builder shall submit an Application for Payment on the of each month.

14.4 The Design/Builder's Proposal includes:
(List below: this Part 2, Supplementary and other Conditions, the drawings, the specifications, and Modifications, showing page or sheet numbers in all cases and dates where applicable to define the scope of Work.)

This Part 2 entered into as of the day and year first written above.

OWNER

____

____

____

BY ______________________________

DESIGN/BUILDER

____

____

____

BY ______________________________

THE AMERICAN INSTITUTE OF ARCHITECTS

AIA Document B901

Standard Form of Agreements Between Design/Builder and Architect

1985 EDITION

THIS DOCUMENT HAS IMPORTANT LEGAL CONSEQUENCES; CONSULTATION WITH AN ATTORNEY IS ENCOURAGED.

TABLE OF ARTICLES

PART 1 AGREEMENT—PRELIMI… …N

…IGN

B901-1985
COVER PAGE

THE AMERICAN INSTITUTE OF ARCHITECTS

AIA Document A491

Standard Form of Agreements Between Design/Builder and Contractor

1985 EDITION

THIS DOCUMENT HAS IMPORTANT LEGAL CONSEQUENCES; CONSULTATION WITH AN ATTORNEY IS ENCOURAGED.

TABLE OF ARTICLES

PART 1 AGREEMENT—PR ... ON

... ON

A491-1985
COVER PAGE

THE AMERICAN INSTITUTE OF ARCHITECTS

AIA Document A401

Standard Form of Agreement Between Contractor and Subcontractor

1987 EDITION

THIS DOCUMENT HAS IMPORTANT LEGAL CONSEQUENCES; CONSULTATION WITH AN ATTORNEY IS ENCOURAGED WITH RESPECT TO ITS COMPLETION OR MODIFICATION.

This document has been approved and endorsed by the American Subcontractors Association and the Associated Specialty Contractors, Inc.

AGREEMENT

made as of the day of in the year of Nineteen Hundred and

BETWEEN the Contractor:
(Name and Address)

and the Subcontractor:
(Name and Address)

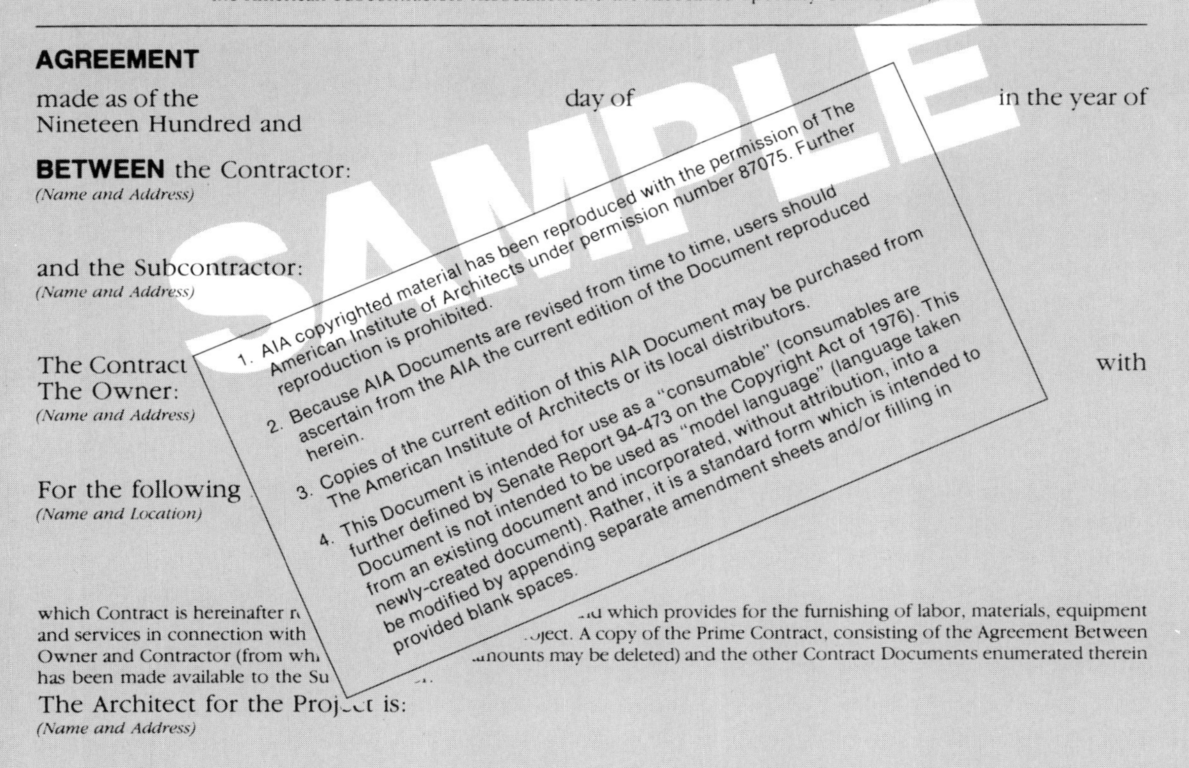

The Contract ... with The Owner:
(Name and Address)

For the following ...
(Name and Location)

which Contract is hereinafter r... ...d which provides for the furnishing of labor, materials, equipment and services in connection with ... ...oject. A copy of the Prime Contract, consisting of the Agreement Between Owner and Contractor (from wh... ...amounts may be deleted) and the other Contract Documents enumerated therein has been made available to the Su... ...r.

The Architect for the Proj...t is:
(Name and Address)

The Contractor and the Subcontractor agree as set forth below.

INDEX

Index

D